MODERN DAIRY TECHNOLOGY

Volume 1

Advances in Milk Processing

MODERN DAIRY TECHNOLOGY

Volume 1

Advances in Milk Processing

Edited by

R. K. ROBINSON
M.A., D.Phil.
Department of Food Science, University of Reading, UK

ELSEVIER APPLIED SCIENCE PUBLISHERS
LONDON and NEW YORK

ELSEVIER APPLIED SCIENCE PUBLISHERS LTD
Crown House, Linton Road, Barking, Essex IG11 8JU, England

Sole Distributor in the USA and Canada
ELSEVIER SCIENCE PUBLISHING CO., INC.
52 Vanderbilt Avenue, New York, NY 10017, USA

WITH 126 ILLUSTRATIONS AND 36 TABLES

British Library Cataloguing in Publication Data

Modern dairy technology.
Vol. 1: Advances in milk processing.
1. Dairying — Great Britain — Technological innovations
I. Robinson, R. K.
338.1'77'0941 S494.5.I5

Library of Congress Cataloging-in-Publication Data

Modern dairy technology.

Includes bibliographies and index.
Contents: v. 1. Advances in milk processing — v. 2. Advances in milk products.
1. Dairy processing. I. Robinson, R. K. (Richard Kenneth) II. Title: Dairy technology.
SF250.5.M63 1986 637 85-20414

ISBN 0-85334-391-8

Phototypesetting by Tech-Set, Gateshead, Tyne & Wear.
Printed in Great Britain by Page Bros (Norwich) Ltd.

Preface

The dairy industry is, in many countries, a major contributor to the manufacturing capacity of the food sector, and as more components of milk are ultilised in processed foods, so this importance is likely to grow. Already dairy operations range from the straightforward handling of liquid milk through to the production of highly sophisticated consumer items, and it is of note that all this activity is based on a raw material that is readily perishable at ambient temperatures.

This competitive, commercial position, together with the fact that the general public has a high regard for dairy products, is an indication of the extent to which milk producers and processors have combined to ensure that retail products are both nutritious and hygienically acceptable. Achievement of these aims, and at reasonable cost, has depended in large measure on the advances that have been made in the handling of large volumes of milk. Thus, factories designed to handle millions of litres of milk per week are now commonplace, and it is the plant and equipment involved that provides the factual background for this two-volume book.

In some instances, the increased capacity has arisen simply from an expansion of a traditional method, but in others a totally new approach has had to be adopted either for the manufacturing process *per se,* or for utilisation of the end products. Success has also depended on the derivation of accurate process controls, both through automation, and through improved procedures for quality control. Together, these diverse facets make up the modern dairy industry, and it is to be hoped that these two volumes will do justice to the innovative genius of those who have been involved in its evolution.

R. K. Robinson

Contents

List of Contributors

Mr F. A. GLOVER

Formerly National Institute for Research in Dairying, Shinfield, Reading RG2 9AT, UK. Present address: 39b St Peter's Avenue, Caversham, Reading, Berkshire, UK.

Dr M. E. KNIPSCHILDT

APV Anhydro A/S, 8, Ostermarken, DK-2860 Søborg-Copenhagen, Denmark.

Dr M. J. LEWIS

Department of Food Science, University of Reading, Food Studies Building, Whiteknights, P.O. Box 226, Reading RG6 2AP, UK.

Mr L.-E. NILSSON

Alfa-Laval, Food & Dairy Engineering AB, P.O. Box 64, S-22100 Lund, Sweden.

Mr C. R. SOUTHWARD

Section Head, Casein and Related Products, New Zealand Dairy Research Institute, Private Bag, Palmerston North, New Zealand.

Mr C. TOWLER

New Zealand Dairy Research Institute, Private Bag, Palmerston North, New Zealand.

Mr R. A. WILBEY

Department of Food Science, University of Reading, Food Studies Building, Whiteknights, P.O. Box 226, Reading RG6 2AP, UK.

Dr J. G. ZADOW

Division of Food Research, CSIRO, Dairy Research Laboratory, P.O. Box 20, Highett, Victoria 3190, Australia.

Chapter 1

Advances in the Heat Treatment of Milk

M. J. Lewis
Department of Food Science, University of Reading, UK

In this chapter milk will refer to bovine milk, either as full cream milk, skim-milk obtained by centrifugal separation, or standardised milk made by combining skim-milk with cream. Currently UK milk is not standardised, although low-fat, homogenised milk containing between 1 and 1·5 per cent fat is becoming popular. Other types of milk which will require heat treatment include flavoured milks, reconstituted milk, filled milk, evaporated milk, milks modified in composition by demineralisation or lactose hydrolysis, and protein-enriched milks produced by ultrafiltration.

Milk can be regarded as a complete food, containing protein, fat, lactose, vitamins and minerals, together with natural enzymes and those derived from micro-organisms within the milk. It has a high nutritional value, but is an excellent medium for microbial growth.

Milk is extremely variable in its composition. There are variations between individual cows in a breed, between breeds and between seasons. Variations between species are also very considerable (Jenness, 1982; Walstra and Jenness, 1984).

Milk is heated for a variety of reasons, the main ones being to remove pathogenic organisms, to increase its shelf-life up to a period of six months, as an aid to further processing, e.g. forewarming prior to separation and homogenisation, or as an essential treatment prior to cheesemaking, yoghurt manufacture and the production of evaporated and dried milk.

When milk is heated, many changes take place, and some of these may lead to protein coagulation. These have been summarised by Fox (1982) as:

decrease in pH;
precipitation of calcium phosphate;
denaturation of whey proteins and interaction with casein;
Maillard browning;
modification of casein: dephosphorylation, hydrolysis of κ-casein and general hydrolysis;
changes in micellar structure: zeta potential, hydration changes, association-dissociation.

The casein fraction of milk is very heat stable, whereas the whey protein fraction is heat labile and almost completely denatured at 100 °C. However, the denatured whey protein complexes with the casein and does not usually precipitate. When cheese whey is heated, the whey proteins start to denature, coagulate and precipitate at about 75–80 °C, illustrating the protective effect of casein toward coagulation; a solution of sodium caseinate can withstand heating at 140 °C for longer than 60 min at pH 6·7.

The extent of any changes will depend upon the time–temperature conditions and other compositional factors, and these reactions will affect the colour, texture and flavour of the milk, and generally reduce its acceptability. Loss of fat, vitamins and minerals will also occur.

It is well known that good quality bovine milk can withstand high processing temperatures without coagulating; a typical bovine milk being stable for several hours at 100 °C, and for about 20 min at 140 °C. The methods for assessing heat stability have been reviewed by Fox (1982) and Walstra and Jenness (1984). These conditions are much more stringent than those used for most milk processing operations, and it would be reasonable to expect no coagulation problems during the processing of good quality milk. However, it is well known that coagulation may occur, and that heat exchangers are susceptible to the deposition of milk solids and associated fouling problems. It must be concluded, therefore, that much of the milk being processed is not of the highest quality, and those properties of raw milk, which are pertinent to its behaviour during heating, will be briefly reviewed.

RAW MILK

Raw milk from healthy cows (sampled from the bulk tank and handled hygienically) should give rise to a low total colony count, and be free of pathogenic organisms. Cousins and Bramley (1981) state that initial counts range between less than 1000 ml^{-1}, indicating minimal contamination, to greater than 10^6 ml^{-1}. Counts of greater than 10^5 ml^{-1} are indicative of serious faults in production hygiene, whereas consistent counts less than 10^4 ml^{-1} reflect good hygienic practices.

There are only three possible sources of contamination, namely from within the udder, from the exterior of the teats and udder, and from milking and storage equipment.

The two most serious human diseases disseminated by the consumption of contaminated raw milk are tuberculosis and brucellosis, as the micro-organisms can be excreted from infected animals. A number of other pathogenic organisms maybe present as a consequence of udder disease (mastitis), such as *Staphylococcus* and *E. coli.* Some species of *Staphylococcus* produce a heat-resistant enterotoxin during storage, which may not be completely inactivated by processing. Very occasionally other more pathogenic organisms may affect the udder. *Salmonella* and *Campylobacter* may also be present from faecal contamination of raw milk, after secretion. Occasionally human carriers have been implicated, and it has been shown also that the udder can also be affected by these organisms.

Pasteurisation will inactivate all of these pathogens, apart from the spore-forming organisms, *Clostridium perfringens* and *Bacillus cereus.* Adams *et al.* (1976) showed that growth of gram-negative psychrotrophs in raw milk led to detectable proteolysis, particularly of κ- and β-casein, even after two days, and it is now well established that several species produce highly heat-resistant enzymes, particularly proteases and lipases. These proteases may survive thermal processing and cause further proteolysis in the stored product, leading eventually to gelation. If the psychrotrophic count is above 10^6 ml^{-1}, significant levels of the proteases may survive (Van den Berg, 1981), whilst Law *et al.* (1977) found that milk containing more than 8×10^6 ml^{-1} of *Pseudomonas fluorescens* AR11, subjected to UHT treatment, gelled about 10–12 days after production. Therefore it is useful to monitor psychrotrophic organisms in raw milk. Evidence of their presence will only become apparent if milk samples stored at 4°C, examined microscopically for counts at one or two successive intervals of 24 h, show a rapid increase in count. Raw milk also contains acid-producing streptococci and bacilli, and excessive activity will result in a fall in pH, and a decrease in the heat stability of the milk.

Badings and Nester (1978) suggest that fresh milk (raw or low pasteurised) has a bland, but characteristic flavour, with many compounds, mostly in sub-threshold concentrations, contributing to its aroma. Off-flavours may arise due to weed taints or feed off-flavours, oxidation reactions, and light induced reactions (Badings, 1984). Connoisseurs of raw milk claim that even a mild form of heat treatment, such as pasteurisation, will significantly change its delicate characteristic flavour. Raw (untreated) milk sold in the United Kingdom must be labelled as 'raw unpasteurised milk'. However, in most processing operations milk is pasteurised as quickly as possible.

PASTEURISATION

The following definition of pasteurisation has been adopted by the Internation Dairy Federation (IDF). 'Pasteurisation is a process applied to a product with the object of minimising possible health hazards arising from pathogenic micro-organisms associated with milk by heat treatment, which is consistent with minimal chemical, physical and organoleptic changes in the product'.

Processing Conditions

The UK heat treatment and labelling regulations (SI 1977/1033) state that the following time–temperature combinations should be used for pasteurised milk. It should be:

(a) retained at a temperature of not less than 62·8 °C and not more than 65·6 °C for at least 30 min, and be immediately cooled to a temperature of not more than 10 °C;
(b) retained at a temperature of not less than 71·7 °C for at least 15 s and be immediately cooled to a temperature of not more than 10 °C; or
(c) retained at such a temperature for such a period as may be specified by the licensing authority, with the approval of the Minister.

A number of other points are detailed. The milk should not be subject to atmospheric contamination. If the processing temperature is greater than 65·6 °C, there should be provision for diverting milk which is under-processed. Suitable indicating and recording thermometers should be installed, and temperature records should be kept for at least one month. The pasteurised milk should also satisfy the phosphatase test, and the milk should be properly sealed and labelled.

The regulations for pasteurising semi-skimmed and skimmed milk (SI 1973/1064) are almost similar, however in addition, these milks must satisfy a methylene blue test. The hygiene and heat treatment regulations (SI 1983/1508) for milk-based drinks were introduced in 1983.

A milk-based drink is defined as a liquid drink (other than a fermented milk) comprising a minimum of 85 per cent milk and other permitted ingredients, a list being supplied. The pasteurisation regulations are the same as for milk, although a coliform test may be necessary if the drink has a colour which interferes with the phosphatase test. The milk regulations, applying at the end of 1983, are summarised by Jukes (1984).

It is interesting to note that the UK regulations permit the use of other time–temperature combinations, and the processing times and temperatures can be plotted on semi-log graph paper. The plot can be used to give an indication of the times required at temperatures between 62·8 °C and 71·7 °C, or extrapolated to give processing times at temperatures greater than 71·7 °C. Pasteurisation requiring very short times is known as 'flash pasteurisation'. For example, at 77 °C the extrapolated time would be about one second. All these interpolated time–temperature combinations would be sufficient to inactivate *Mycobacterium tuberculosis* (Society of Dairy Technology, 1983). If data for the disappearance of the cream line (Society of Dairy Technology) (SDT) is considered, these pasteurisation conditions will not lead to a loss of cream line. This is very important with bottled milk, but less so for milk in non-transparent packaging.

Pasteurised milk still accounts for greater than 85 per cent of liquid milk consumed in the UK, and homogenised milk accounts for a further 6 per cent. Statistics for the sales of liquid milk over the period 1975 to 1982, classified according to their heat treatment, are given in Table I. One interesting effect of homogenisation is that it appears to give the milk a creamier texture.

Pasteurisation conditions are fairly uniform worldwide, although slight variations may occur in time–temperature combinations, cooling temperatures and testing procedures. In the USA, the following time–temperature combinations apply to Grade A milk (Busse, 1981):

63 °C	30 min	94 °C	0·1 s
77 °C	15 s	96 °C	0·05 s
89 °C	1 s	100 °C	0·01 s
90 °C	0·5 s		

TABLE I
Types of heat treated milk sold for liquid consumption in England and Wales

	Pasteurised	*Sterilised*	*Homogenised*	*UHT*
		(million litres)		
1975	5643 (85·3)	500 (7·6)	437 (6·6)	35·9 (0·5)
1978	5453 (85·9)	464 (7·3)	394 (6·2)	41·0 (0·6)
1982	5125 (86·2)	386 (6·5)	389 (6·5)	47·0 (0·8)

Figures in parenthesis represent the percentage of the total.
Data from: United Kingdom Dairy Facts and Figures (1983).

It is interesting to note that such short residence times were used, because there are difficulties in measuring and controlling short residence times. This milk will have a shelf-life of 18 days or longer at a maximum of 7 °C. In Central Europe, most dairies use 74 °C for about 30–40 s, but such harsh conditions are likely to result in a cooked flavour. There is little evidence that harsher processing conditions will increase the shelf-life, as this is more likely to be influenced by storage temperature and the extent of post-processing contamination (Busse, 1981). In fact, in the absence of post-processing contamination, Kessler and Horak (1984) found that more stringent conditions, i.e. 85 °C/15 s and 40 s and 78 °C/40 s reduced the shelf-life slightly. This loss was attributed to the germination of spores stimulated by the harsher processing conditions.

Pasteurisation is a relatively mild form of heat treatment, and most consumers would probably find difficulty distinguishing between raw and pasteurised milks. Pasteurised milk has no readily apparent cooked flavour; no active sulphydryl groups are found. Whey protein denaturation is low (between 5 and 15 per cent), and there is relatively little loss of the heat sensitive nutrients. The rennet coagulation properties are not affected. Aboshama and Hansen (1977) observed a slight loss of amino acids, approximately 6 per cent. More recently Scott *et al.* (1984 a and b) have surveyed the vitamin contents of milk as delivered to dairies, bulk pasteurised milk, and packaged milk as delivered to the home and after storage in the refrigerator. Pasteurisation was found to have hardly any effect on vitamins, but considerable changes occurred during subsequent transportation and storage. For example, an average value for vitamin C in pasteurised milk was 19·4 μg ml^{-1}, whereas the range for bottles sampled as delivered was 0·1–18·4 μg ml^{-1}. However, some of these reactions may well be more seriously affected if the temperature and residence times are not properly controlled. Enzymes are not normally a problem with pasteurised milk, mainly because the milk is used fairly soon after processing.

Excessive agitation and turbulence may cause damage to the fat globule membranes and stimulate the naturally-occurring, lipoprotein lipase in raw milk, so causing hydrolysis and a slight impairment of the flavour (Solberg, 1981). Homogenisation down-stream of the holding-tube usually eliminates any problems of lipase activity.

One of the more recent faults with pasteurised milk is 'sweet curdling' caused by bacterial proteinases, due mainly to *Bacillus cereus*; this is best controlled by close supervision of cleaning and disinfection procedures, and the use of nitric acid which has been found to be the best material for eliminating these spores.

Post-processing contamination (PPC) has a decisive influence on the keeping quality of pasteurised milk. It may arise from balance tanks, pipelines, filling and capping machines, and containers downstream of the pasteuriser, and may reduce its keeping quality from two weeks down to only a few days (Ashton and Romney, 1981). Schroeder *et al.* (1982) and Schroeder (1984) have reviewed the effects of psychrotrophic PPC on the keeping quality of milk, and the origins and levels of PPC in commercial dairies in the UK.

Pasteurised milk should be stored at as low a temperature as possible, and kept in the cold-chain at all points during its distribution and sale. Under these conditions, psychrotrophic gram-negative bacteria proliferate and control shelf-life, but as products are stored at warmer temperatures, thermoduric organisms begin to assume dominance. Solberg (1981) makes the novel suggestion of incubating raw milk with certain lactic starter cultures with the aim of lowering the oxidation–reduction potential and inhibiting the growth of psychrotrophic organisms. These starters would be inactivated during subsequent pasteurisation.

The predominant, surviving thermodurics are *Streptococcus thermophilus*, certain micrococci, *Arthrobacter* and *Microbacterium.* Spores of bacilli and clostridia are not inactivated by pasteurisation, but clostridia will not develop in milk due to its high oxygen content, whereas aerobic spore-formers will, and may well, cause spoilage even in refrigerated milk. Spoilage by thermodurics is most likely to occur when the temperature of the milk rises.

Micro-organisms arising from post-pasteurisation contamination may initially be present in small numbers. Ashton and Romney (1981) suggest that bacterial counts on the product, together with a repasteurised sample of the product, be undertaken, or that use be made of enrichment techniques for psychrotrophic organisms. Close scrutiny of cleaning procedures will help to reduce PPC. Solberg (1981) suggests that long runs on pasteurisers may give rise to serious bacteriological problems; plant should be efficiently cleaned every 8 h to avoid these. PPC can also be avoided by HTST pasteurisation followed by hot filling into plastic or laminate cartons between 70 and 80 °C in an aseptic, protective gas atmosphere, fairly rapid cooling to 50 °C followed by slower cooling to either ambient or refrigerated temperature.

Pasteurisation Equipment

The Holder process

This batch heating process involves heating the milk to a temperature between 62·8 °C and 65·6 °C, holding it at that temperature for 30 min, and rapidly cooling it to below 10 °C. It is generally favoured by processors with low throughputs. The equipment is relatively cheap and simple, and can be readily adapted for different volumes and other fluids. However, it is more labour intensive than HTST processes, and energy costs are higher as it is not easy to incorporate regeneration.

The heating time in a jacketted, heating vessel is given by:

$$t = \frac{MC}{AU} \ln \frac{\theta_h - \theta_I}{\theta_h - \theta_F}$$

where

t = heating time (s)
M = mass (kg)
C = specific heat (J kg^{-1}K^{-1})
θ = temperature.
A = heat transfer area (m^2)
U = overall heat transfer coefficient (W m^{-2}K^{-1})

Subscripts h, I and F refer to the heating medium, the initial temperature of the milk, and its final temperature respectively. Therefore, the time required to heat 3000 kg of milk from 4 °C to 65 °C using steam (100 °C) in a vessel (A = 3 m^2; U = 730 W m^{-2}K^{-1}) is given by

$$t = \frac{3000}{3} \times \frac{3900}{730} \ln \frac{100 - 4}{100 - 63}$$

and equals approximately 27·6 min (note the heat transfer coefficient value is taken from Kessler (1981). Further details about batch pasteurisers are given by the Society of Dairy Technology (1983).

For higher throughputs, the high temperature–short time continuous process (HTST) is favoured. HTST pasteurisers are available in a range of sizes from 400 to 50 0001 h^{-1}. They can be extremely efficient in terms of energy utilisation by incorporating a regeneration section into the heat exchanger; this is shown schematically in Fig. 1.

Raw milk enters the plant through the regeneration section, where it is heated by the hot milk, which in turn cools. After regeneration, it enters a final heating section, where hot water or electricity is used to bring the milk up to the desired temperature. It then passes through a holding tube, back into the regeneration section, followed by the mains water and chilled water cooling sections.

The regeneration efficiency (RE) is defined as:

$$\text{RE} = \frac{\text{Amount of heat supplied by regeneration}}{\text{Total heat load, assuming no regeneration}} \times 100$$

$$= \frac{MC(\theta_2 - \theta_1)}{MC(\theta_3 - \theta_1)} \times 100$$

M = mass flow rate (kg s^{-1}) θ_1 = inlet temperature (°C)
C = specific heat θ_2 = temperature leaving regeneration
θ_3 = pasteurisation temperature.

Manufacturers are now claiming regeneration efficiencies in excess of 95 per cent. Therefore, if milk enters at 10 °C, it will leave the regeneration section at 68·6 °C. If it is assumed that there is no heat loss over the regeneration section, then

$$(\theta_2 - \theta_1) = (\theta_4 - \theta_3)$$

i.e. the cooled milk will leave the regeneration section at a temperature of 13·1 °C. In this case, the mains water cooling section would be dispensed with, as it is highly unlikely that it would be below this temperature. The product is then cooled from 13·1 °C, to below 10 °C in the chilled water (refrigerated) section. Consequently, regeneration saves energy in terms of both heating and refrigeration costs. Regeneration efficiencies are increased by the use of larger heat exchanger surface areas, but capital costs are higher, and the fluid has a longer residence time within the heat exchanger. This latter point is extremely important in connection with the performance of UHT plants.

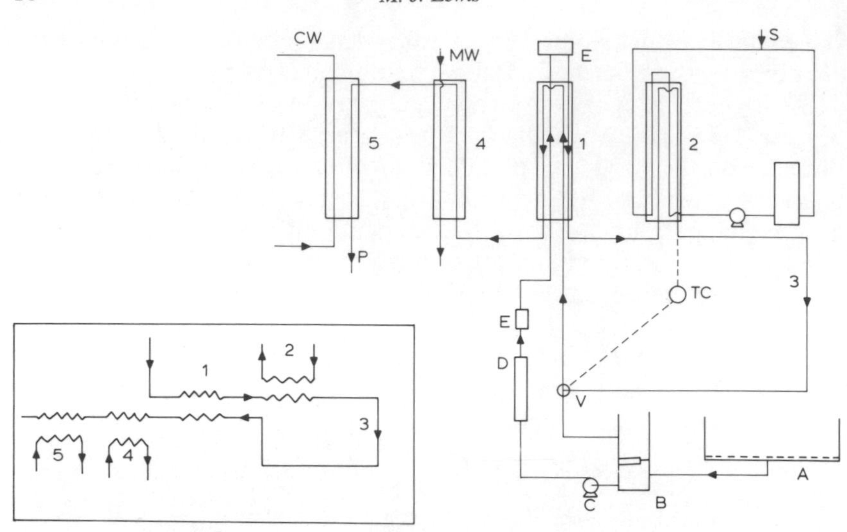

Fig. 1a. Layout of HTST pasteuriser (the insert shows a schematic diagram of the heat exchange sections). (A: feed tank; B: balance tank; C: feed pump; D: flow controller; E: filter; P: product; S: steam injection (hot water section); V: flow diversion valve; MW: mains water cooling; CW: chilled water; TC: temperature controller; 1: regeneration; 2: hot water section; 3: holding tube; 4: mains cooling water; 5: chilled water cooling).

The layout of a typical HTST pasteuriser incorporating regeneration is shown in Fig. 1. The milk is fed from the main feed tank (A), which contains a coarse filter, to the balance tank (B) which keeps a constant head. It is then pumped by either a centrifugal or positive displacement pump through a flow controller (D) and additional filter (E), into the regeneration section. A fine cloth filter may also be included in the regeneration section. From regeneration the milk goes to the heating section, where energy is supplied by a hot water 'set'. Steam is injected into circulating hot water in a closed system until the temperature reaches the set-point temperature. Steam consumption can be monitored by measuring the amount of steam which condenses. The hot water, set-point temperature is fixed, so that the milk leaves this section at the correct pasteurisation temperature. The milk then enters the holding section, which is normally tubular, and designed to ensure that all the milk resides in the tube for at least 15 s. A temperature probe is located in the holding tube, and is attached to a recorder and a flow diversion valve. If the temperature falls below the desired temperature, the flow

Fig. 1b. A side view of a typical plate heat exchanger (reproduced by courtesy of Stork-Amsterdam Int., Middx., UK).

diversion valve will operate and direct the under-processed milk back to the balance tank. Under normal processing conditions, the milk will pass into the regeneration section, then through the mains water and chilled water sections before being stored in tanks or packaged into bottles and cartons.

Most HTST pasteurisers are of the plate heat exchanger type, and these are described in more detail by Scott (1970), Kessler (1981) and Society of Dairy Technology (1983). They offer the advantage of providing a large surface area for heat transfer in a compact space. The gaps between the plates are narrow to induce turbulence, and the plates are designed with this, and pressure drop considerations, in mind. The system is extremely suitable for fluids of low viscosity which are reasonably sensitive to heat, but it cannot cope with particulate matter or fluids of high viscosity, as pressure drops would be too high. It is also unsuitable for liquids which will cause substantial fouling, and milk is not normally susceptible to this under pasteurisation conditions. Although steam and hot water are the main heating fluids, there are some designs which rely upon electrical heating, with the heating wire either wrapped round the outside of quartz tubes, through which the milk flows, or with the heating element inside the tube in direct contact with the milk. Most milk is still pasteurised in large dairies, but the last few years has seen a large increase in the number of producer-processors who pasteurise their milk on the farm. Several manufacturers now produce HTST pasteurisers specifically for these locations, most of which are plate heat exchangers.

Normally the diversion valve is positioned at the end of the holding tube, although there is an interesting debate as to the optimum position for the temperature probe controlling diversion. If it is positioned at the beginning of the holding tube, it is possible to detect under-processed milk quickly enough to be able to operate the flow diversion valve before the milk reaches the end of the tube, even if the response time is fairly slow. However, it would not indicate the extent of heat loss as the milk passed through the tube, and hence the fall in temperature. A temperature probe positioned at the end of the holding tube would monitor the lowest temperature in the tube, but the instrumentation would need a fast response-time to prevent under-processed milk going forward. Most holding tubes are not thermally lagged, and there is little information presented in the literature about temperature drops in holding tubes. If temperature drops are high, then the location of the temperature probe would be very important.

Residence Times

It has long been recognised that there is a distribution of residence times when liquids flow through pipes. This distribution can be measured and analysed by a variety of methods, which are described in more detail by Levenspiel (1972). One common method, which is easily applied, involves the injection of a pulse of some suitable tracer material into the tube inlet at time zero, followed by sampling and analysis of the outlet stream at regular time intervals. Figure 2 shows the traces obtained for (a) plug flow, (b) streamline flow and (c) turbulent flow.

Plug flow is an idealised situation, where there is no velocity profile across the tube. The fluid passes through the tube in a similar fashion to a piston in a cylinder, and consequently there is no spread of residence times. However, due to internal friction and/or viscosity of the fluid,

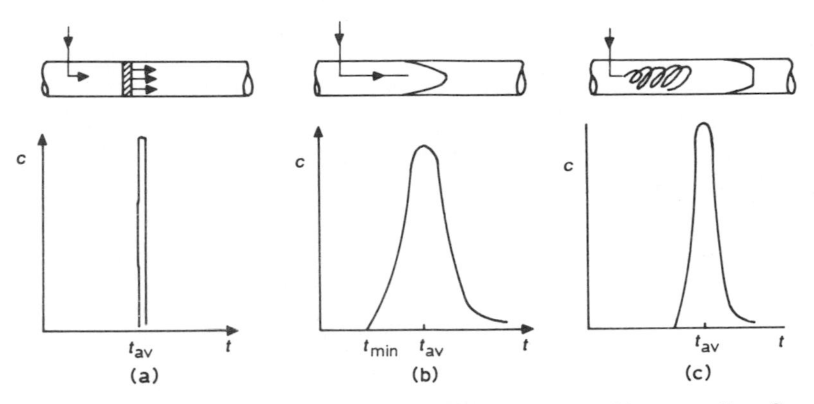

Fig. 2. Residence time distributions for, (a) plug flow, (b) streamline flow, (c) turbulent flow.

there is a velocity distribution across the tube, the nature of which depends upon whether the flow is streamline or turbulent. The type of flow can be determined by evaluating a dimensionless constant known as the Reynolds Number (*Re*), where:

$$Re = \frac{VDP}{\mu} \quad \text{or} \quad \frac{4Qp}{\pi \mu D}$$

V = average velocity (m s^{-1})
D = tube diameter (m)
p = fluid density (kg m^{-3}).
μ = dynamic viscosity (N s m^{-2})
Q = volumetric flow rate (m^3 s^{-1})

If the Reynolds Number is less than 2000, the flow is streamline, if it is greater than 4100, the flow is turbulent. With streamline flow, there is a parabolic velocity profile across the tube with no bulk movement across the tube. The maximum velocity occurs at the centre of the tube, and is approximately twice the average velocity, where:

$$V = \frac{\text{volumetric flow rate}}{\text{cross sectional area of tube}} = \frac{4Q}{\pi D^2}$$

and $V_{max} = 2V_{av}$.

The average residence time (t_{av}) is based on the average velocity, and is equal to L/V_{av}.

It can also be shown that

$$t_{av} = \frac{\text{volume of tube}}{\text{volumetric flow rate}}$$

In the case of the injection of a pulse of tracer, the average residence time is the time for 50 per cent of the material in the pulse to pass through the tube. The average residence time is determined from the volumetric flow rate and the dimensions of the holding tube, both of which are easily measured. However, the minimum residence time, which is based on the maximum velocity, will equal $t_{min} = 0{\cdot}5\, t_{av}$. Therefore, a holding tube designed to give an average residence time of 20 s would give a minimum time of 10 s, under these flow conditions, and result in some of the milk being under-processed. Thus, holding tube design should be based upon the minimum residence time when the eradication of micro-organisms is the main criterion.

When the flow is turbulent, the velocity profile is much flatter, and there is good bulk mixing across the tube. In this case

$$V_{max} = 1{\cdot}2 V_{av}$$

$$t_{min} = 0{\cdot}83 t_{av}$$

The distribution of residence times is also much narrower, thereby reducing the chance of over-processing.

Typical results from a 900 litre h^{-1} HTST pasteuriser, using an injection technique with water, gives the following data

$$Re = 13\,000, \quad t_{min} = 16 \text{ s}$$

t_{av} (evaluated from holding tube size and volumetric flow rate) = 21·5 s. The distribution shown in Fig. 2 is not quite symmetrical, but there is a close agreement between the average residence time (evaluated) and the time corresponding to maximum concentration of tracer in the outlet stream. However, as the distribution is asymmetrical, a long tail often results, and, although the area under this tail is usually small, a small portion of fluid will have a residence time significantly longer than the average residence time. As the flow in the tube becomes more turbulent, the residence time distribution tends to be reduced. In practice, this can be achieved by making holding tubes longer and thinner, provided the pressure drops are not excessive.

Dickerson *et al.* (1968) describe a technique for measuring the minimum residence time in holding tubes of HTST pasteurisers. Using identical flow rates minimum residence times for raw milk (15·8 s) and chocolate milk (15·5 s) agreed fairly closely with those for water (16·6 s). However for more viscous products, such as condensed skim-milk (40 per cent TS), ice cream mix (10 per cent milk fat) and cream (40 per cent fat), the discrepancies were larger; the size of the discrepancy increasing as the viscosity increased. This was attributed to the development of streamline flow with its more parabolic velocity profile and higher maximum velocity.

It may be necessary to lengthen holding tubes when processing more viscous materials to ensure that all the material is sufficiently processed; residence time distribution data become difficult to obtain at short residence times.

STERILISED AND UHT MILKS

Pasteurisation conditions are not sufficient to inactivate the thermoresistant spores in milk, but elimination of these spores would give a milk with much better keeping qualities, and one whose shelf-life would be limited by non-microbial spoilage reactions. Milk subjected to temperatures in excess of 100 °C and packaged in air-tight containers goes under the general name of 'sterilised milk'. Packaging can take place either before or after heat treatment, but if after heat treatment, aseptic packaging is essential. Milk, thus processed, is termed 'commercially sterile'; i.e. it is not necessarily free of micro-organisms, but those which survive the heat treatment are unlikely to proliferate during storage and cause spoilage of the product. Since the primary objective is the inactivation of spores, this will be covered in more detail.

Effects of Heat on Micro-organisms

The destruction of vegetative organisms and spores follows first order reaction kinetics. When the log of the population (N) is plotted against time, a straight line relationship results (at constant temperature)

$$\frac{dN}{dt} = kN \quad \therefore \quad \frac{dN}{N} = k\,dt$$

If this is integrated between the final number (N_F) and the initial number (N_0)

$$\ln\frac{N_0}{N_F} = kt$$

or $2{\cdot}303 \log_{10}\dfrac{N_0}{N_F} = kt$

$\log_{10} \frac{N_0}{N_F}$ = number decimal reductions

k = reaction velocity constant (s^{-1}) or (min^{-1})

t = heating time (s) or (min).

An alternative way of expressing the heat resistance is using the decimal reduction time (D_T). This is defined as the time required for the population to be reduced by 90 per cent (see Fig. 3) or alternatively 1 log cycle.

The number of decimal reductions is given by:

$$\log_{10} \frac{N_0}{N_F} = \frac{\text{heating time}}{D_T} = \frac{t}{D_T}$$

combining these equations: $D_T = \frac{2{\cdot}303}{k}$

These relationships are useful for evaluating sterilisation effects at constant temperature. It is common practice in canning of low-acid products (pH 4·6) to achieve at least 12 decimal reductions ($10^{12} - 1$) for the spores of *Clostridium botulinum,* because they are the most heat resistant of the major food poisoning organisms. However, in milk there are spores more resistant than *Clostridium botulinum,* and which are likely to cause spoilage over a long storage period; the most heat resistant being *Bacillus stearothermophilus.* Unfortunately, most sterilisation procedures do not take place at constant temperature, and in many cases, the heating and cooling periods may make a substantial contribution to the total lethal effect.

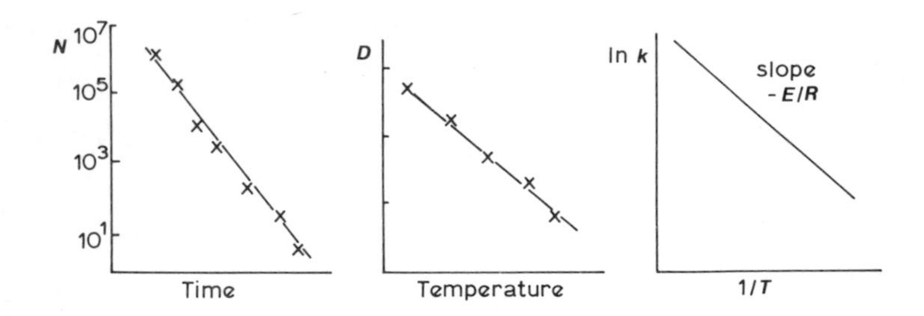

Fig. 3. Representation of kinetic data for the inactivation of micro-organisms.

Variation of Heat Resistance with Temperature

There are three ways of representing the variation of reaction rate with temperature.

Activation energy (E)

When the logarithm of reaction velocity constant ($\log k$) is plotted against the reciprocal of the absolute temperature ($1/T$), a straight line relationship is often observed (Fig. 3).

This is described by the Arrhenius Equation:

$$k = Ae^{-E/RT} \quad \text{or} \quad \ln k = \ln A - \frac{E}{RT}$$

E = activation energy, A = collision frequency.

Z value

Over a limited temperature range, a straight line relationship also results when the logarithm of decimal reduction time ($\log D$) is plotted against temperature. This is used to define the Z value for that reaction.

The Z value is the temperature difference which results in a 10 fold change in the D value (see Fig. 3). Burton (1983) reviews Z value data for heat resistant spores found in milk, and concludes that $Z = 10\,°C$ reasonably describes the data for mesophilic and thermophilic spores. The equation relating D values at different temperatures is

$$\frac{\log D_1 - \log D_2}{T_1 - T_2} = -\frac{1}{Z}$$

Chemical reaction rates (low Z, high E) are much less sensitive to changes in temperature than those for the inactivation of vegetative organisms and spores.

The relationship between E and Z is given by:

$$E = \frac{2{\cdot}303\,RTT_1}{Z\,(°C)}$$

E = activation energy (cal $mole^{-1}$)
T = temperature (K)
T_1 = reference temperature (K)
R = gas constant (cal K^{-1} $mole^{-1}$).

A reference temperature is quoted because most experimental data are determined about a reference temperature, but the equation should only be used if the actual temperature is within one Z value of the reference temperature.

Q_{10} value

The Q_{10} value provides a quick check as to how sensitive reaction rate is to temperature. It is defined as follows

$$Q_{10} = \frac{\text{Rate of reaction at } (T + 10)\,^{\circ}\text{C}}{\text{Rate of reaction at } T\,^{\circ}\text{C}}$$

The relationship between the Q value and Z value is

$$Z = \frac{10}{\log_{10} Q}$$

For most chemical reactions, Q lies between 2 and 4, whereas values for the inactivation of micro-organisms are usually between 10 and 30. Kinetic parameters for a wide range of microbial and chemical reactions occurring in foods are summarised by Lund (1975).

A bacteriologically effective residence time can be evaluated from the residence time distribution function, as described by Burton *et al.* (1977). It is defined as the effective heating time for that process; the sterilisation effect resulting from the process would be the same, provided that the sample was held at the same temperature for the effective heating time. For example, a direct steam injection plant was found to give an average residence time of 3·26 s, and an effective bacteriological holding time of 2·31 s. The effective time was then used to evaluate decimal reduction times for heat resistant spores directly from data from a UHT plant, and there was a reasonable agreement between data obtained by this method and by capillary tube experiments.

The bacteriologically effective residence time falls between the average and the minimum residence times. Therefore, sterilisation efficiencies, based on average residence times, will give an over-estimate of the lethal effect (Brown and Ayres, 1982). More detailed analyses of residence time distribution functions are given by Levenspiel (1972), Loncin and Merson (1979) and Kessler (1981).

Most thermal processes do not take place at constant temperature, and some means of evaluating the sterilisation effect is required:

$$2{\cdot}303 \log_{10} \frac{N_0}{N_F} = \int k \, dt = \int A e^{-E/RT} \, dt$$

If the relationship between temperature and time is known, the expression can be integrated directly, but otherwise it can be evaluated by plotting $Ae^{-E/RT}$ against time, and determining the area under the curve. However, a more popular approach has been to adopt a 'total integrated lethal effect value', known as the F_0 value, for the process.

The equation describing two sets of conditions along the thermal death line is:

$$\frac{\log t_1 - \log t_2}{\theta_1 - \theta_2} = -\frac{1}{Z} \quad \text{or} \quad \log\left(\frac{t_1}{t_2}\right) = \frac{\theta_2 - \theta_1}{Z}$$

The lethality (L) at any experimental temperature (θ_2) is defined as the number of minutes at the reference temperature (θ_1) that would have the same sterilisation effect as one minute at the experimental temperature.

$$t_1 = L, \quad t_2 = 1, \quad \theta_1 = 121{\cdot}1\,^\circ\text{C}, \quad \theta_2 = \text{experimental temperature}$$

$$\therefore \log L = \frac{\theta_2 - 121{\cdot}1}{Z}$$

Values of L are tabulated from this equation, and summarised in standard lethality equations. The temperature–time heating profile can be converted to a lethality–time profile where the area under the curve is equal to the F_0 value

$$\underset{(F_0)}{\text{Total lethality}} = \int L \, dt$$

Many investigators ignore the heating and cooling sections in UHT plants, but this approach is much more valid for direct steam injection processes. The F_0 value is evaluated from the following equation

$$F_0 = 10^{\frac{\theta - 121{\cdot}1}{10}} \cdot \left(\frac{t}{60}\right)$$

θ = processing temperature (°C)
t = time (s).

Most of the work on the heat resistance of spores at UHT conditions has been done with *Bacillus stearothermophilus.* Problems arise because decimal reduction times are very low at UHT temperatures, and difficulties have also arisen when trying to relate data obtained from capillary tube experiments to those from the pilot-plant. Burton *et al.* (1977) found that corrections for temperature distributions within the tube were necessary at temperatures above 135 °C. The determination of the total number of viable spores remaining after heat treatment is not straightforward, due to a number of factors that affect spore resistance, such as the age and storage conditions of the spores, the recovery of the spores and the presence of inhibitory substances produced by the heat treatment. Shew (1981) suggested that a maximum spoilage rate of 1 in 10^4 containers was acceptable. Assuming that raw milk contains an average of one heat resistant spore per ml and UHT processing brings about eight decimal reductions, the likely survival would be one organism in every 10^5 one-litre packages. If the raw milk were more contaminated, the spoilage rate would increase. Methods for determining the heat resistance of spores at UHT conditions are reviewed by Brown and Ayres (1982), whilst Burton (1983) reviews the most recent data.

Distinction Between In-bottle Sterilised and UHT Milk

The traditional way of producing sterilised milk involves filtering and homogenising the milk, filling into bottles, sealing them and heat treating them in batch or continuous retorts. More recently milk has been processed by using higher temperatures for shorter times, followed by aseptic packaging; when temperatures exceed 135 °C, the holding times become short enough to allow a continuous processing operation. The production of sterilised milk under these conditions has been known as ultra-high temperature (UHT) processing.

The distinction between sterilised and UHT milk can best be seen by plotting reaction kinetic data for the destruction of spores, the inactivation of enzymes and various chemical reactions on the same graph, using semi-logarithmic paper (Fig. 4). Examples are shown for the destruction of thermophilic and mesophilic spores, the inactivation of protease and a chemical reaction — superimposed on this are the conditions for sterilised and UHT milks.

Kessler (1981) presents similar reaction kinetic data for the discoloration of whole milk, evaporated milks, skimmed milk and cream containing 10, 30 and 40 per cent fat respectively, loss of vitamin B_1, the formation of hydroxymethyl furfural and losses of lysine. It can be seen that both types of process produce a commercially sterile product. However, the UHT process offers other advantages, which arise from the fact that chemical reaction rates are less sensitive to changes in temperature than the thermal inactivation of spores. Therefore, the much shorter

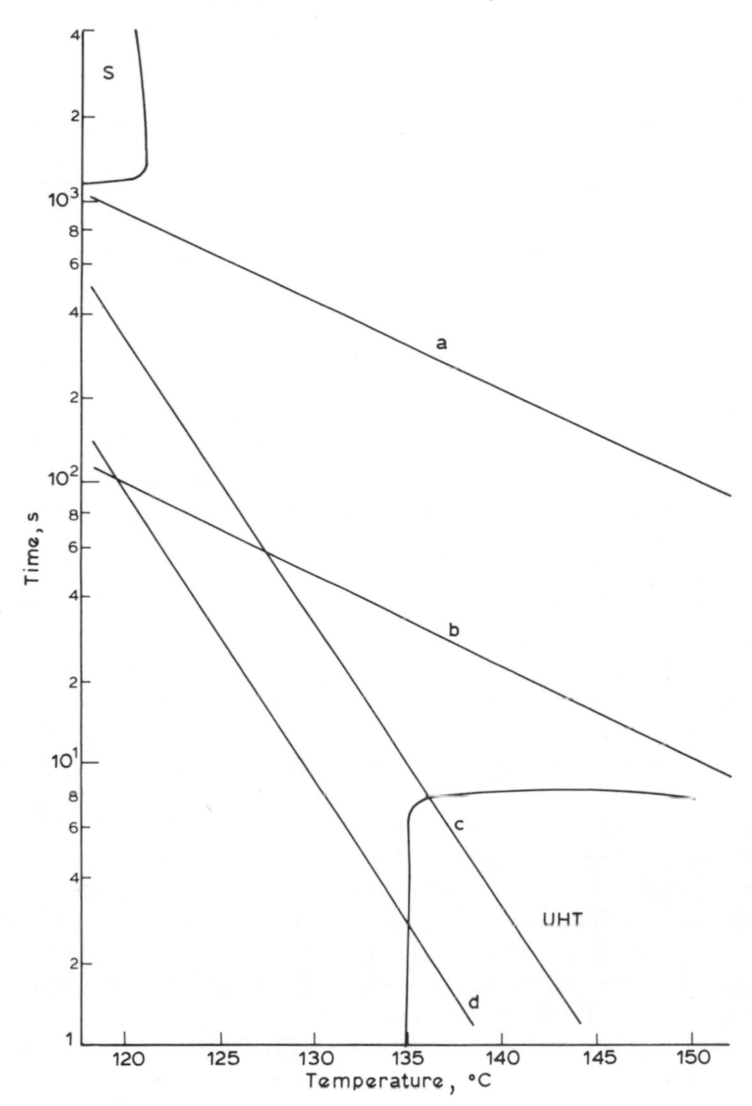

Fig. 4. Reaction kinetic data for the following reactions. (a) Inactivation of 90% of protease from *Pseudomonas* (Cerf, 1981). (b) Destruction of 3% thiamin (Kessler, 1981). (c) Inactivation of thermophilic spores (Kessler, 1981). (d) Inactivation of mesophilic spores (Kessler, 1981). (S) Sterilised milk region. (UHT) UHT milk region.

processing times required to achieve commercial sterility result in much less chemical reaction occurring, as well as allowing a continuous processing operation.

In comparison to sterilised milk, UHT milk is whiter, tastes less caramellised, and undergoes less protein denaturation and loss of heat sensitive vitamins. Consequently, the two products are different in overall quality, and specifically in sensory, chemical and nutritional terms; note that the quality will also change during subsequent storage. However, one drawback of UHT processing may be the incomplete inactivation of the more heat-resistant, proteolytic enzymes, particularly when initially present at high concentrations. More recently it has been realised that there are considerable differences between UHT milks produced not only by the methods of indirect heating and direct steam injection, but also from different equipment using the same methods.

Some countries now distinguish between sterilised milk and UHT milk, whilst in other countries, the term sterilised milk refers to milk produced by both methods.

Sterilised Milk (Heat Treatment Regulations, United Kingdom)

Initially, the regulations applied to milk which was heat-treated in the container after it had been sealed, but they have now been amended to incorporate milk which has been treated by continuous flow methods and then filled into sterile bottles; additional heat treatment after filling being optional.

The sterilised milk regulations (Statutory Instruments, 1977) are much less specific, in terms of times and temperatures, than those for pasteurisation. They state that the milk should be filtered or clarified, homogenised, filled into containers of less than 1 gallon (UK) capacity and maintained at a temperature above 100 °C for sufficient time to ensure that a negative result is obtained when the turbidity test is performed, and labelled sterilised milk.

Additional regulations for sterilised milk produced by continuous flow methods state that the milk should be filled into sterile bottles, in a clean environment, and immediately after processing. Temperature records should be kept for the continuous heat exchanger, and a flow diversion valve should be provided. Sterilised milk, thus produced, must also satisfy the colony count for UHT milk.

The turbidity test involves the addition of 4 g of ammonium sulphate to 20 ml of milk. At this concentration, this salt will precipitate the casein and denatured whey protein, but any undenatured protein will be left in solution. The mixture is then filtered, and the resulting, clear filtrate is heated. Any undenatured whey protein present in the filtrate flocculates and precipitates on heating, resulting in a positive turbidity. Therefore, sterilised milk would contain no denatured whey protein, whereas most UHT milk should give a positive turbidity reading.

Conventional, in-bottle sterilised milk is subjected to processing temperatures of 110–116 °C for 20–30 min, depending upon the degree of caramellisation required; batch or continuous retorts are used. Sterilised milk produced by continuous processing is normally given an additional retorting period after filling, but one which is substantially lower than for in-bottle sterilisation. This additional heat treatment is generally sufficient to fully denature the whey proteins, and hence satisfy the requirement for negative turbidity. Prior to heat treatment, the milk is preheated and filled into the glass bottles at 80 °C, or plastic bottles at 55 °C (Ashton and Romney, 1981).

It may appear that the turbidity test could be used to distinguish between UHT and sterilised milks. However, it is possible to find some UHT milks which will give negative turbidity, and this arises from the use of equipment with high regeneration efficiencies, and long heating and cooling periods.

In the United Kingdom, sterilised milk is a traditional product, and in 1982 still commanded 6·5 per cent of the liquid milk market (Table I). The brown colour and caramellised flavour are still popular, particularly with the elderly, and it sells well in industrial areas. Its production is described in more detail by Ashton and Romney (1981). The use of continuous retorts, e.g. hydrostatic cookers or rotary retorts, is favoured for larger processing units, i.e. greater than 10 000 units per day. When using plastic bottles, consideration has to be paid to the barrier properties of the material, the processing temperatures used, and pressure regulation during processing to ensure that the internal pressure within the bottle is not significantly different to that in the retort.

UHT Regulations

The UK heat treatment and labelling regulations for UHT milk state the following.

> The milk should be treated by the ultra-high temperature method, that is to say retained at a temperature of not less than 132·2 °C, for not less than 1 s.

Flow diversion facilities should be provided if the temperature falls below 132·2 °C; suitable indicating and recording thermometers should be provided and records preserved for three months. The milk should be packaged aseptically, securely fastened and labelled either ultra-heat treated milk or UHT milk. A sample of the milk should satisfy the UHT colony count, which is detailed in the regulations.

UHT milk can be produced using a variety of heat exchangers, in particular the plate and tubular types, where the heat transfer medium and the milk are separated. More recently, heating by mixing the milk with steam, either by injecting steam into the milk (injection) or by injecting milk into a steam chamber (infusion), has been introduced. Consequently, the UHT treatment is referred to as either indirect or direct.

The UHT regulations for direct methods require that there should be no dilution of the milk, and how this is achieved is described by Perkin and Burton (1970). The injected steam should be dry and saturated, free from foreign matter, and readily removed for sampling, and the boiler water should only be treated with certain listed, permitted compounds. The UHT regulations for semi-skimmed and skimmed milk are similar.

The minimum conditions for milk of 132·2 °C for 1 s corresponds to an F_0 value of 0·21 (assuming no contribution is made from the heating and cooling periods); this is well below the normal value of 3 which is recommended for low-acid products. Consequently milk, thus processed, will show a high level of microbial spoilage, and in practice, temperatures between 140 °C and 145 °C and times between 2 and 4 s are used. In addition, the heating and cooling periods will contribute to the total lethality of the process. The regulations do not require that direct and indirect UHT processes should be distinguished.

It is also interesting to note that both the terms ultra-heat treatment and ultra-high temperature are used. My preference is for the ultra-high temperature, because this gives a more accurate description of the process. Ultra-heat treated implies that the product has been very severely treated: this is not so, since relatively less chemical change takes place in UHT milk, in comparison to sterilised milk.

More recently, regulations for the UHT treatment of milk-based drinks and cream have been introduced. These stipulate minimum processing times and temperatures of 2 s and 140 °C respectively, these being considerably in excess of those for milk; this corresponds to an F_0 value of 2·59.

TABLE II
UHT regulations for milk in other countries

	Temp (°C) (minimum)	*Time (seconds) (minimum)*	*Other comments*
Germany	135–150	'short time'	D, I; should keep for 6 weeks at room temperature
Denmark	135	1	D, I; should give +ve turbidity; stable for 14 days at 30 °C and 7 days at 55 °C; should be stored at 5 °C
Finland	135	2–3	D, I; incubate at 30 °C for 14 days; 100 ml^{-1}; 90 day shelf-life
France	140	1	D, I; considerable labelling requirements; should be stable after 21 days at 31 °C and 10 days at 55 °C
Switzerland	130–150	'few seconds'	Storage up to 1 to 4 months (depending upon packaging conditions)

D: direct; I: indirect.
Compiled from Staal (1981).

Several countries now distinguish between UHT and sterilised milks, and these differences have been discussed by Staal (1981), and are summarised in Table II. Other countries where there are specific regulations for UHT products are Sweden, Australia and, more recently the USA, once aseptic packaging facilities were granted approval by USDA. In many countries, UHT products are subject to the same regulations as sterilised milk.

Despite the low popularity of UHT milk in the United Kingdom, it is much more acceptable in many of the other EEC countries; in France and Germany, UHT milk commands over 40 per cent of the liquid milk market. Table III gives information for the amounts of pasteurised, sterilised and UHT milk sold in the EEC countries, together with the milk consumption (kg/person/year).

TABLE III
Sales of liquid milk by heat treatment

	Pasteurised	*Sterilised*	*UHT*	*Total*	*UHT* (%)	*Total milk consumption* (*kg/person/ year*)
	(*million litres*)					
Germany	1832	43	1471	3346	43·9	70·4
France (1979)	1416	540	1454	3427	42·4	76·4
Italy (1974)	1288	1126	286	2726	10·5	93·3
Netherlands	777	— 185 —		962	(19·2)	94·0
Belgium	34	521	157	712	22·1	75·0
Luxembourg	19	2	4	25	16·0	84·3
UK	5795	413	46	6254	0·7	132·9
Denmark	551	— very small —		551	-	132·0
Irish Republic	465	-	-	465	-	196·9
			3418			

Source: EEC Dairy Facts and Figures (1981).

UHT Processing Plant

The short times necessary for UHT processing allow the use of continuous processes, and the equipment used for indirect UHT processing is not dissimilar to HTST pasteurisation plant (Fig. 1). Plate heat exchangers or tubular heat exchangers are used for low viscosity fluids, but for more viscous products or materials containing particulate matter, scraped-surface heat exchangers can be used. Milk is heated by regeneration, and finally by steam or pressurised hot water. Homogenisation occurs either between the heating or the cooling stages, and the sterile product is pumped to either an aseptic storage tank, or directly to an aseptic filling system.

The major differences between a pasteuriser and UHT plant are as follows:

(1) Higher operating pressures are required in order to prevent the milk boiling at the processing temperatures.

(2) An homogenisation step would be required for milk containing an appreciable amount of fat, to prevent separation and hardening of the fat during storage.
(3) The plant would need to be sterilised down-stream of the holding tube prior to processing, and maintained sterile throughout processing. The product would need storing in an aseptic tank, and packaging into cartons aseptically. Sterilisation is achieved by circulating hot water through the plant, and ensuring that all points down-stream from the holding tube reach a temperature of 130 °C for 30 min; aseptic packaging equipment is discussed later. The more recent trends in design have been to use high regeneration efficiencies, and a low temperature differential in the final stage of heating.

Direct Steam Injection

One of the most interesting developments is that of direct steam injection. The mixing of saturated steam and milk results in almost instantaneous heating to the final UHT temperature, which is regulated by a combination of steam pressure and the pressure in the holding tube. The milk is normally heated (by regeneration) in a plate or tubular heat exchanger to between 70 and 80 °C, prior to steam injection (see Fig. 5). After injection, there is a between 10 and 15 per cent dilution

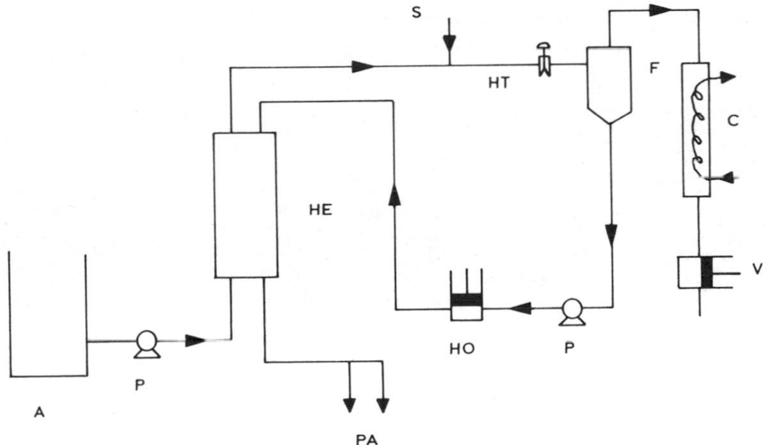

Fig. 5. Schematic layout for a direct UHT plant. (A: feed tank; C: condenser; F: separator; HE: heat exchanger; HO: homogeniser; HT: holding tube; P: pump; PA: aseptic packaging; S: steam injection; V: vacuum pump).

of the milk, and the resulting solids content will fall; this water is subsequently removed in the flash-cooling stage. The milk then passes through the holding tube, where it is held for the desired holding time; similar comments apply to residence time distributions as were made earlier.

After the holding tube, the milk passes into a vessel held under vacuum conditions. The milk is well above its boiling point at these conditions, and consequently there is a rapid fall in temperature, and the resulting loss of energy causes some of the water to vaporise. This process is known as flash-cooling. The two most important effects are virtually instantaneous cooling, and the removal of water vapour and other volatile components. The instantaneous heating and cooling of the milk is very important, because it means that the milk suffers very little chemical thermal damage, since it is at a high temperature for only a very short time. The microbiological damage will depend upon the UHT temperature and holding time. One disadvantage of the flash-cooling is that energy is lost from the system and cannot be recovered by regeneration. Therefore, the energy costs of direct systems tend to be higher than for indirect systems.

The pressure (vacuum) in the chamber will control the boiling temperature, the degree of cooling and the amount of water removed. As the pressure is reduced, the temperature falls, and the amount of water evaporated increases. The pressure should be adjusted so that the amount of water removed is the same as the amount of steam condensed during the heating process. The heat balance and a detailed analysis of the process is given by Perkin and Burton (1970). A mathematical solution relies on a knowledge of the heat loss from the plant, but unfortunately this is not easy to evaluate and will vary from one plant to another. Therefore, direct UHT plants require calibrating, and the temperature conditions which give the correct water removal are determined by processing the fluid, and then altering the vacuum conditions until no dilution is observed. This desired point can be checked by recirculating the product, and noting whether there is any accumulation or loss as it goes through the steam addition and flash-cooling process. As a first approximation, the temperature of the milk in the vacuum chamber should be approximately 2 °C higher than the temperature of the milk prior to steam injection. Once these conditions have been determined, the temperatures are recorded during the processing run together with the temperature in the holding tube. This procedure is necessary to avoid dilution of the milk products. With other formulated products, e.g. milk-type drinks, reconstituted filled milks, creams and ice cream mixes, there may not necessarily be the same requirement to remove the exact quantity of water, and conditions may not be so strict.

A process could also be envisaged in which the formulated mix is made to a higher solids content than is required, and brought down to its intended solids content by steam injection; in this case, flash-cooling could be replaced by conventional cooling. A whole range of possible combinations are possible, and as the amount of flash-cooling is reduced, the regeneration efficiency would increase.

A further advantage of flash-cooling is the loss of volatile components in the milk. Oxygen and low molecular weight volatile components are removed along with the water vapour; of particular importance are the low molecular weight sulphur components which arise as breakdown products of the sulphur-containing amino acids. Their presence in milk gives rise to a typical 'cabbage' flavour, and a pungent odour; it is this group that makes UHT-indirect milk unacceptable at this stage. However, it is hard to distinguish milk produced by direct steam injection from pasteurised milk, and the low level of oxygen is beneficial in retarding oxidation reactions during storage; these will be discussed in further detail in a later section.

After flash-cooling, the milk is homogenised. Homogenisation is preferred after steam injection as it results in a more stable product, but the homogeniser needs to be used under aseptic conditions; normally this is accomplished by incorporating a steam chest through which the piston rods pass (sterile block).

COMPARISON OF DIRECT AND INDIRECT PROCESSES

There have been a whole series of papers devoted to comparing direct and indirect heating processes. Most of the differences arise because of the different time–temperature profiles the milk is subject to and, to some extent, to the inclusion or not of the flash-cooling process.

The capital cost of a direct steam injection plant is usually higher than that of an indirect plant, mainly because the direct plant is slightly more complex. The energy costs of direct steam injection plants are also higher, because it is not possible to obtain such high regeneration efficiencies; the energy lost in the flash-cooling process is not easily recoverable. Virtually all results indicate that direct steam injection is a milder form of heat treatment as far as chemical reactions are concerned, that it is more suitable for milks of poor microbiological quality, and that the plant is less susceptible to fouling. It is claimed that it is possible to process milk for up to 20 h on a direct plant, compared to a maximum of 4–5 h on an indirect plant. Perkin *et al.* (1973) showed that direct heat treated milk produced twice as much sediment in the package after storing for 100 days. Both types of process reduced the rates of clotting with pepsin and rennin, but the effect of indirect processing was more severe. Lyster *et al.* (1971) used lactoglobulin denaturation to monitor the severity of the heat treatment, and while direct steam injection resulted in between 65·1 and 71·1 per cent denaturation, indirect heating resulted in between 78·8 and 83·5% denaturation. The temperature profiles for the two processes were used, in conjunction with reaction kinetic data for denaturation of β-lactoglobulin (Lyster, 1970), to calculate these levels. The calculated values were always lower than those found experimentally, and agreement between the results was only fair. Jelen (1982) presented some further results, and some of these are given in Table IV.

However, difficulties arise when comparing results when only holding times and temperatures are recorded.

Kessler (1981) has adopted an alternative procedure for assessment of UHT plants. He recognised that all UHT plants would give different time–temperature profiles as the product passed through, and that in some cases, considerably more chemical damage would result from excessive heating and cooling periods, than for the short time in the holding tube. Some typical time–temperature profiles are shown in Fig. 6.

He introduced a microbiological parameter (B^*) and a chemical parameter (C^*); a temperature of 135 °C was adopted as the reference temperature. The microbiological parameter was based on the fact that

TABLE IV
Concentration of components after direct and indirect processing

		Direct	*Indirect*
Whey protein nitrogen (mg/100 g)		38·8	27·6
Loss of available lysine (per cent)		3·8	5·7
Vitamin loss %	B_{12}	16·8	30·1
	Folic acid	19·6	35·2
	Vitamin C	17·7	31·6
Hydroxymethyl furfural (μmol litre^{-1})		5·3	10·0

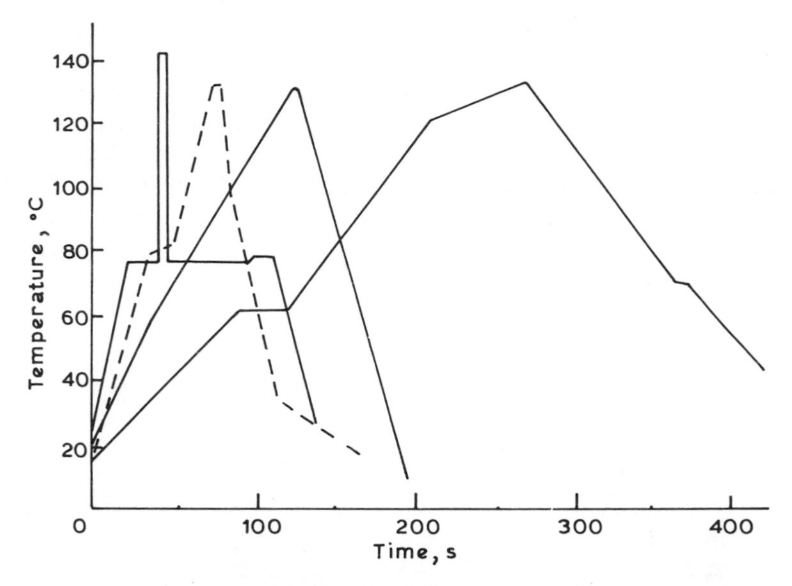

Fig. 6. Temperature–time profiles for different UHT plants. (Adapted from data in Reuter (1984).)

commercial sterility could be achieved at 135 °C for 10·1 s, the corresponding Z value being 10·5 °C. Such a process was given a B^* value of 1·0. If the temperature–time profile is known for the product passing through the plant, B^* is evaluated in a fashion similar to F_0:

$$B^* = \int \frac{10^{(135-\theta)/10 \cdot 5}}{10 \cdot 1} \, \mathrm{d}t$$

θ = temperature
t = time (s).

Thus, a process with a B^* of 4·0 would be equivalent to 135 °C for 40·4 s.

The chemical parameter (C^*) was based on the conditions required for a 3 per cent destruction of thiamin. This was chosen as it represented a typical chemical reaction in milk, and corresponded to a change which would only just be detectable. The conditions required to bring this about were 135 °C for 30·5 s; the Z value being 31·4 °C.

$$C^* = \int \frac{10^{(135-\theta)/31 \cdot 4}}{30 \cdot 5} \, \mathrm{d}t$$

Conventional thermal processes use 121·1 °C as a reference temperature, but on this occasion 135 °C has been selected because the extent of extrapolation required for typical UHT processing conditions is greatly reduced. It is highly desirable that B^* should be greater than one and C^* should be as low as possible.

Kessler (1981) gives examples of B^* and C^* calculations for typical direct and indirect UHT processes. The following values were recorded.

Indirect UHT — $B^* = 1{\cdot}25$ $C^* = 0{\cdot}49$
Direct UHT — $B^* = 2{\cdot}18$ $C^* = 0{\cdot}30$

This approach is extremely useful as it allows a direct comparison of all UHT processes. Reuter (1984) summarises the situation, to date, by describing three types of UHT plant:

First generation
Direct steam injection with rapid heating and cooling, and low overall chemical damage.

Second generation
Older, indirect processing plant with low regeneration efficiency (70–80 per cent), and shorter, non-stop operating times (6–10 h).

Third generation
Indirect processing with high regeneration efficiency (85–90 per cent), broad heating profiles and longer, non-stop operating times (14–18 h). However, compared to second generation plants, the milk is subject to a much higher degree of chemical damage.

Since the regulations in some countries differentiate between sterilised and UHT milks, there have been many attempts to distinguish between these by chemical methods. These are reviewed by Burton (1983), and some results are summarised in Table V. However, the range of values is considerable, and results for these different types of milk overlap.

Other changes involving hydroxymethyl furfural (HMF), lysinoalanine, furosine and pyridosine were also examined. Andrews (1984) found that lactulose levels could be used to distinguish between direct and indirect UHT processes, but there was an overlap between indirect UHT milk and sterilised milk. It was observed that lactulose increased during storage in sterilised milk, but not in UHT milk. It is now obvious that some UHT milk is being more severely heat treated than some in-bottle sterilised milk.

TABLE V
Chemical changes in heat treated milk

	Whey protein nitrogen (mg/100 g mean)	*Turbidity*	*Lysine loss (%)*	*Lactulose (mg/ml)*
Raw	95·5			
Pasteurised	80·8	771	0·7–2·0	0·1
UHT direct	38·8	181	0–4·3	0·3
UHT indirect	27·6	14·2	1·7–6·5	
Sterilised (glass)	21·9	0·8	3·3–13	2·9

'COOKED MILK' FLAVOURS

It is widely accepted that consumer acceptance of, and/or preference for, a certain type of milk is influenced more by its flavour than by any other attribute. Flavour is a property detected by the senses, in particular taste and smell, and thereby requires taste panel work for its evaluation. Flavour changes in milk arise because of changes in its chemical constituents, and the various types of flavour defect in milk have been reviewed by Badings (1984) and Shipe *et al.* (1978). Therefore, researchers studying flavours use a combination of taste panel work and chemical analysis; the sensory aspects have been described in more detail by Piggott (1984).

The flavours of particular interest in heated milk are the 'cooked' flavour, and the bitter and oxidised flavours which develop during long-term storage of sterilised and UHT milk. The picture is further complicated because the 'cooked' flavour, which is developed on heating, changes rapidly during the early days of storage. The vocabulary used for description is also not straightforward, and terms, such as 'cooked', 'cabbagey', 'sulphury' and 'caramellised', are all frequently used. It is this 'cooked' flavour which makes it unpopular with the consumer, and is probably the principle reason why sales of UHT milk remain low (see Table I). The 'cooked' flavour is easily detectable both in liquid milk and in drinks, such as tea and coffee, where only small quantities of milk are used. One of the best accounts of the flavour changes in milk on heating and during storage is given by Ashton (1965), and he recognised two phases, each with a number of distinct stages. These are summarised as follows.

Primary Phase

(a) Initial heating flavour accompanied by a strong sulphydryl or 'cabbagey' smell.
(b) Weaker sulphydryl or cabbage odour with residual 'cooked' flavour.
(c) Residual 'cooked' flavour with normal, acceptable, agreeable flavour.

Secondary Phase

(d) Normal, acceptable to agreeable, flat, acceptable flavour.
(e) Flat, acceptable to mild, oxidised flavour.
(f) Incipient, oxidised flavour (or rancidity) to pronounced rancidity.

UHT milk is best consumed whilst in stages (c) and (d).

There have been a number of examinations of the 'cooked' flavour problem, such as:

(i) Measuring the amount of sulphur components in the milk; this portion includes active sulphydryls, total -SH and disulphide (S-S) groups; total (-SH + S-S) expressed as -SH groups, sulphur-containing amino acids, and the low molecular weight volatile sulphur components, e.g. H_2S and CH_3SH.
(ii) Examining the denaturation levels of the proteins in milk, in particular β-lactoglobulin and α-lactalbumin, as these are usually implicated as the major source of sulphur-containing components.
(iii) Adding substances to milk, prior to heat treatment, which will reduce the intensity of the 'cooked' flavour.

Some of these approaches will now be discussed in more detail.

Josephson and Doan, as early as 1939, made some observations on the source and significance of the 'cooked' flavour in milk heated above temperatures of 170 °F, and they concluded that sulphydryl groups were wholly responsible for the 'cooked' flavour. These sulphydryl groups also reduced the oxidation–reduction potential of heated milk and acted as antioxidants, preventing both the oxidation of ascorbic acid and the development of tallowy and oxidised flavours. They suggested that the lactalbumin of milk, and some of the proteins associated with the fat globule membrane, were the most likely sources of the sulphydryl groups.

The role of sulphydryl groups was recognised at an early stage, although later work has drawn attention to active and total sulphydryl groups, disulphide groups and low molecular weight, sulphur components.

The presence of high concentrations of oxygen in the packaged milk will accelerate all the relevant reactions, i.e. the disagreeable 'cooked' flavour will disappear more quickly, but the 'cardboardy', oxidised note will appear more quickly.

Milk produced by the indirect process has been found to have a very intense initial sulphydryl or 'cabbagey' smell, due, most likely, to a high concentration of hydrogen sulphide, which disappears rapidly within a few days. However, unless this milk is specially de-aerated, the oxygen content immediately after packaging is very high, and the oxidation reactions proceed quickly.

Milk produced by direct steam injection has a much reduced 'cabbagey' smell, resulting from the flash-cooling process which removes most of the volatile sulphur components and oxygen. This milk (I would suggest) is already at stage (c) in Ashton's list, but if the oxygen content is still reasonably high, it may proceed to an oxidised stage reasonably quickly. Nevertheless, many investigators have reported that it is quite difficult to distinguish direct UHT milk from pasteurised milk.

This reaction scheme is accelerated by increasing the storage temperature, and Thomas *et al.* (1975) examined the effects of high, medium and low oxygen levels on UHT milk during storage at room temperature for 150 days. High initial levels of oxygen led to rapid depletion of -SH groups, and thereafter rapid depletion of ascorbic acid and folic acid; losses of these vitamins were much reduced at lower initial oxygen levels. Milks with the higher initial oxygen content were preferred for up to 13 days, but thereafter, acceptability was independent of initial oxygen content. They concluded that the beneficial effect of high initial oxygen on flavour appears to be so slight, and confined to a such short period in the early life of the milk (probably before it even reaches the consumer), as to be completely outweighed by the adverse nutritional effects.

Blanc and Odet (1981) found similar results at refrigerated storage (5 °C) and ambient storage (25 °C). At 5 °C, the milk loses its sulphur flavour three weeks after production, is at its optimum period between 4½ and 7 weeks, and starts to become stale after 8 weeks. At 25 °C, the sulphur flavour is lost after 2 weeks, the milk is at its optimum between

3 and 5 weeks, and the stale flavour becomes evident after 6 weeks.

Many other reactions which will affect the acceptability of the milk, such as browning, caramellisation and the onset of gelation, are also accelerated by increased storage temperatures.

Sulphydryl and Disulphide Group Determination

The occurrence of 'cooked' flavours in milk has been related to the changes in sulphydryls and disulphides in milk and, in particular, to the presence of free sulphydryl groups. Sulphydryl groups in the whey proteins, particularly β-lactoglobulin, become exposed as the molecules unfold. These may themselves contribute directly to the 'cooked' flavour, or further react to form low molecular weight compounds, such as hydrogen sulphide and dimethyl disulphide.

Total sulphydryl groups, free or active sulphydryl groups and disulphide groups can be determined by the same procedure.

(1) For active sulphydryl groups, milk is used directly.
(2) For total sulphydryl groups, the milk is reacted with a protein unfolding, denaturing agent, such as urea (Lyster, 1964).

Disulphide groups are determined by reducing the disulphide group, using sodium borohydride, and measuring total -SH. Patrick and Swaisgood (1976) expressed this as 'half cystine'; this is a useful approach, because it gives a measure of total sulphur. However, several chemical methods are available, and while the most popular is based on Ellman's reagent (5,5′-dithiobis 2-nitrobenzoic acid), others based on a spectrofluorometric determination (Pofahl and Vakaleris, 1968) and a silver nitrate titration (Dill *et al.,* 1962) have been described.

Lyster (1964) distinguished between active and masked sulphydryls, and measured both in skim-milk heated at temperatures between 40 °C and 100 °C for 15 min. Active sulphydryl groups remained low until a temperature of 70 °C was achieved, after which they increased sharply to a maximum at 85 °C, followed by a gradual fall. Total sulphydryl groups fell throughout the temperature range, the fall being greatest from 85 to 100 °C. No information was given on disulphide activity. Milk from a UHT plant (no conditions given) had similar concentrations of free and total -SH groups.

Pofahl and Vakaleris (1968) suggested that whey proteins in their native form contain a significant number of sulphydryl groups, and that this is increased on heating, probably at the expense of disulphide groups. Overall, the total amount of disulphide and sulphydryl decreased, probably due to volatilisation, and no distinction was made between active and total sulphydryl groups. Casein was found to contain hardly any sulphydryl groups, and only very small amounts were generated on heat treatment.

Koka *et al.* (1968) determined that the production of activated sulphydryl groups followed first order reaction kinetics, with reaction rate constants (min^{-1}) of 0·078 (75 °C), 0·142 (80 °C), 0·384 (85 °C) and 0·805 (90 °C) over the initial heating period, when skim-milk was heated for 15 min at different temperatures. The activated -SH groups showed a similar pattern to that observed by Lyster, with the maximum level of free -SH being observed at 100 °C, followed by a small decrease at higher temperatures.

Dill *et al.* (1962) found that, in direct steam injection, heat-activated sulphydryl groups increased as temperature increased over the temperature range 190–300 °F for a time of 2 s. However, at holding times of 20 s and 150 s, the heat-activated sulphydryl groups went through a maximum, followed by a decrease, which was associated with volatilisation of sulphur compounds.

Patrick and Swaisgood (1976) performed work on milks heated by the UHT direct steam injection method, and they measured free sulphydryl, total sulphydryl and half cystine (total sulphydryl + reduced disulphide). For direct steam injection, the concentration of reactive sulphydryl groups, after heat treatment, approached or occasionally exceeded the concentration of total sulphydryl groups in the raw milk. Concomitant with the production of reactive sulphydryl groups was a decrease in measurable half cystine, which was about 6 per cent for milk heat treated at 150 °C for 1·5 s, compared to 17 per cent for milk heated at 100 °C for 30 min. These results suggest that both scission of disulphide bonds and volatilisation of sulphydryl groups could be occurring during heating. A similar pattern of events was observed for milk heated at 90 °C for up to 60 min. Blanc *et al.* (1977), by comparing results with Lyster's (1964), also concluded that direct steam injection caused less protein denaturation than indirect processing. The loss of sulphydryl groups during storage was accelerated by increasing the temperature — a finding in agreement with Lyster (1964) — but at 4 °C, the number of buried sulphydryl groups increased during storage, suggesting that the proteins may fold up during refrigerated storage.

VOLATILE SULPHUR COMPONENTS IN MILK

Dill *et al.* (1962) showed the existence of hydrogen sulphide in heated milk by flushing it out through lead acetate, precipitating lead sulphide and measuring it qualitatively. Thomas *et al.* (1975) described a rapid method for determining small amounts of H_2S in the head-space of milk products. Although H_2S concentration showed a good relationship to 'cooked' flavour intensity, it was not established that it was responsible.

Jeon (1976) showed the existence of dimethylsulphide and isopropyl sulphide in UHT milk, and observed that these disappeared less rapidly in UHT milk to which ascorbic acid had been added.

Jaddou *et al.* (1978) isolated flavour volatiles in UHT milk, by a low temperature distillation technique, and identified them using gas chromatography and mass spectrometry. 'Cabbagey' defects in heated milks correlated very well with total volatile sulphur, the main components implicated being hydrogen sulphide, carbonyl sulphide (COS) methanethiol (CH_3SH), carbon disulphide (CS_2) and dimethyl sulphide $(CH_3)_2S$. Burki and Blanc (1978) suggested that volatile sulphur components were more responsible than reactive -SH groups for the early 'cabbagey' flavour of UHT milk.

Jeon *et al.* (1978) analysed volatiles produced in UHT milk heated at 145 °C for 3 s, aseptically packed and stored for up to 150 days. Twenty-six compounds were identified, most of which were carbonyl compounds, and it was believed that aldehydes were the most important in contributing to off-flavours in UHT milk; no mention was made of sulphur-containing components.

WHEY PROTEIN DENATURATION

The formation of activated sulphydryls has been associated with the denaturation of whey proteins, in particular β-lactoglobulin. This fraction is implicated because each dimer (36 000 Daltons) contains two -SH groups and four -S-S groups, and in addition, it accounts for over 50 per cent of the whey proteins. However, α-lactalbumin (0, 4) and blood serum albumin (1, 17) also contain sulphydryl and disulphide groups (SH, S-S), but caseins appear to make little contribution.

β-lactoglobulin denaturation was found to follow second order reaction kinetics (Lyster, 1970). Denaturation levels were measured by immunodiffusion, and the variation in reaction velocity constant (k)

with temperature was found to be as follows:

$$\log k = 37{\cdot}95 - 14{\cdot}51\left(\frac{10^3}{T}\right) \quad \text{(Temperature} = 68\text{–}90\,°\text{C)}$$

$$\log k = 5{\cdot}98 - 2{\cdot}86\left(\frac{10^3}{T}\right) \quad \text{(Temperature} = 90\text{–}135\,°\text{C)}$$

T = temperature (K)
k = reaction velocity constant ($\text{l g}^{-1}\,\text{s}^{-1}$).

Hillier and Lyster (1979) using quantitative polyacrylamide gel electrophoresis evaluated whey protein denaturation in heated skim-milk, and were able to evaluate the reaction velocity constants for β-lactoglobulin variants A and B.

$$\text{lactoglobulin A}: \log k_A = 4{\cdot}25 - 1{\cdot}91\left(\frac{10^3}{T}\right): \text{(Temp. range 100–150 °C)}$$

$$\text{lactoglobulin B}: \log k_B = 3{\cdot}48 - 1{\cdot}67\left(\frac{10^3}{T}\right): \text{(Temp. range 95–150 °C)}$$

At temperatures below 95 °C, β-lactoglobulin (*B*) was more stable to heat, whereas above 95 °C it was less stable, but differences between the two variants were small. The irreversible step in this process was believed to involve a second order reaction between an S-S bond and an -SH group to form a new disulphide link.

The second order reaction equation is

$$kt = \frac{1}{C} - \frac{1}{C_0}$$

where C = final concentration.
C_0 = initial concentration.

Similar data were also presented for the denaturation of α-lactalbumin and blood serum albumin.

Blanc *et al.* (1977) used several different methods for assessing β-lactoglobulin denaturation in pasteurised and UHT treated milks. These include electrophoresis, immunoelectrophoresis, immunodiffusion, rocket electrophoresis gel filtration, differential scanning calorimetry and analysis of -SH and -S-S groups. There was reasonable agreement between the two quantitative methods employed (see Table VI).

TABLE VI
Concentration of β-lactoglobulin in heated milk

	Electrophoresis	*Immunodiffusion*
	(mg/ml)	
Untreated	3·0	3·0
Pasteurised (72 °C)	3·0	2·9
Pasteurised (92 °C)	2·1	1·4
UHT direct	1·8	1·25
UHT indirect	0·75	0·96

Differential scanning calorimetry (DSC) was used to measure the denaturation temperature and heat of denaturation for β-lactoglobulin in salt solutions of different composition, and the effects of the presence of protein on denaturation reactions. From the results, it was concluded that the presence of α-lactalbumin and κ-casein would exert a protective effect on β-lactoglobulin denaturation, and that the denaturation temperature of β-lactoglobulin was dependent upon the ionic environment. Considerable differences were observed in the DSC thermograms for the different milks.

Melo and Hansen (1978) found evidence for a complex between α-lactalbumin and β-lactoglobulin in model systems subject to UHT direct heating (143 °C for 8–10 s). They suggested that aggregation was due to the formation of disulphide bonds; evidence for the formation of this complex at lower temperatures (70–90 °C) has been shown by Elfagm and Wheelock (1977).

FLAVOUR IMPROVEMENT

There have been several attempts to improve the flavour of UHT milks. Potassium iodate (10–20 ppm) has been used to reduce the amount of 'cooked' flavour in milk, by causing the oxidation of any exposed sulphydryl groups (Skudder *et al.,* 1981), and it was also found that the presence of potassium iodate reduced denaturation of α-lactalbumin, probably by not allowing sulphydryl disulphide exchange reactions to occur. The presence of free -SH groups encourages the formation of protein aggregates on heated surfaces, and the addition of L-cysteine, prior to UHT processing, resulted in a massive increase in the amount of deposit, as well as a most unacceptable 'cabbagey' or 'sulphury' flavour. Unfortunately, the addition of iodate (at these levels) results in the formation of bitter components about 14 days after processing, due possibly to increased instability of proteases within the milk, or suppression of natural protease inhibitors.

Badings (1977) reported that the 'cooked' flavour could be reduced by adding L-cystine to milk prior to heating, both in direct and indirect processes. For milk treated by the indirect process, the amount of hydrogen sulphide produced immediately after processing was 82·5 μg/kg, which was approximately eight times greater than that produced by direct steam injection. The addition of 30 and 70 mg cystine/kg milk reduced the hydrogen sulphide level to 9·5 and 1·7 g ml^{-1} respectively for indirect milk. Hydrogen sulphide disappeared very quickly during the first 24 h of storage, even at 3 °C, with changes being quicker in the indirect milk, presumably due to its higher oxygen content. They also concluded that addition of L-cystine produced no inclination toward oxidation or other flavour defects; it was later proposed that the hydrogen sulphide was removed by L-cystine with L-cysteine as the reaction product.

Swaisgood (1977) patented a process for removing the 'cooked' flavour from milk (30–35 °C) by immobilised sulphydryl oxidase attached to glass beads, and he suggested placing the reactor down-stream of the UHT holding tube. At present, the flow rate is limited to 40 ml h^{-1}, and there may also be problems of sterilising the reactor contents prior to processing.

Another patent application claimed to improve flavour by preheating milk to between 70 and 90 °C, centrifuging it at high speed to remove many of the micro-organisms and spores, and then holding the milk at 35–40 °C for a period of 10–20 min. Thereafter, the milk was UHT treated at 100–140 °C for short periods of time up to 10 s. It is claimed to improve the keeping quality by removing the vast proportion of bacteria by centrifugation prior to heat treatment, so preventing them deteriorating in the milk during storage.

Most attention has been paid here to the 'cooked' flavour, but Mehta (1980) has reviewed the factors affecting the onset of the stale or oxidised flavour, which appears after the 'cooked' flavour has disappeared, together with the volatile components responsible for it. Methyl ketones were the largest class of compounds isolated, although aldehydes were thought to make the most significant contribution to this off-flavour (Blanc and Odet, 1981).

NUTRITIONAL VALUE

Most of the loss in nutritional value is concerned with the vitamin and, to a lesser extent, mineral fractions. Although whey proteins are considerably denatured, there is very little loss in biological value or available lysine; Burton (1982) lists the biological values of raw milk, UHT milk and in-bottle sterilised milk as 0·90, 0·91 and 0·84 respectively.

Losses of vitamins during sterilisation processes have been reviewed by Mehta (1980) and Burton (1983), and they may occur as a result of processing, or during storage. On the whole, they are reasonably stable to sterilisation processes, although losses are greater during in-bottle sterilisation than in UHT production.

The fat-soluble vitamins (A, D and E) and the water-soluble vitamins of the B group (pantothenic acid, nicotinic acid and biotin) are hardly affected by UHT processes; riboflavin is stable to heat, but not to light. Up to 20 per cent of thiamin and 30 per cent of vitamin B_{12} is lost in UHT processing, whereas losses during in-bottle sterilisation are up to 50 per cent for thiamin and 100 per cent for B_{12}. Losses of vitamin C occur during thermal processing, but are small in comparison to losses which can occur during storage; some values for vitamin C contents of heated milk are presented in Table VII (Scott *et al.* 1984 a and b).

Losses of vitamin C, folic acid and vitamin B_{12} during storage are inter-related; folic acid starts to be lost once all the ascorbic acid is oxidised, and vitamin B_{12} losses are also accelerated at high oxygen levels.

COLOUR OF MILK

Blanc and Odet (1981) reviewed the effects of heat on the colour of milk. Changes in casein size and the denaturation of whey protein both increase the amount of light scattering (reflectance), making milk appear whiter. However, this improvement is balanced by browning,

TABLE VII
Vitamin C content of heat-treated milk

	Vitamin C (μg/ml)
Fresh milk	23·3
72°C/16 s	19·4
80°C/16 s	18·3
110°C/3·5 s	17·3
140°C/3·5 s	16·4

which lowers the degree of reflectance and gives a striking increase in the green and yellow components on the Hunter or Munsell solids colour system (Francis, 1975; Clydesdale, 1975).

Homogenisation is an additional, complicating factor, as most UHT milk is homogenised at some stage. In the absence of heat, homogenisation of skim-milk makes the milk very slightly lighter, and gives a slight decrease in the green component suggesting, perhaps, a slight modification to the size of the casein micelles. Whole milk shows a marked increase in lightness and a very striking decrease in the yellow component, attributed to a considerable increase in the number of fat globules, and the dispersal of their coloured, fat-soluble vitamins. The resulting effect is that UHT milk is usually whiter than raw milk, whereas sterilised milk is slightly browner. Non-enzymatic browning is linked to parameters, such as pH, extent of heat treatment, storage temperature, as well as milk composition. In general, direct processes cause less browning than indirect heat; in UHT milk, significant browning occurs during storage, particularly at higher temperatures.

TEXTURE OF UHT MILK

The texture of milk is related to mouth-feel. Blanc and Odet (1981) summarised the factors affecting texture, and the methods of assessing some of these factors. The main defects are the separation of fat, and the formation of a sediment which contains protein, fat, lactose and minerals in varying proportions. Fat may separate due to inefficient homogenisation, and when it does, it rises and eventually hardens. Aggregation increases as the severity of the heat treatment increases, and may be reduced, to some extent, by homogenisation after heat treatment. The concensus of opinion is that more sediment is produced in milk from the direct heating process, but the use of chemicals which provide additional anions, such as citrate, bicarbonate, hydrogen phosphate, can be beneficial. The total amount of sediment increases during storage, but the protein content decreases over the first five weeks and then rises again.

AGE THICKENING — GELATION

UHT milk shows a greater tendency to thicken and coagulate during storage than sterilised milk; this effect being related to the severity of the heat treatment. The gel structure is believed to be caused by casein micelles, and these will also trap fat globules and whey proteins in a three-dimensional network. The UHT treatment of milk leads to a much larger proportion of small-sized casein micelles compared to raw or pasteurised milk. When viewed under the electron microscope, the surface of the micelles appears rough, which results in a viscosity increase of about 0·1–0·2 mPa s (cp). Casein micelle size also changes during storage, more so at conditions of fluctuating temperature, and particularly prior to gelation.

The physico-chemical and biochemical processes involved in age-gelation are summarised in Table VIII.

Adams *et al.* (1976) concluded that the growth of gram-negative psychotrophs in raw milk led to detectable proteolysis, particularly of κ- and β-casein, even after two days. This breakdown had, subsequently, a deleterious effect on the proteins during UHT treatment, with decreased levels of the κ-, β- and α-caseins and β-lactoglobulin.

It was further claimed that coagulation during, or shortly after, heating increased with severity of heat treatment and size of psychrotrophic population. These findings were confirmed by Law *et al.* (1977) who found that the level of psychotropic bacteria determined whether UHT milk would gel during subsequent storage. Samples containing greater than 8×10^6 colony forming units of *Pseudomonas fluorescens* AR11 gelled about 10–12 days after production when stored at 20 °C; below this count no gelation was observed after 20 weeks storage. The protease from this organism caused extensive protein breakdown of

TABLE VIII
Processes involved in age-gelation

Physico-chemical

- Dissociation of the casein/whey protein complexes
- Cross linking due to Maillard reaction
- Removal or binding of calcium ions
- Conformational changes of casein molecules: breakdown of micelle structure; interaction of β-lactoglobulin and κ-casein; S-S exchange reactions; pH change; dephosphorylation of casein, and interaction of casein and carbohydrate

Biochemical

- Heat resistance and reactivation of natural and bacterial proteinases
- Survival of bacterial spores

κ-casein in a way similar to rennet action; β-casein was also broken down rapidly, while α_{s1}-casein was degraded only slowly.

Chilled storage of raw milk may lead to high counts of psychrotrophic bacteria, and some of these are spore forming (Van den Berg, 1981). Several species also produce significant amounts of highly heat-resistant enzymes, particularly proteases and lipases. In pasteurised milk, the effects of these enzymes is not usually noticeable, but if the count is above 10^6 ml^{-1}, significant amounts of proteases may survive UHT treatment and cause problems during storage. The clotting properties of UHT and sterilised milks also appear to be altered. Thus, while clotting times are increased and the coagulum is not so firm in hard cheese manufacture, heat treatment has been reported to increase activity of the starter, reduce clotting times, and increase yield in cottage cheese manufacture using skim-milk. Chemical and physical changes during processing and storage are accentuated in concentrated milks, and Muir (1984) reviews some practical methods for the production of UHT sterilised milk concentrates.

HEAT EXCHANGER FOULING

Milk starts to form deposits at 80 °C, and the extent of the deposition can be monitored by weighing, or indirectly by monitoring the pressure drop. In practice, there is a time period over which there is no significant loss of pressure, the length of which can vary significantly. The pressure drop then increases in a parabolic fashion up to a value where the limiting pressure drop is achieved; the plant is then closed down and cleaned.

Deposit formation is also dependent upon the temperature of the milk, and the temperature for maximum deposit formation is 110 °C (Skudder, 1984). The deposit that forms between 80 °C and 105 °C is a white voluminous precipitate, which is predominantly proteinaceous, and tends to block the flow passages. Above 110 °C, the deposit is finer, more granular, and predominantly mineral in origin, and because the build-up is in the final section, it tends to reduce the efficiency of heat transfer; consequently temperatures begin to fall. The use of pressurised hot water sets tends to induce less fouling than steam heating, and high velocities also help to reduce deposits; this latter approach is limited by holding time and pressure drop considerations.

In addition, fore-warming of the milk at 80–85 °C for between 5 and

10 min will reduce both the amount and nature of the deposit, which is then predominantly a mineral one. Such fore-warming may be ideal prior to evaporation for the production of high heat powders, or prior to UHT processing followed by an in-container sterilisation process. It is also important to stop the air coming out of the milk, as the presence of bubbles increases deposit formation. The application of a reasonable back pressure, approximately one atmosphere (14·7 psi) above that required to prevent boiling, should keep the air in solution.

The quality of the milk plays an important role in the formation of deposits. Ageing the milk at 4°C for 10–24 h improves the heat stability, perhaps due to lipolysis, because the presence of carreic acid, and later stearic acid, inhibits the fouling process. Milks with high contents of β-casein are more prone to deposit formation, but the addition of pyrophosphates to milk brings about a significant improvement.

Alcohol can be used to quickly assess the heat stability of raw milk prior to processing. Equal volumes of milk and alcohol solutions are mixed together, and if the milk flocculates in less than 74 per cent alcohol, it is likely to cause fouling problems during processing. Zadow (1971) reported that milk started to become unstable to UHT treatment, using direct steam injection, below a pH of 6·62; at a pH below 6·50, virtually complete precipitation was observed. Reduction in the calcium ion concentration through the addition of EDTA, phosphate or citrate salts had a slight effect on stability, but very little sediment was observed above pH 6·7. Skudder (1984) observed that it is extremely useful to measure pH, as a slight increase in pH improves processing times, and similar results were obtained with reconstituted milk powder (Zadow, 1978) with a plate-type UHT plant. The natural range of pH values is probably not greater than 0·15 units, but a change in initial pH from 6·67 to 6·54 vastly reduces the processing time attainable; note that during UHT processing, the pH of milk falls to well below 6·0. At pH 5·3–5·4, the milk becomes grossly unstable, but the addition of NaOH to increase the pH by 0·1 unit, prior to heat treatment, improves stability. The addition of trisodium citrate (4 mM) raises the pH by 0·7, increases processing times, and appears to have a slightly beneficial effect on quality, due to an increase in viscosity of the product.

Reducing the pH increases the amount of fat in the deposits formed from whole milk. The surface finish and temperature differentials across the plate are also important. The presence of cold spots results in calcium phosphate deposits remaining on the plate after in-place cleaning with sodium hydroxide followed by orthophosphoric acid.

Tissier *et al.* (1984) found two major deposit peaks, one at 90 °C (predominantly protein-50%) and one at 130 °C (predominantly mineral-75%). The major protein contributing to the lower temperature peak was β-lactoglobulin (62%), while in the second peak, β-casein (50%) and α_{s1}-casein (27%) were dominant. The effects of plant design have been evaluated by Lalande *et al.* (1984).

ASEPTIC PACKAGING

UHT technology relies on the ability to be able to transfer the sterile product into a container under aseptic conditions. Obviously the container itself will need sterilising prior to filling; cans (Dole) are sterilised by superheated steam, whilst plastic pots (Fresh-Fill) and laminates (Tetrapak and Combiblok) are sterilised by hydrogen peroxide. Care should be taken to ensure that all hydrogen peroxide is removed, as it is a strong oxidising agent. Recently, a large volume aseptic package has been introduced, making possible the bulk transportation of UHT products.

The advantages and disadvantages of some of these systems are described by Cerf and Brissendon (1981) and Kosaric *et al.* (1981).

CONCLUDING REMARKS

One interesting development is concerned with the direct conversion of electrical energy to heat, by using the fluid as an electrical resistance and conducting medium (Ohmic heating). The fluid is placed in a non-conducting tube which has electrodes at each end, and as an alternating current is passed through the fluid, heat is generated; the conversion efficiency being greater than 90 per cent. The major advantages are that even heating results, there are no temperature gradients, and none of the usual limitations due to conduction and convection arise. In addition, it is suitable for continuous processing, and there is no requirement for a hot, heat-transfer surface, which will reduce fouling. Liquids containing particulate matter can be processed, and will not be subjected to high shear rates found in scraped-surface heat exchangers. In principle, the technique is similar to microwave heating, over which it is claimed to have a lower capital cost and a higher conversion efficiency (Hasting, 1985).

Other developments will be aimed towards obtaining longer processing runs, and improving the quality of heat-treated products, by a better understanding of the physical, chemical and biochemical reactions which occur during heating and subsequent storage.

REFERENCES

Aboshama, K. and Hansen, A. P. (1977). *J. Dairy Sci.,* **60**, 1374.
Adams, D. M., Barach, J. T. and Speck, M. L. (1976). *J. Dairy Sci.,* **59**, 823.
Andrews, G. R. (1984). *J. Soc. Dairy Technol.,* **37**, 92.
Ashton, T. R. (1965). *J. Soc. Dairy Technol.,* **18**, 65.
Ashton, T. R. and Romney, A. J. D. (1981). I.D.F. Document No. 130.
Badings, H. T. (1977). *Nordeuropaeisk Mejeri-tidsskrift,* **43**, 379.
Badings, H. T. (1984). *Dairy Chemistry and Physics* (Eds P. Walstra and R. Jenness), John Wiley, New York.
Badings, H. T. and Nester, R. (1978). *XX Int. Dairy Congress,* Paris, 1978, 77ST.
Blanc, B. and Odet, G. (1981). I.D.F. Document No. 133.
Blanc, B., Baer, R. and Ruegg, M. (1977). *Schweizerische Milkwirtschaftliche Forschung,* **6**, 21.
Brown, K. L. and Ayres, C. A. (1982). In: *Developments In Food Microbiology — 1* (Ed. R. Davies), Applied Science Publishers, London.
Burki, C. and Blanc, B. (1978). *XX Int. Dairy Congress,* Paris, 1978, Brief Communications.
Burton, H. (1982). In: *CRC Handbook of Processing and Utilization in Agriculture* (Ed. I. A. Wolfe), Chemical Rubber Company, New York.
Burton, H. (1983). I.D.F. Document No. 157.
Burton, H., Perkins, A. G., Davies, F. L. and Underwood, H. M. (1977). *J. Fd Technol.,* **12**, 149.
Busse, M. (1981). I.D.F. Document No. 130.
Cerf, O. (1981). I.D.F. Document No. 130.
Cerf, O. and Brissendon, C. H. (1981). I.D.F. Document No. 133.
Clydesdale, F. M. (1975). In: *Theory, Determination and Control of Physical Properties of Food Materials* (Ed. C. Rha), D. Reidel, Dordrecht.
Cousins, C. M. and Bramley, A. J. (1981). In: *Dairy Microbiology, Volume I* (Ed. R. K. Robinson), Applied Science Publishers, London.
Dickerson, R. W. Jr, Scalzo, A. M., Read, R. B. Jr and Parker, R. W. (1968). *J. Dairy Sci.,* **51**, 1731.
Dill, C. W., Roberts, W. M. and Aurand, L. W. (1962). *J. Dairy Sci.,* **45**, 1332.
EEC Dairy Facts and Figures (1981). Economics Division, Milk Marketing Board.
Elfagm, A. A. and Wheelock, J. V. (1977). *J. Dairy Res.,* **44**, 367.
Fox, P. F. (1982). In: *Developments In Dairy Chemistry — 1* (Ed. P. F. Fox), Applied Science Publishers, London.
Francis, R. J. (1975). In: *Theory, Determination and Control of Physical Properties of Food Materials* (Ed. C. Rha), D. Reidel, Dordrecht.
Hasting, A. P. M. (1985). Private communication.

Hillier, R. M. and Lyster, R. L. J. (1979). *J. Dairy Res.,* **46**, 95.
Jaddou, H. A., Pavey, J. A. and Manning, D. J. (1978). *J. Dairy Res.,* **45**, 391.
Jelen, P. (1982). *J. Food Protection,* **45**, 878.
Jenness, R. (1982). In: *Developments in Dairy Chemistry — 1* (Ed. P. F. Fox), Applied Science Publishers, London.
Jeon, I. J. (1976). Ph.D. Thesis, University of Minnesota, St. Paul, Minnesota.
Jeon, I. J., Thomas, E. L. and Reineccius, G. A. (1978). *J. Agric. Fd Chem.,* **26**, 1183.
Josephson, D. V. and Doan, F. J. (1939). *The Milk Dealer,* **29**, 35.
Jukes, D. J. (1984). *Food Legislation of the U.K. — A Concise Guide,* Butterworths, London.
Kessler, H. G. (1981). *Food Engineering and Dairy Technology,* Verlag A. Kessler, Freising.
Kessler, H. G. and Horak, F. P. (1984). *Milchwissenschaft,* **39**, 451.
Koka, M., Mikolajcik, E. M. and Gould, I. (1968). *J. Dairy Sci.,* **51**, 217.
Kosaric, N., Kitchen, B., Pandial, C. J., Sheppard, J. D., Kennedy, K. and Sargant, A. (1981). *CRC Critical Reviews in Food Science and Nutrition,* **14**, 153.
Lalande, M., Tissier, J. P. and Corriev, G. (1984). *J. Dairy Res.,* **51**, 557.
Law, B. A., Andrews, A. T. and Sharpe, M. E. (1977). *J. Dairy Res.,* **44**, 144.
Levenspiel, O. (1972). *Chemical Reaction Engineering,* John Wiley, New York.
Loncin, M. and Merson, R. L. (1979). *Food Engineering, Principles and Selected Applications,* Academic Press, New York.
Lund, D. (1975). In: *Principles of Food Science, Part 2, Physical Principles of Food Preservation* (Ed. O. Fennoma), Marcel Dekker, New York.
Lyster, R. L. J. (1964). *J. Dairy Res.,* **31**, 41.
Lyster, R. L. J. (1970). *J. Dairy Res.,* **37**, 233.
Lyster, R. L. J., Wyeth, T. C., Perkin, A. G. and Burton, H. (1971). *J. Dairy Res.,* **38**, 403.
Mehta, R. S. (1980). *J. Food Protection,* **43**, 212.
Melo, T. S. and Hansen, A. P. (1978). *J. Dairy Sci.,* **61**, 710.
Muir, D. D. (1984). *J. Soc. Dairy Technol.,* **37**, 135.
Patrick, P. S. and Swaisgood, H. E. (1976). *J. Dairy Sci.,* **59**, 594.
Perkin, A. G. and Burton, H. (1970). *J. Soc. Dairy Technol.,* **23**, 147.
Perkin, A. G., Henschel, M. J. and Burton, H. (1973). *J. Dairy Res.,* **40**, 215.
Piggot, J. R. (Ed.) (1984). *Sensory Analysis of Foods,* Elsevier Applied Science Publishers Ltd, London.
Pofahl, T. R. and Vakaleris, D. G. (1968). *J. Dairy Sci.,* **51**, 1345.
Reimerdes, E. H. and Mehrens, H. A. (1978). *Milchwissenschaft,* **33**, 345.
Reuter, H. (1984). In: *Engineering and Food,* Vol. 1 (Ed. B. M. McKenna), Applied Science Publishers, London.
Schroeder, M. J. A. (1984). *J. Dairy Res.,* **51**, 59.
Schroeder, M. J. A., Cousins, C. M. and McKinnan, C. H. (1982). *J. Dairy Res.,* **49**, 619.
Scott, K. J., Bishop, D. R., Zechulko, A., Edwards-Webb, J. D., Jackson, P. A. and Scuffam, D. (1984a). *J. Dairy Res.,* **51**, 37.
Scott, K. J., Bishop, D. R., Zechulko, A., Edwards-Webb, J. D., Jackson, P. A. and Scuffam, D. (1984b). *J. Dairy Res.,* **51**, 51.
Scott, R. (1970). *Proc. Biochem.* **5** (5), 39.
Shew, D. I. (1981). I.D.F. Document No. 133.

Shipe, W. F., Bassette, R., Deane, D. D., Dunkley, W. L., Hammond, E. G., Harper, W. J., Kleyn, D. H., Morgan, M. E., Nelson, J. H. and Scanlan, R. A. (1978). *J. Dairy Sci.,* **61**, 855.
Skudder, P. J. (1984). Private communication.
Skudder, P. J., Thomas, E. L., Pavey, J. A. and Perkin, A. G. (1981). *J. Dairy Res.,* **48,** 99.
Society of Dairy Technology (1983). *Pasteurising Plant Manual,* Huntingdon.
Solberg, P. (1981). I.D.F. Document No. 130.
Staal, P. F. J. (1981). I.D.F. Document No. 133.
Statutory Instruments (1973). 1064, HMSO, London.
Statutory Instruments (1977). 1033, HMSO, London.
Statutory Instruments (1983). 1508, HMSO, London.
Suhren, G. (1983). I.D.F. Document No. 157.
Swaisgood, H. E. (1977). U.S. Patent No. 4053644.
Thomas, E. L., Burton, H., Ford, J. E. and Perkin, A. G. (1975). *J. Dairy Res.,* **42**, 285.
Thomas, E. L., Reineccius, G. A., De Waard, G. J. and Slinkard, M. S. (1976). *J. Dairy Sci.,* **59**, 1865.
Tissier, J. P., Lalande, M. and Corriev, G. (1984). In: *Engineering and Food,* Vol. 1 (Ed. B. M. McKenna), Applied Science Publishers, London.
United Kingdom Dairy Facts and Figures (1983). The Federation of U.K. Milk Marketing Boards.
Van den Berg, M. G. (1981). I.D.F. Document No. 130.
Walstra, P. and Jenness, R. (1984). *Dairy Chemistry and Physics,* John Wiley, New York.
Zadow, J. G. (1971). *J. Dairy Res.,* **38**, 393.
Zadow, J. G. (1978). *XX Int. Dairy Congress,* Paris, Brief Communications 711.

Chapter 2

Developments in Cream Separation and Processing

C. Towler
New Zealand Dairy Research Institute, Palmerston North, New Zealand

Cream consists of a concentration of the fat in milk, with the fat mainly existing as globules protected by a membrane. As such, cream can have a variety of compositions and is normally defined according to fat content or function. Fat content may range from 10 per cent (half-cream) to 80 per cent plus (plastic cream). Cream for butter manufacture would normally contain approximately 40 per cent fat. The United Nations Food and Agricultural Organization and World Health Organization (1977) have suggested the following standards for Market Cream:

Pasteurised, sterilised and ultra-high temperature (UHT) treated cream > 18 per cent milkfat.
Half-cream, 10–18 per cent milkfat.
Whipping cream > 28 per cent milkfat.
Heavy whipping cream > 35 per cent milkfat.
Double cream > 45 per cent milkfat.

The physico-chemical properties of cream are very much influenced by the state of dispersion of the milkfat globules and the globule membrane which surrounds them. Mulder and Walstra (1974) have prepared a comprehensive treatise on the chemistry of the milkfat globule, including the properties of various systems which incorporate milkfat. The fat globules in milk or cream are not of uniform size, and vary from 0·5 μm to 10 μm in diameter. Figure 1 shows a typical Gaussian distribution curve for milkfat globules in cow's milk. As the concentration of milkfat globules is altered in cream, changes take place which have a marked effect on the rheology and physical state. Temperature changes also have a marked effect as the different lipid components undergo changes of state. Not only is the magnitude of change important, but the rate of change, as this affects crystallisation patterns. The non-fat milk solids also play an important part in the properties of cream, and additives such as salts, proteins, emulsifiers and hydrocolloid stabilisers all affect cream properties. The properties of cream are also affected by physical handling such as pumping, aeration and agitation as they affect the disintegration and agglomeration of the globules. It will be the function of this chapter to explain the principles of cream manufacture from milk, along with further treatments which may be applied to give cream products for direct consumption, or cream which may be used for manufacture of other dairy products.

SEPARATION AND DEVELOPMENTS IN SEPARATORS

The separation of milk is a process whereby an essentially fat-free portion (skim-milk) is separated from a fat-rich portion (cream). The

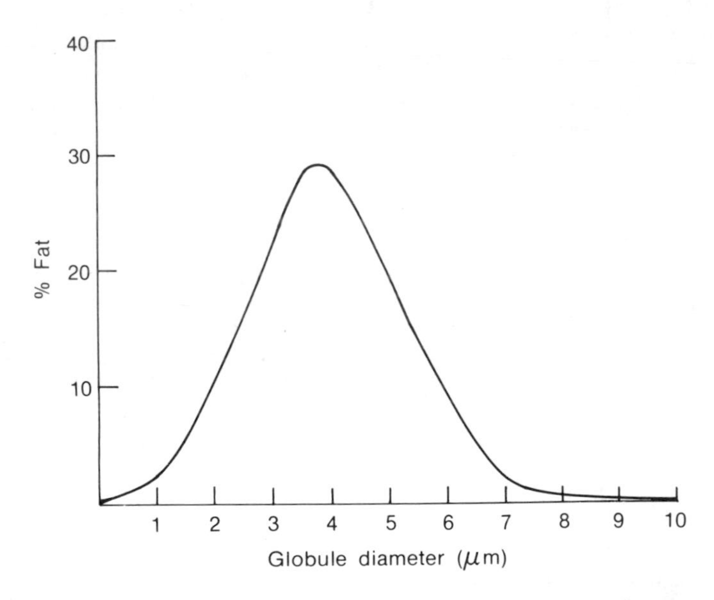

Fig. 1. Typical distribution curve of fat globules in milk expressed as percentage of total fat taken up by globules with a particular diameter.

process is a physical process relying on the density difference between the milkfat in the globules and the aqueous phase in which they are dispersed. If milk is allowed to stand, fat rises, and the familiar process of 'creaming' is observed with a fat-rich fraction collecting at the surface. The upward gravitational force (f_u) on a fat globule is given by;

$$f_u = (4/3)\pi r^3 g(\rho_s - \rho_f)$$

where r = radius of globule
g = acceleration due to gravity
ρ_s = density of serum (skim-milk)
ρ_f = density of fat globule.

The rise of the globule is inhibited by frictional force (f_f) which is given by Stokes' Law.

$$f_f = 6\pi\eta rv$$

where η = fluid viscosity of serum
v = velocity of globule.

Thus, when the fat globule is rising at a constant terminal velocity, then

$$f_f = f_u$$

$$6\pi\eta rv = (4/3)\pi r^3 g(\rho_s - \rho_f)$$

$$v = \frac{2r^2 g(\rho_s - \rho_f)}{9\eta}$$

Thus, the velocity with which a fat globule rises is directly proportional to the square of its radius and the density difference between the globule and the serum, and is inversely proportional to the viscosity of the serum. The densities of the milkfat and serum and the viscosity of the serum can be manipulated by altering the temperature, but the radius of a globule is fixed. Table I gives the theoretical velocities of fat globules of 1, 2 and 5 μm diameter at different temperatures, with the relevant time it would take to rise 200 mm; this is the approximate depth of milk in a 600 ml bottle. These figures are only of academic interest as small globules tend to remain dispersed through thermal currents and Brownian motion, whilst larger globules tend to coalesce leading to more rapid separation. The figures do illustrate, however, that separation by gravity is most inefficient.

TABLE I
Velocity of different diameter fat globules at different temperatures with the time taken to rise 200 mm

Temperature (°*C*)	*Density of serum* (*kg* m^{-3})	*Density of fat* (*kg* m^{-3})	*Viscosity of serum* (*N* m^{-2} *s*)	*Diameter of globule* (*μm*)	*Velocity of globule* (*μm* s^{-1})	*Time to travel 200 mm* (*days*)
5	1037	961	$2{\cdot}96 \times 10^{-3}$	1	0·014	165
				2	0·056	41
				5	0·350	6·6
20	1034	930	$1{\cdot}79 \times 10^{-3}$	1	0·032	73
				2	0·127	18
				5	0·791	2·9
35	1029	908	$1{\cdot}17 \times 10^{-3}$	1	0·056	41
				2	0·225	10
				5	1·410	1·6
50	1022	898	$8{\cdot}5 \times 10^{-2}$	1	0·080	29
				2	0·318	7·3
				5	1·99	1·2
65	1015	888	$6{\cdot}5 \times 10^{-2}$	1	0·106	22
				2	0·426	5·4
				5	2·66	0·9

Centrifugal Separators

The other factor which may be manipulated to increase sedimentation is the force acting on the globules. This increase can be achieved through the use of centripetal force in a rotating vessel. The resultant globule velocity can then be given as;

$$v = \frac{2r^2(\rho_s - \rho_f) \times R\omega^2}{9\eta}$$

where R = radial distance of globule from axis of rotation
ω = angular velocity

or

$$v = \frac{2r^2(\rho_s - \rho_f) \times 4\pi^2 RN^2}{9\eta}$$

where N = rotational frequency in revolutions s^{-1}.

Table II gives the velocity of fat globules 0·1 m from the axis of rotation in a centrifuge rotating at different speeds with milk at different temperatures. Figures 2 and 3 illustrate the effect of the various parameters on the velocity. These indicate that the velocity of globules can be increased substantially by increases in centrifuge speed or milk temperature, but the size of the fat globules is a critical factor, and small globules have only a limited velocity at high rotational speeds and high milk temperatures.

Gustaf de Laval devised the first continuous centrifugal separator, the principle of which is illustrated in Fig. 4. Whole milk was fed in through the top of the bowl to a distributor in the base, which brought it to rotational speed while being channelled into the bowl itself. The fat globules moved in toward the axis of rotation to form a cream layer, whilst the denser skim-milk flowed to the outside of the bowl and was channelled out. The cream was taken out of the inner bowl as overflow, and control of separation was achieved by altering the flow of incoming milk. In later models, the fat content of the cream was controlled through restriction in the skim-milk flow. Allowing more skim-milk to flow out reduced cream flow, and the fat content was increased. Such separators had limited capacity, as fat globules had a reasonably large distance to travel before reaching the cream layer. If high milk flows were used, then separation was inefficient with fat globules escaping with the skim-milk flow.

TABLE II
Velocity of fat globules of various diameters in a centrifuge rotating at different speeds at different temperatures

Temperature (*°C*)	*Globule diameter* (*μm*)	*Rotational frequency of centrifuge* (*rpm × 10³*)	*Velocity of globule* (*mm s⁻¹*)
5	1	3	0·014
		5	0·039
		7	0·077
	2	3	0·056
		5	0·156
		7	0·307
	5	3	0·352
		5	0·978
		7	1·916
35	1	3	0·057
		5	0·158
		7	0·309
	2	3	0·227
		5	0·630
		7	1·235
	5	3	1·418
		5	3·938
		7	7·718
65	1	3	0·107
		5	0·298
		7	0·583
	2	3	0·429
		5	1·190
		7	2·333
	5	3	2·678
		5	7·440
		7	14·582

This difficulty was resolved by a German inventor von Bechtolsheim (1888), who provided a separator with a number of separation zones through a 'disc stack'. The discs were conical in shape (cone angle approximately 60°) with holes in them to channel milk through. Identical discs were stacked one on top of the other with spacers (caulks) fitted on the upper surfaces to provide a gap between adjacent discs. Such discs are still features of modern separators, and Fig. 5(a) shows how fat globules have only a limited distance to move before being captured on the upper surface of a disc to be channelled inwards, with

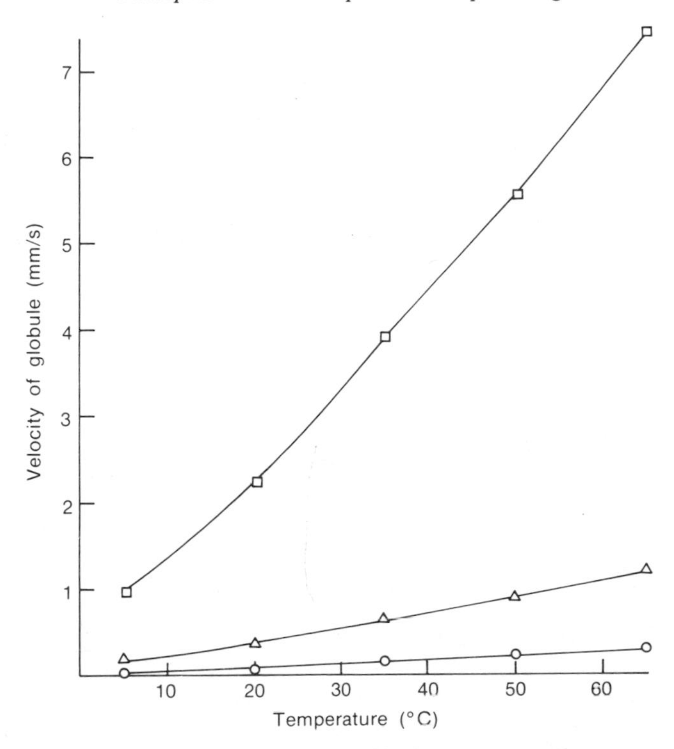

Fig. 2. The dependence of the velocity of fat globules of different diameter on temperature in a centrifuge rotating at 5000 rpm, the globules being 0·1 m from the axis of rotation. ○: globule 1 μm diameter; △: globule 2 μm diameter; □: globule 5 μm diameter.

skim-milk being captured and channelled out on the lower surface of the adjacent disc. Figure 5(b) shows the arrangement of a complete disc stack with relevant flows, and Fig. 5(c) is a photograph of an actual disc showing the holes for channelling milk, and the spacers (caulks) which maintain the required separation between the discs.

Paring Discs

A further development came with provision of 'paring discs' at the outlet of separators. The paring disc converts the rotational energy of exiting milk and cream into linear kinetic energy and, in fact, is a stationary centripetal pump. Figure 6 shows a cutaway diagram to illustrate the

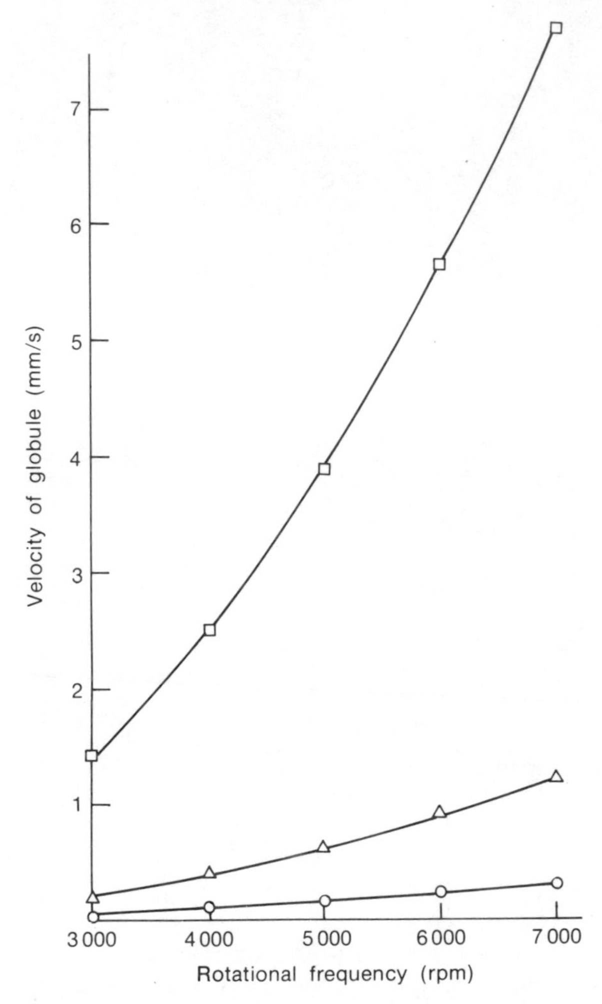

Fig. 3. The dependence of the velocity of fat globules of different diameters on the speed of rotation of a centrifuge at 35°C, the globules being 0·1 m from the axis of rotation. ○: globule 1 μm diameter; △: globule 2 μm diameter; □: globule 5 μm diameter.

action of a paring disc. The advantages of the paring disc are;

1. The generated pressure can be used to push the exiting cream or skim-milk through heat exchangers.
2. Variable flow restrictive devices on the outlets can be used to

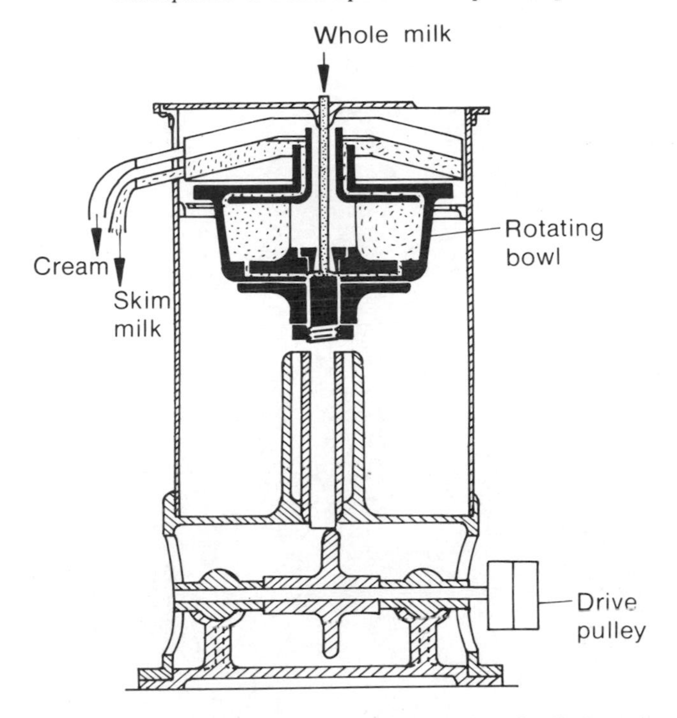

Fig. 4. An early model separator (courtesy of Alfa-Laval).

generate back-pressures which will control flow. In this way, a fairly accurate control of fat content in the cream can be achieved.

Separators with milk feed at atmospheric pressure and paring discs at the outlet are known as semi-open types, as opposed to the previous ones with straight discharges which are open types.

Hermetic Separators

Entrainment of air in milk inhibits separation. Although it might be expected that low density air would migrate rapidly in towards the cream line, the presence of naturally occurring surface active agents in the milk provides a somewhat stable structure to the air bubbles. This led to the next development in separators, the air-tight or hermetic separator. The development was based on the provision of hermetic seals which effectively isolated the separator from the atmosphere. In contrast to open or semi-open separators, the milk is introduced into the

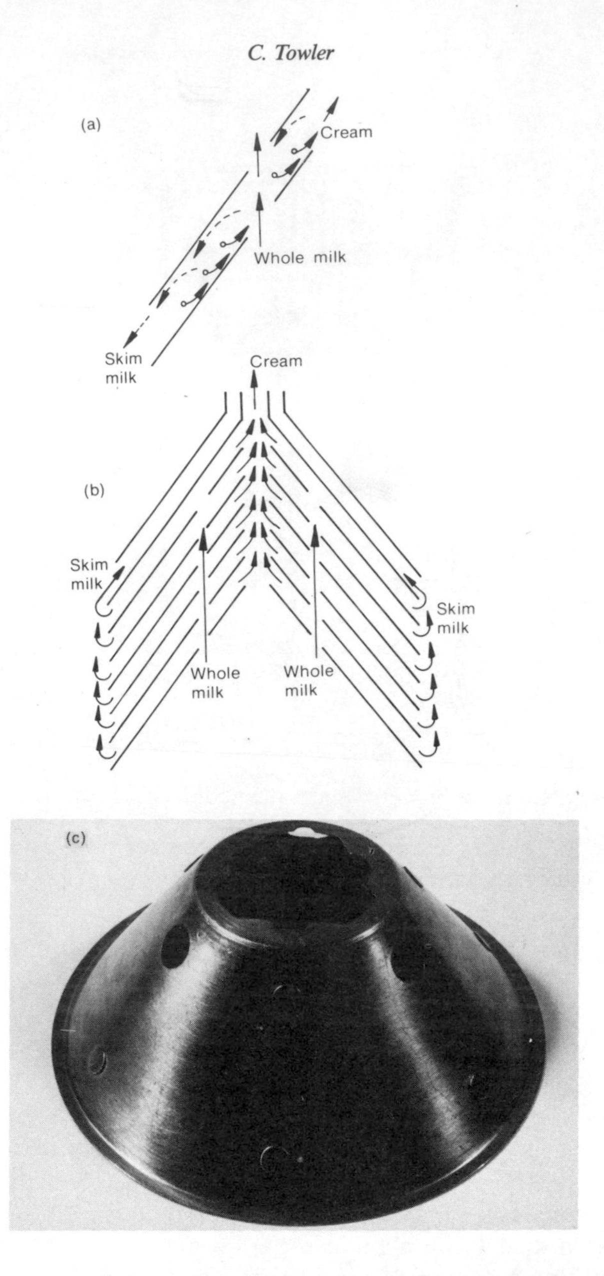

Fig. 5. (a) Flow of cream and skim-milk in the space between discs in a centrifugal separator. (b) A disc stack. (c) Photograph of a separator disc showing holes for channelling of milk and spacers (caulks).

separator from below via a hollow spindle in the central shaft. The milk gradually reaches the rotational speed of the separator in contrast to the open types when rapid acceleration takes place on reaching the distributor. Separation takes place as in a normal separator through the disc stack. Efflux of cream and skim-milk takes place through hermetic seals under a moderate pressure. If a higher pressure discharge is required to feed heat exchangers, then pumps or paring discs may be built into the outlets. The feeding of hermetic separators is important, as both correct flow and discharge pressure must be achieved. The effectiveness of a hermetic separator does depend on the whole milk being air-free as it is fed to the separator, and it is important that air entrainment be avoided. The importance of hermetic sealing is that air will not be mixed into the product as a result of the separation process.

Self-desludging Separators

Milk contains solid particles, including contaminating dirt and cellular material from blood and bacteria. This dense, solid material collects on the outside of the spinning bowl and, if left, inhibits the efflux of skim-milk and stops the separation process. This limited running time of

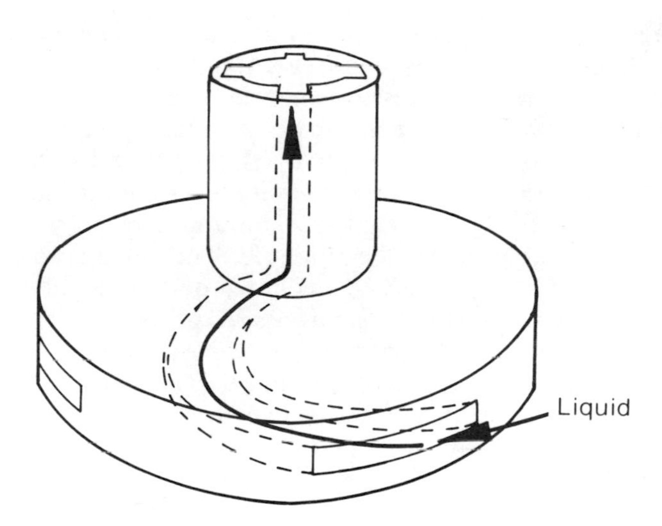

Fig. 6. A paring disc.

separators led to the development of a mechanism for automatically removing the solid material without having to interrupt the operation of the machine. Such separators are variously called 'self-desludging' or 'self-cleaning' separators. The diagram of a self-desludging separator (see Fig. 7) shows the principle of operation. Slots are cut in the outside of the bowl, and are normally kept closed by a sliding bowl bottom (piston) which is elevated by the hydraulic pressure of running water in a reservoir underneath. When desludging is required, an hydraulically operated valve opens up and the reservoir drains to allow the bowl bottom to fall, thus opening the slots. The outward pressure on the sediment forces it outwards into an outer bowl. The waste material has a rotational motion which allows it to be taken tangentially out of the outer bowl into a cyclone for suitable disposal. The valve then closes to allow the reservoir to refill, and the slots are closed-off. The action is very rapid with the slots being open for less than one fifth of a second. Discharges are normally programmed to occur at regular, predetermined intervals, which depend on the volume of the sediment space in the separator bowl and the condition of the milk. The above mechanism is not exclusive, and several other methods of operating the sliding bowl have been devised.

Factors Affecting Separation

The previous sections have covered the major advances in separator design and principles involved in separation. At this stage it is pertinent to look at the different processing conditions which have an influence on separation, and how these should be viewed in a practical situation.

Normally the object of separation is to attempt to get all the fat in the whole milk within the cream fraction, with the minimum amount of fat retained in the skim-milk fraction. Skimming efficiency is normally assessed as the fat content of the skim-milk. The statement 'normally' is used as, in fact, it is dependent very much on the fate of the various fractions. If the skim-milk is merely going to be remixed with cream to produce milk with a standardised fat content, then the skimming efficiency is not so vital. If the skim-milk is to be converted into skim-milk powder or casein, then it is important that the fat content be low to meet the various specification and functional requirements of these products: particularly if a casein whey is to be subjected to ultrafiltration or other membrane processing, as the fat can have an inhibiting effect on the flux rate through a membrane.

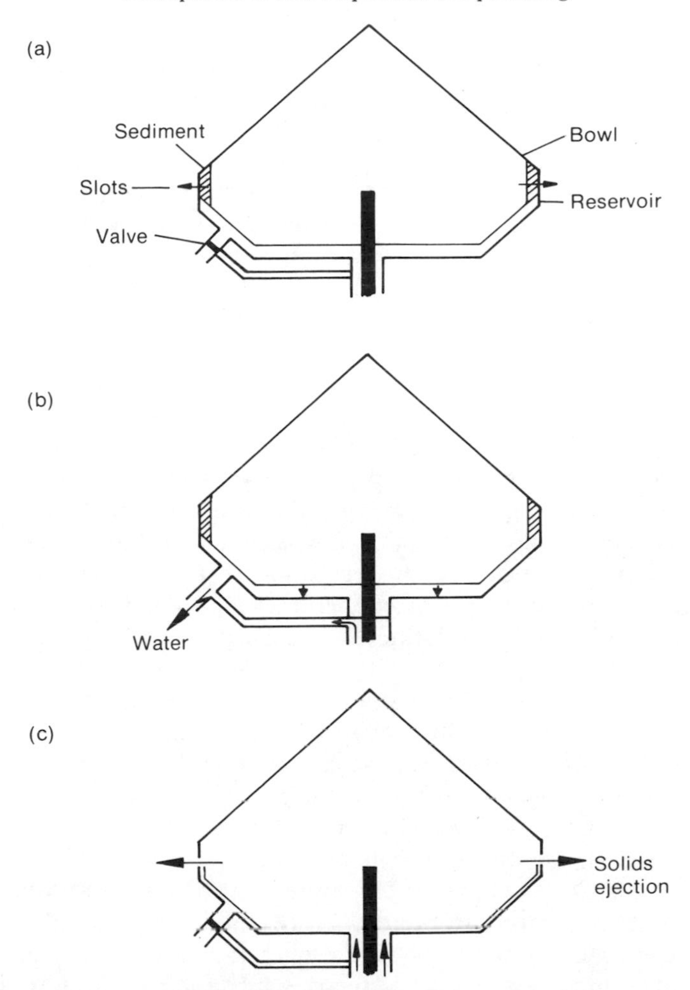

Fig. 7. (a) Self-desludging separator in normal operation with bowl bottom elevated by hydraulic pressure of water in reservoir. (b) Activation of hydraulically operated valve causes reservoir to empty and bowl bottom moves down. (c) Bowl bottom depressed opens up slots and solid sediment is ejected. Valve closes to allow reservoir to refill moving bowl bottom up again.

Figures 2 and 3 illustrate the theoretical effects of globule size, milk temperature and rotational speed of a centrifuge on the velocity of fat globules, but, for practical purposes, other factors have to be taken into account. The most comprehensive treatises on separation are generally of Russian or German origin, Lipatov (1976) having written a book on the subject. Other pertinent literature is to be found in Lang and Thiel (1955), Rothwell (1975) and Alfa-Laval (1980).

Temperature

An increase in milk temperature leads to both an increase in density difference between skim-milk and milkfat, and a reduced viscosity of the skim-milk, so that, in theory, an increase in temperature should lead to a rise in separation efficiency. In practice, a higher temperature can lead to disruption of fat globules and, as noted earlier, this has a very major effect on rate of separation. King *et al.* (1972) showed that fat losses in skim-milk were slightly higher with separation at 72 °C than at 54·5 °C, but separation at 32 °C gave a higher fat loss than separation at 54·5 °C or 72 °C. The temperature history of the milk has an important bearing on separation efficiency, and the modern practice of farmers cooling to 5 °C and holding for collection leads to some disruption of globule agglomerates. In practice, it would appear that most plants in New Zealand have settled on temperatures around 50 °C as being optimal in terms of skimming efficiency, and this is in agreement with manufacturer's recommendations (Westfalia, personal communication). Separation temperatures can also have an effect on free fat levels. Te Whaiti and Fryer (1975) found greater increases of free fat levels in cream with separation at 70 °C than with separation at 35 °C. However, separation at low temperature (less than 10 °C) will give little increase in free fat levels as the fat is mostly in the solid form. Thus, free fat formation is less when the fat is either all in liquid form or all in solid form, and the range 20–40 °C is the most critical. Free fat formation is, however, largely a factor of mechanical treatment, and can be minimised by preventing air incorporation and taking care in pumping the cream.

If separation is practised at high temperature, then the separation step can be included as part of the pasteurisation holding time. This approach can reduce the heat exchange requirements as the separate cream and skim-milk streams need not be pasteurised, although it will mean a somewhat more complex regenerative cooling system to get the cream and skim-milk to suitable final temperatures. The choice will depend very much on the fate of the cream and the skim-milk. The major effects of a high separating temperature are protein denaturation and phospholipid migration to the skim-milk; this latter point is of great importance in cream properties, and will be referred to later. Phospholipids in whey also inhibit flux during membrane processing.

Cold milk separators are available which will operate at 4–5°C. These allow separation of milk as it is received at the factory, and, although fat losses to skim-milk are somewhat higher, they do allow substantial savings in energy and capital costs in certain situations. In some cheesemaking operations, heat treatment of the milk is undesirable, and cold milk separators offer some advantage. In addition, cold milk separators give cream with a greater phospholipid content which will give better whipping properties. The major modification in a cold milk separator is a wider disc spacing than with a conventional model (approximately double) to allow adequate flow of the more viscous cold cream.

Bowl speed

The velocity of a fat globule is proportional to the square of the rotational speed, and thus an increase in bowl speed will have a very major effect on separation efficiency. An increase in bowl speed, however, requires an increase in energy input, and a more robust design to withstand the large forces at the bowl periphery. The separator also generates more noise. For these reasons bowl speeds have not increased significantly, as skimming efficiency is quite adequate at moderate speeds of 4000–5000 rev min^{-1} (rpm). Early design separators operated with bowl speeds of 3000 rpm. It is important, however, that the bowl speed be maintained during operation, and many separators are fitted with tachometers to ensure that the rotational velocity is consistent. Momentary deceleration does take place during a desludging operation, so it is important that it be carried out within a minimum time period, as separation efficiency is affected.

Disc separation

Lang and Thiel's (1955) review on Centrifugal Separators in the Dairy Industry summarised several studies on the effect of disc spacing on separation efficiency. In theory, the smaller the spaces between discs, the higher should be the efficiency of separation as the fat globules have less distance to travel before being 'captured' on the disc surface. However, flow patterns must be taken into consideration, and it is important that laminar flow conditions exist for maximum efficiency of separation, as any turbulence will result in fat globules being remixed with the milk stream, so increasing the possibility of their being swept out with the skim stream. For relatively narrow spacings (less than 0·2 mm), it is found that separation efficiency is independent of disc spacing, because the disc spacings, in fact, take equal volumes of milk in a given time. Thus, the tangential force moving milk through the disc space is created by the friction of the disc on the milk, and with a narrow space, there is more surface contact with the milk giving a proportionately greater flow outwards than with a wide disc spacing. Thus within certain limits, separation efficiency is independent of the disc spacing. These limits are dependent on flow factors (i.e. rate of milk flow and viscosity) and bowl rotational speed; for example, a cold milk separator requires a larger disc spacing than a hot milk separator because of the higher viscosity of the cream.

It is important that discs are not distorted through physical damage from stress, or dismantling and cleaning operations, and that the disc spacers (caulks) ensure that excessive turbulence does not occur as fluid passes by. Thus, the assembly of a disc stack containing the necessary multitude of discs makes it necessary to use a large amount of physical pressure to force the discs into the necessary configuration, with the discs in intimate contact with spacers to get the required interdisc spaces. An O-ring locking device is also used to ensure that this spacing is maintained. The placement of channels in the discs is also important, and should reflect the likely proportion of flow between light and dense phases. For milk separation, the channels are displaced toward the axis of rotation as the cream flow is less than the skim. In contrast, a separator for anhydrous milkfat (AMF) production from molten butter has channels that are very close to the disc periphery. The number of discs in a separator is related to the capacity.

Flow

The flow of incoming milk and the relative flows of exiting skim-milk and cream are factors which must be carefully controlled in order to achieve good separation efficiency with adequate throughput. Much more efficient separation can be achieved with lower input feeds, as the milk has more time within the separator to allow fat globules to segregate. However, it is important that the flow should not be so low as to allow significant air entrainment to fill available bowl space, as separation efficiency will be adversely affected.

In practice, as high a flow rate as possible will be aimed for, i.e. minimum processing time, but separating efficiency cannot be compromised as this will lead to high fat losses. In extreme cases, if incoming milk flows are too high, 'flooding' of the separator occurs and virtually no separation takes place. The flux of cream and skim-milk must naturally equal the flow of incoming whole milk. In a fully open separator with no control on discharge flow, the fat content of the cream must be controlled by the flow of the incoming milk. In commercial separators, however, the flow of skim-milk and cream can be controlled by back-pressure, and this will control the fat content of the cream. The fat content of the cream is dependent on its final use, as will be referred to later on standardisation. Modern separators are normally supplied with variable back-pressure valves on the cream and skim-milk lines to adjust for maximum separating efficiency, and maintain the desired fat content in the cream.

Raw milk

The state of the raw milk has a large bearing on separation efficiency. Figures 2 and 3 illustrate that, whatever conditions prevail in separation, the velocity of a fat globule is very dependent on its size, and a high proportion of small fat globules will lead to a reduced separation efficiency. In practice, separation of fat globules less than approximately 1 μm represents the lowest limit of separation; in normal milk that would represent a level of approximately 0·04 per cent fat in the skim-milk. Fat contents in skim-milk in commercial operation are normally in the range 0·04–0·06 per cent with cream being taken-off at approximately 40 per cent fat, so this represents a good separation efficiency. The factors which may affect the fat globule size distribution are:

1. Breed of cow.
2. Stage of lactation.
3. Temperature history of the milk.
4. Handling of the milk through agitation, pumping, aeration, etc.

As mentioned previously, violent treatment of the milk will lead to high free fat levels as the globule membranes become ruptured, and the presence of large amounts of free fat can cause gelation of cream (Te Whaiti and Fryer, 1975).

Conclusions on Separation

Figures 8, 9 and 10 show diagrams of commercial separators of various types. More recent advances in separators have been confined to increases in capacity to fit the needs of larger processing units and refinements to cleaning-in-place (CIP) procedures, whereby they can be washed and rinsed without stripping down at regular intervals.

Mechanical advances have concentrated on energy efficiency and noise reduction. The latter has been achieved through the production of double-walled skins with water in the gap. This water, together with the water used in the self-desludging mechanism, also exerts a cooling action.

The major advances are currently taking place in the automation and control of the separation process. The process is normally included as an integral part of milk reception and pasteurisation processes, with the whole being subject to microprocessor control. The microprocessor controls the operation of valves and pumps with information being fed back from devices which measure temperature, pressure, flow and level in tanks. With such controls, milk can be received, preheated, separated, pasteurised and cooled, with cleaning to follow, using a variety of conditions and a number of modes of operation through different programmes which can be fed into the microprocessor from keyboard, disc or tape. The microprocessor controls which particularly apply to the separator are the incoming milk flow, back-pressures on skim-milk and cream, and the desludging frequency. More sophisticated control systems will be referred to in the next section on standardisation.

Special separators have been designed to perform functions other

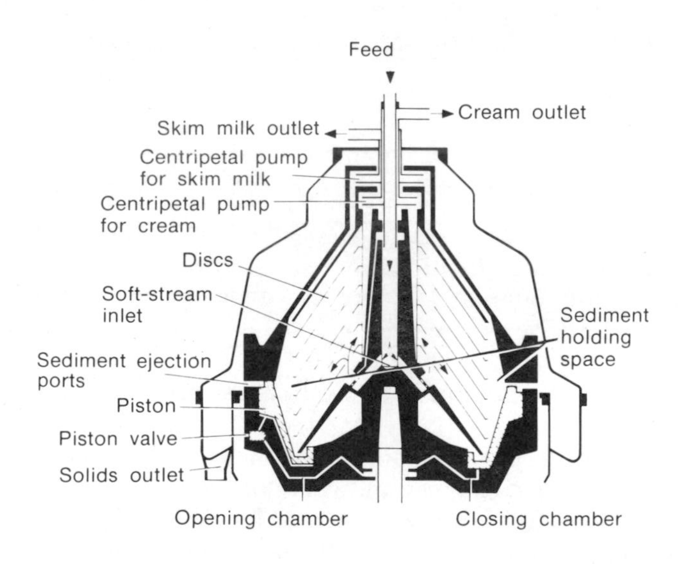

Fig. 8. Cross-section of a semi-open separator (courtesy of Westfalia Separator).

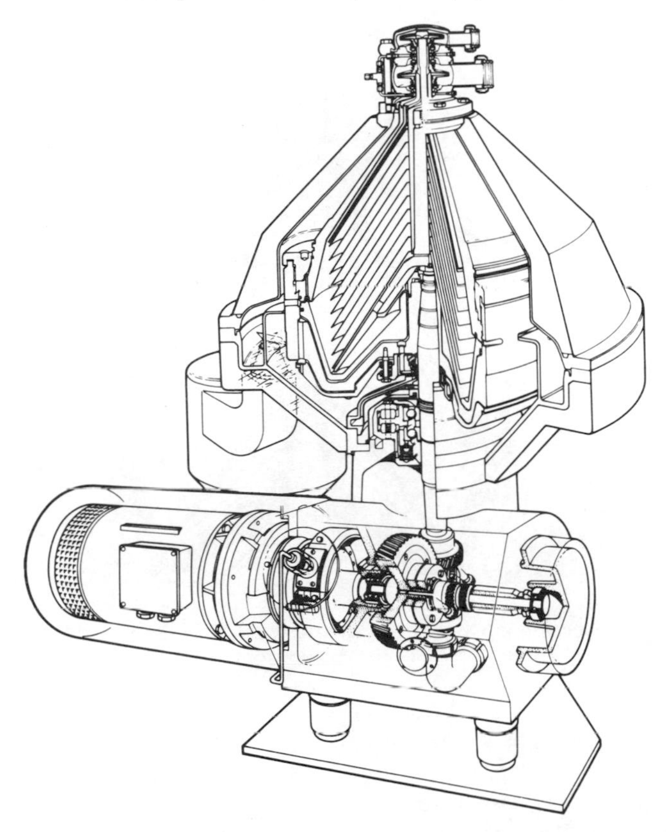

Fig. 9. Cutaway diagram of a modern hermetic separator including drive system (courtesy of Alfa-Laval).

than skim-milk/cream separations. These include:

1. Cream concentrators to produce plastic cream from normal cream of approximately 40 per cent fat content.
2. Clarifixators for inverting phase and separating aqueous serum from liquid milk fat in anhydrous milkfat (or butteroil) manufacture.
3. Clarifiers to merely remove solid material from milk or whey.
4. Bactofuges which decrease the bacterial count in milk by centrifugal action — such centrifuges have a high rotational speed.
5. Nozzle bowl centrifuges for continuously ejecting dense phase concentrates.

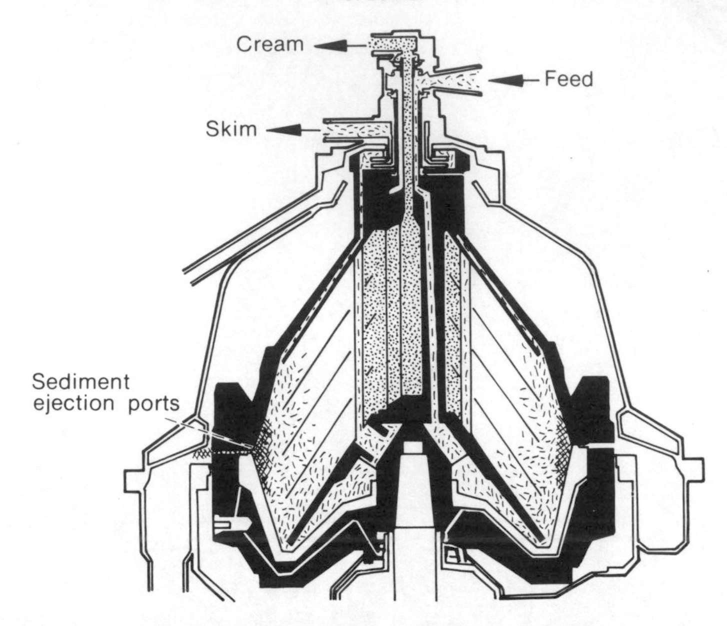

Fig. 10. Cross-section of a 'semi-hermetic' cold milk separator (courtesy of Westfalia Separator).

6. Desludging type separators for separating fine particulate matter from slurries, e.g. lactalbumin manufacture.
7. Cream cheese separators for separating cream cheese (which forms the light phase) from whey.

These modified separators have different disc designs and various systems of feeding and ejecting components dependent on consistency and relative flows of material. The large desludging separators are normally belt-driven to cope with the extra mechanical strain associated with frequent desludging operations.

STANDARDISATION

The important factor in standardisation of cream is the fat content. In market cream, if the fat content is higher than the specified requirement, then financial returns will suffer. If the fat content is low, then it may not meet regulatory requirements, and it may also lack desired functional properties, such as viscosity or whipping properties. In cream for butter manufacture, although accurate standardisation is not so important, too high a fat content may lead to difficulties in churning. Too low a fat content will result in a high buttermilk volume, and this will not give as good a return as the skim-milk which would have resulted from better separation control. Standardisation is also an important facet of market milk production, and much of the technology is applicable to cream and milk. At the present time, the technology of standardisation is changing at a rapid rate, so that it is only pertinent to touch on general principles.

To achieve accurate standardisation, the fat content should, ideally, be controlled as it leaves the separator. For small scale operations, however, standardising in bulk is the norm. The separator is adjusted to give a slightly higher fat content than required, and a suitable diluent, such as skim-milk, is added to get the required fat content. Such dilutions can be calculated from first principles, or by using Pearson's Square. A very important consideration in this operation is the measurement of fat content. Normally, a reasonably quick measurement is required which precludes the accurate Werner–Schmidt or Rose–Gottlieb methods. The Babcock or Gerber methods, which rely on volumetric measures of extracted fat, are quicker, if a little less accurate. Most dairy factories, however, possess Milkotesters (A. N. Foss Electric, Denmark) or similar instruments which will give rapid measurements of good accuracy, provided they are set in the correct mode with a properly prescribed diluent in the cream. Once the fat content of the cream has been determined, it is necessary to add the correct proportion of diluent to get the required fat content. Tanks are normally fitted with volumetric measures of contents, and as such the density of the contents should be taken into account as higher fat contents give lower densities. For creams in the vicinity of 40 per cent fat content, a density of 1 kg l^{-1} can be assumed. For bulk standardisation, it is more appropriate to consider some examples.

Example 1. A 5300 litres quantity of cream has a fat content of 42·3 per cent. A fat content of 40 per cent is required. How much skim-milk should be added? If we assume that the skim-milk contains 0 per cent fat, then

$$\text{Quantity of fat in cream} = 5300 \times 0{\cdot}423$$
$$= 2241{\cdot}9 \text{ kg}$$

The total amount of cream which would have this quantity of fat as 40 per cent of its content would be

$$2241 \times \frac{100}{40} = 5605 \text{ kg} \triangleq 5605 \text{ litres}$$

$$5605 - 5300 = 305$$

Thus, 305 litres of skim-milk should be added.

If, in the above example, whole milk with 4·2 per cent fat content were to be used as a diluent, then the Pearson Square method becomes much easier to use. The essentials of the method are shown in Fig. 11. The fat content of the cream used is set in the top left hand corner of a square (42·3 per cent), with the fat content of the diluent at the bottom left hand corner (4·2 per cent). The required fat content is then placed at the intersection of the square's diagonals (40 per cent), and the proportions of the cream and the whole milk are simply obtained by subtraction along the diagonals,

i.e. cream required = 40 − 4·2 = 35·8
whole milk required = 42·3 − 40 = 2·3

Thus, the amount of added whole milk to 5300 kg of cream should be

$$5300 \times \frac{2{\cdot}3}{35{\cdot}8} = 341 \text{ kg}$$

i.e. the cream should be made up to 5641 litres with whole milk.

These are only approximations requiring some assumptions, and a final analytical check of the fat content is advisable.

In high volume factories, it is more convenient to produce cream of a required fat content directly from the separator. This can be done purely

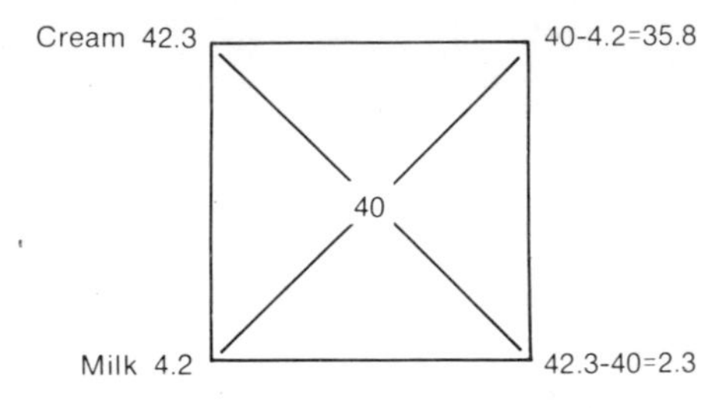

Fig. 11. Pearson's Square as a means of calculating proportions of components for standardising.

by adjusting the flow of skim-milk and cream via back-pressures for a fixed flow of incoming milk. Provided the fat content of the milk is known, the back-pressures on the cream and skim-milk lines can be adjusted with special valves to provide the required proportional flow of skim-milk and cream. A further refinement is a branch from the cream line, with a flow adjuster, to feed a proportion back into the skim-milk and produce a milk with a standard fat content. The important factors in such an operation are that standard conditions prevail during separation, and that the flow regulators are accurately calibrated.

More sophisticated operations require an accurate measure of milkfat content as the cream leaves the separator. A convenient method of ascertaining the fat content of cream is to measure its density, which is a function of the fat content. For example in the temperature range 40–80 °C, cream density (kg m^{-3}) can be computed as

$$D = 1038{\cdot}2 - 0{\cdot}17T - 0{\cdot}003T^2 - \phi(133{\cdot}7 - 475{\cdot}5/T) \text{ (Phipps, 1969)}$$

where T = temperature in °C
ϕ = fractional fat content.

As cream density is largely influenced by fat content, then use of an in-line densitometer can provide a constant monitor of fat content. For standardisation, a constant density is required, and in this regard the unsophisticated principle of maintaining a constant volume at a constant weight can be used. A U-shaped tube of liquid is counter-balanced by weights, and any movement of the tube through losing balance can provide a signal to a controller to restore the balance. A diagram of such an instrument can be found in the Alfa-Laval Dairy Handbook (1980), and Pato (1978) has described the incorporation of density measurement as a means of direct standardisation.

The densitometer is sensitive to vibration, and the incorporation of air into the cream will greatly affect its accuracy; other mass flow meters are available which will relate density and total flow of liquid. Less sophisticated systems which incorporate a Milko-tester (A. N. Foss Electric, Denmark) are available, and although not as accurate as a densitometer, are much less sensitive to environmental changes. The current advances are in the control systems, which incorporate micro-processors to receive signals from the various measuring devices and transmit signals to operate the various controls. Figure 12 illustrates a typical arrangement for producing standardised cream, with Fig. 13 showing an extended system for producing standardised milk and cream. The major perturbations for such systems are changes of milk feed silos, which may change milk composition as well as feed pressure to the separator, and desludgings of the separator. The efficiency of a system depends very much on the rapidity of response to such changes, and the re-establishment of the required conditions through design and tuning of the feedback control loops.

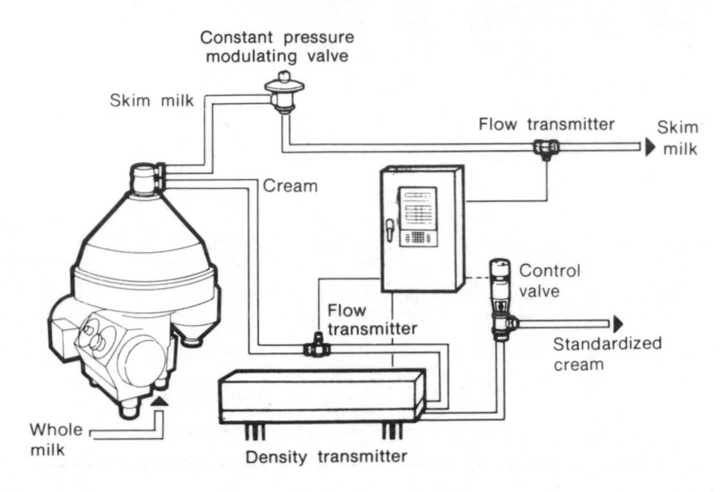

Fig. 12. System for automatic production of standardised cream (ALFAST courtesy of Alfa-Laval). Key: ———, electrical signals; — — —, pneumatic signals.

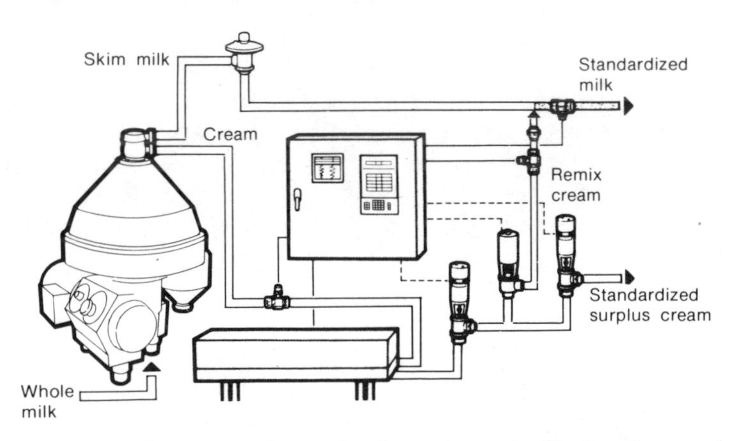

Fig. 13. System for automatic production of standardised cream and milk (ALFAST courtesy of Alfa-Laval). Key as for Fig. 12.

VACREATION

Heat treatment of cream is necessary to destroy organisms and enzymes which may be pathogenic and cause spoilage. Lactic acid producing bacteria, in particular, will cause souring and coagulation of cream. Lipase enzymes will break down the lipids to produce free fatty acids which give a rancid flavour. Cream can be pasteurised by conventional means using heat exchangers of various designs, and Chapter 1 has given accounts of these.

The cream from pasture fed animals can have flavour taints derived from the herbage that animals feed on. This has been particularly prevalent in New Zealand, where all milk for manufacturing is derived from pasture fed animals. As most of the tainting substances are relatively volatile, a process was devised to both pasteurise the cream and remove the volatiles through what is essentially a steam distillation process. The piece of equipment is known as a Vacreator, which was the trade name adopted for the Murray Vacuum Pasteurizer (present manufacturers and agents Protech Engineering, Auckland, New Zealand), and the process as vacreation. Vacreation is a treatment which is normally reserved for cream used in the manufacture of butter as the tainting substances are fat soluble. In the Vacreator, steam is intimately mixed with cream and the condensed vapour plus volatiles are removed by flash evaporation under vacuum. Figure 14 shows a diagram of a modern Vacreator consisting of 5 vessels. The typical pressure and temperature conditions pertaining to each vessel are shown on the diagram. Raw cream is preheated by vapours exiting from vacuum vessels 3 and 4 in a tubular heat exchanger. The cream is mixed with steam and vapours exiting from vessel 1, and passes into vessel 3 whence it is subjected to a vacuum and the cream and vapour are separated. The cream is then mixed with steam and vapour exiting from vessel 2, and the mix is passed into vessel 4 for separation. The vapours from vessels 3 and 4 are combined and passed through the preheater, before passing to a water jet condenser which provides vacuum and condenses the remaining condensible vapours. A spring-loaded baffle valve applies a back-pressure to vapour from vessel 3, so that the pressure difference required to transfer cream between vessels is maintained.

The cream from vacuum vessel 4 passes into an internal cream pump and is pumped to vessel 1, where it meets fresh in-coming steam. The cream separated in vessel 1 is mixed with fresh steam again before

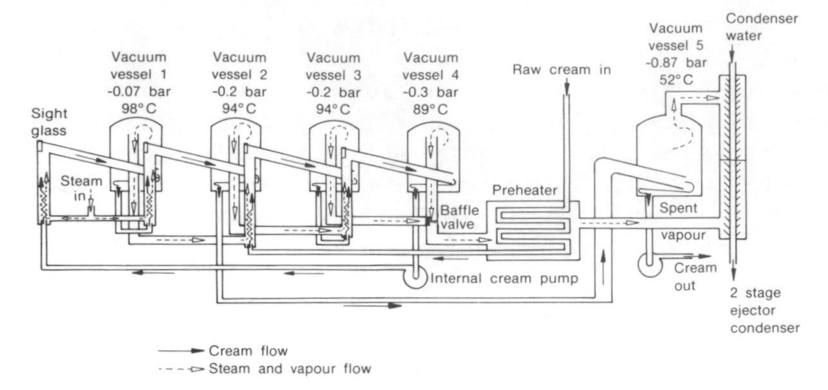

Fig. 14. Schematic diagram of a Vacreator.

passing into vessel 2. The cream exiting from vessel 2 passes into vessel 5 which acts purely as a flash cooler, with the vacuum removing water vapour and the associated latent heat. The cream exits at approximately 52 °C, and the vapours are removed in the condenser. The somewhat complicated system is called 'weaving flow', and is essentially counter-current with the cleanest cream meeting the cleanest steam.

The complex nature of flows has been found to be necessary because of the tendency of cream to foam under the vacuum treatment, so that the separation of the liquid and vapours becomes difficult. The liquid and vapours are separated by a cyclonic centrifugal action with the cream being fed tangentially into the vessel at a slight downwards angle. The vapours are taken out through a centrally mounted pipe. If foaming is severe, then liquid gets mixed with the vapour stream resulting in product loss. The flow of steam assists in providing the cream with sufficient kinetic energy to flow through the system, but loss of energy occurs in the separation process which necessitates the use of the internal cream pump to push the cream through the final two stages. The flow from vessel to vessel is also controlled through the pressure differential between the vessels, but the high operating temperatures (and thus low vacuums) mean that transfer of cream by vacuum differences is limited to two vessels in series. The vacuum levels, temperatures and flows of cream and steam thus require very careful control to ensure that excessive product loss through foaming or flooding does not occur. The modern Vacreator is now equipped with microprocessor control to assist in achieving optimum temperature conditions during operation.

The amount of taint removed is proportional to the quantity of steam used. In the spring or prolonged periods of wet weather, weed growth and the proportion of green feed in the diet result in high levels of taints, and consequently high steam flows during vacreation are required (typically 0·25–0·3 kg steam/kg cream). During drier parts of the season, less green feeds are consumed so that less steam is required (typically 0·18 kg steam/kg cream) to remove the lower levels of taint. Taints resulting from poor quality cream may also be removed by vacreation, but high steam flows are required.

The major disadvantage of vacreation is its energy usage through the relatively large quantities of steam required, as the design of the Vacreator is such that vessel flooding will prevent operation at steam flows less than 0·15 kg steam/kg cream. Some heat is recovered through the preheater, and thermo-recompressors are available which will generate low pressure steam from waste heat recovered from the vapours; this can be fed to the raw steam entering the Vacreator, but the cost only warrants their use with large, high-throughput units. The high energy usage of the Vacreator has led to some companies preferring flash pasteurisation incorporating a limited vacuum treatment. Such a process is acceptable for treating cream with a low taint level, but is generally unsuitable for cream with a high taint level that is to be used for producing butter. Also, experience tends to indicate that butter made from flash pasteurised cream has poorer keeping quality than that made from vacreated cream.

More recent work in New Zealand has indicated that steam : cream ratios can be reduced if the proportion of steam entering vessels 1 and 2 is more carefully controlled through use of a valve which limits the quantity of steam passing to vessel 2 which is at lower pressure. With such control, the differential pressure between the two vessels can be maintained, and flooding of vessel 1 is eliminated even at low steam flows (Cant, personal communication).

Steam quality is of the utmost importance in Vacreator treatment. It must be of culinary standard because of the intimate contact with the cream, and this limits the use of certain chemicals for the treatment of boiler feed-water.

Steam is injected into cream during vacreation at a velocity of approximately 140 m s^{-1} (500 km h^{-1}), and this violent treatment causes disruption of fat globules with an increase in the proportion of fat present as small globules ($< 2\,\mu m$). Vacreation will also increase the number of large fat globules ($> 10\,\mu m$) due to agglomeration resulting from foaming or flash-boiling (Dolby, 1953, 1957). The increase in the number of small fat globules can lead to higher losses of fat to buttermilk in butter manufacture. The introduction of low velocity steam diffusers should alleviate this, and this is possible when the steam split into vessels 1 and 2 is carefully controlled (Cant, personal communication).

A photograph of a Vacreator is shown in Fig. 15. The four primary vacuum vessels lie in a line with the preheater lying horizontally just underneath and to the rear of these. The flash vacuum and condenser are situated behind the preheater.

Fig. 15. Photograph of a Vacreator.

CONSUMER CREAM PRODUCTS

Towler (1982a) has presented a review of consumer cream products. The variable composition of cream, and the properties relating to its emulsified state, give a wide variety of forms which have different functions for use as food. The following presents a resumé of important consumer cream products.

1. Half-cream and single cream (10–18 per cent fat): These creams would normally be used as pouring creams for use in desserts and beverages.
2. Coffee cream (up to 25 per cent fat): The function of cream is to provide an attractive appearance to the coffee with an appropriate modification in flavour. The hot, acid conditions in coffee provide an alien environment for cream, and protein precipitation may occur (feathering) with the release of free fat and a reduction in 'whitening' effect. The incidence of feathering and methods of alleviating it will be referred to later.
3. Cultured (sour) cream (up to 25 per cent fat): Inoculation of cream with lactic bacteria will induce souring through the conversion of lactose to lactic acid with subsequent coagulation of the protein. The solids-not-fat (SNF) content of the cream is thus as important as the fat dispersion in determining the texture of the final product. Preheating given to the cream will modify the protein in such a way as to alter the protein coagulum, and also a higher protein content will lead to a firmer coagulum. Homogenisation of the cream will also increase its viscosity, and will inhibit creaming during the period of fermentation. The flavour of the finished products is very much a factor of the type of cultures used and the acidity of the final product. Sour cream is used in a number of meat dishes, vegetable dishes and confectioneries.
4. Whipping cream (30–40 per cent fat): When cream is subjected to mechanical beating, air is incorporated and continued agitation results in rupture of the milkfat globule membrane. If there is sufficient solid fat present, a firm structure will develop through bridging of adjacent fat globules. The factors which are of importance for whipping cream are:

   (a) The amount of beating required to form a stable aerated structure (whipping time). This is dependent on the apparatus used to whip the cream, and the state of the cream itself. If whipping is continued beyond the optimum time, phase inversion and separation of fat will take place as in churning.

(b) The overrun which expresses the percentage volume increase of the cream due to air incorporation. Thus, if a certain quantity of cream is whipped

$$\text{Overrun} = \frac{\text{Volume of whipped cream} - \text{Volume of unwhipped cream}}{\text{Volume of unwhipped cream}} \times 100\%$$

i.e. if a quantity of cream doubles in volume on whipping, then the overrun is 100 per cent.

(c) The stiffness of the whipped cream. A stiff whip is particularly important for cream fillings in cake or for 'piped' cream decorations.

(d) Serum leakage from the cream. This will occur with partial churning (overwhipping), and leads to an unattractive 'pool' around the whipped cream or 'sogginess' if applied to cakes.

Variables which affect the whipping properties of cream are:

(e) Fat content of the cream. A high fat content will lead to a whipped cream with a low overrun but a firm structure and little tendency to synerese (provided it is not overwhipped). Conversely, a low fat content cream will form a high overrun whip with a soft structure. A certain minimum fat content is required to form a whip, and this would normally be about 30 per cent fat for a cream without specific additives or treatment.

(f) Temperature of cream. The formation of a stable whipped structure depends on sufficient quantity of solid fat, and cream must be below 10 °C to whip satisfactorily; though there will be differences according to the lipid make-up of the milkfat in the cream. If cream is cooled to below 10 °C, warmed and then cooled again, then an increase in viscosity results and the whipping time of the cream is reduced. The process is known as rebodying, and was patented by Bergman and Svedberg (1934). The exact mechanism of the change has not been elucidated, but it is postulated to be due to changes in the milkfat globule membrane, although fat crystallisation changes could also be involved as marked changes in butter hardness can occur with the same temperature cycling.

(g) The fat globule size distribution and membrane structure. Homogenisation of cream has a deleterious effect on whipping properties resulting in a marked increase in whipping time and a considerable reduction in whip firmness. Homogenisation results in a net increase in surface area of the fat globules, and the new membranes incorporate proportionately more protein from the serum. The new membrane is more stable to mechanical beating, and it is thus much harder to form a whipped structure.

Addition of natural membrane material (phospholipid) or emulsifiers can lead to an improvement in whipping properties. Separation of milk at low temperatures gives a cream with a high phospholipid content which will lead to better whipping properties (Thomé and Eriksson, 1973). Acidity in cream will reduce the solubility of protein in the membrane and this leads to reduced whipping times. Kammerlehner (1973, 1974) has presented much useful information on factors which affect the whippability of cream.

The handling of milk and cream through transportation, separation, pasteurisation and other movements is quite critical in obtaining a cream which is of suitable quality as a whipping cream. Any disruption or homogenisation of globules can lead to a deterioration in whipping properties, and temperature changes have a critical bearing on the distribution of phospholipids which are also important in the whipping properties of the resultant cream.

Aerosol cream represents a variation on whipped cream whereby cream is packed in a can with nitrous oxide gas under pressure. Release of pressure through a valve causes the cream to be propelled out, and the dissolved gas volatilises to form a high overrun whipped cream structure, depending on the volume ratio of gas:cream. Such a whipped cream does not have the permanency of a mechanically whipped cream and shrinkage occurs. Such aerosol creams are thus more suited for desserts and confectioneries which will be consumed immediately.

Machines are also manufactured for bulk dispensing of whipped cream. Such machines normally consist of a refrigerated tank for holding the cream. A pump with an air-bleed will then take the cream and pass it through an insert designed to provide a tortuous path for the cream, resulting in much turbulence and shear which whips the cream before it is dispensed out of a nozzle. The extent of whipping is controlled through the proportion of air which is bled into the cream.

5. Double cream (>48 per cent fat): Double cream is marketed predominantly in Europe, and represents an extra rich product for addition to desserts. Whipping will produce a very dense whipped cream which can be used in gateaux.
6. Clotted cream (>55 per cent fat): This is traditionally produced in S.W. England. The traditional process involves batch heating of gravitationally separated cream. The heating process induces rapid fat rise, and the fat agglomerates on the surface; protein denaturation also takes place. Superficial skimming gives a thick spreadable product with a distinct cooked flavour. More modern methods for high volume manufacture use heat exchangers to get the cream to the required temperature, followed by mechanical separation to the required fat content.
7. High fat creams: Several high fat spreadable creams are found as indigenous products around the world. Examples are Gammer cream (Iraq) and Kajmak (Yugoslavia). High fat creams can be simply produced by passing normal cream through a second separation. Such a separator should have a relatively wide disc spacing, and the distribution channels should be approximately half way down the disc faces. As the fat content in cream rises and the globules pack together more tightly, the stability of the emulsion is reduced and phase inversion takes place very easily. A very high fat cream (70–80 per cent fat) is known as plastic cream.
8. Confectionery, butter or mock creams: Such products are low moisture products containing high concentrations of sugar or other sweeteners, and are used as cake and bun fillings. They are, in fact, phase inverted creams with the aqueous phase emulsified in the fat. The fat base is aerated by mechanical beating with sweetener being added later.

PACKAGING AND PRESERVATION OF CREAM

Cream, being a high moisture product, is perishable and, without special care to preserve it, enjoys only a limited life without spoilage. Pasteurisation of cream will extend the shelf-life to some extent, and this can be accomplished with conventional equipment. Pasteurised cream is packaged in cartons and bottles for local consumption.

For extended shelf-life there are several options, each presenting different technical difficulties in terms of maintaining the required properties of the cream during processing and storage.

Drying

Removal of water from cream will give a product with an extended shelf-life, and commercial production has been practised on a limited scale for some years. The commonest method of producing dry products from liquids is through spray-drying, and Hedrick (1967) has described the manufacture and utilisation of spray-dried cream. The particular problem in the production of spray-dried cream is the high fat: SNF ratio in the finished product, with the fat being in a liquid state at the temperature at which it would exit a spray-drying chamber. It is essential that the fat globules are encapsulated and protected by the non-fat solids, and this normally requires the addition of extra protein, usually sodium caseinate, and the addition of a suitable carbohydrate (e.g. lactose, dextrose, or sucrose) to act as a carrier. Cooling of the powder is necessary to solidify the fat and prevent caking, which would occur if the thin protective membranes are ruptured. Snow *et al.* (1967) have described a suitable method of removing such powders from a drier via a continuous belt, followed by cooling in a fluidised-bed system. Cyclone collection of powder results in mass caking of powder on the cyclone walls, and bag filters soon become impervious if fat is deposited. The continuous belt collection and cooling system is now commercially available with the Filtermat drier (Damrow Co., Wisconsin, USA), which is used extensively for producing powders with high fat contents, these powders usually being based on vegetable fats.

The high fat content of cream powder renders it susceptible to deteriorations which impair the flavour. It is essential that the cream is given sufficient heat treatment to destroy lipases. Oxidation of the fat is also a potential problem, and the addition of antioxidant should be a requirement. Storage of cream powders should be at low ambient temperatures, as an elevated temperature will give high proportions of liquid fat resulting in caking problems as well as more rapid flavour deterioration. The addition of a free-flow agent is recommended to assist in preventing caking.

Cream powder provides a milkfat concentrate in a free-flowing form, and as such has a number of potential uses. It can be used as an ingredient for dried soup, dessert, ice cream or packet cake mixes.

Dried cream has limited functionality when reconstituted as the globules do not reform as in the natural product. To obtain a functional product, incorporation of emulsifiers is necessary, and homogenisation conditions are of particular importance (Griffin *et al.*, 1970; Cooper and Peacock, 1979; Kieseker *et al.*, 1979).

Sterilisation

Applying sufficient heat to destroy all microbes and enzymes in cream, with steps taken to prevent further contamination, will render a product with a shelf-life dependent only on physico-chemical changes which may take place as a result of temperature changes or time. The simplest method of sterilisation is to first package the material, and then heat the complete package and material to render sterility. The can or glass bottle have been the traditional containers for such operations, but other retortable materials are now becoming available for packaging. Sterilisation takes place in a retort or hydrostatic steriliser using temperature/time regimes of 110–120 °C/10–20 min. This severe heating induces gross changes in the cream, with protein denaturation, Maillard browning and fat agglomeration all taking place to modify texture and flavour. Normally cream of approximately 23 per cent fat content ('reduced cream') is the base cream for in-can sterilised cream manufacture, and it enjoys a substantial market for use as a dessert adjunct, or as an ingredient in a number of food items, such as dressings or sauces. Probably its best known function is as a base for party dips. Such a low fat cream will not whip. In-bottle sterilised coffee cream has been marketed in Europe for many years. Higher fat creams are more difficult to deal with as their higher viscosity reduces heat transfer, and it is difficult to stabilise them against coagulation. Kieseker and Zadow (1973a) have described the effects of various parameters on the properties of in-can sterilised whipping cream.

The adoption of ultra-high temperature (UHT) sterilisation followed by aseptic packaging has led to the availability of a wider range of functional creams which can be stored for some months. The high temperature (135–150 °C) short time (3–5 s) treatment does not induce the same amount of chemical change that in-container sterilisation does, but other changes, such as creaming and fat agglomeration, will take place on storage, and steps must be taken in the processing of UHT creams to alleviate these problems.

Coffee cream will tend to feather more after a period of storage, and

Anderson *et al.* (1977) showed that the index of feathering was related to a progressive increase in the proportion of calcium and casein associated with the fat phase of the cream. An increase in casein content (to provide more buffering capacity to the acid in coffee) and a reduction in calcium content markedly increases resistance to feathering during storage (Cheeseman *et al.*, 1978). The gravitational separation of fat in the stored cream is inhibited through homogenisation, and the extent of homogenisation has a marked effect on the whitening effect of the coffee cream (Towler, 1982b).

The production of UHT whipping cream presents problems which require compromises in order that a long shelf-life product with adequate functional attributes will ensue. Homogenisation, which inhibits creaming, has a deleterious effect on whipping properties, although Graf and Müller (1965) showed that adjustment of conditions to give fat globule clusters of 15–20 μm would result in a good whipping cream. Additives can markedly improve the stability and properties of UHT whipping cream. Stabilisers, such as hydrocolloids, gums and gelatin, inhibit fat rise and agglomeration of fat. Australian processors produce thickened whipping cream with a high level of such stabilisers, and the cream may, in fact, be semi-solid. Such a thickened cream has to be pseudoplastic, so that the cream will become more fluid on beating and the requisite quantity of air will be incorporated. Emulsifiers aid the whipping properties of the creams by substantially increasing overrun. Their incorporation is limited, however, by the off-flavours they impart. Kieseker and Zadow (1973b) have reported the effect of several factors on the properties of UHT whipping cream. They showed that separation at low temperature, or the addition of calcium, improved whipping but gave poor storage stability, whereas separation at high temperature and the addition of calcium sequestrants led to creams with good storage stability, but poor whipping properties.

The storage temperature of UHT whipping cream is particularly important in determining its shelf-life. Storage at 30 °C will induce considerable agglomeration of fat within a relatively short time, dependent on the amount of homogenisation and added stabilisers. Storage of a stabilised cream at 5 °C will give a cream with a shelf-life in excess of 12 months. The rebodying process is particularly effective in improving the whipping properties of UHT homogenised whipping cream, but excessive temperature cycling is deleterious to shelf-life.

UHT cream is available in other forms such as single, reduced or double creams. A number of different packaging options are available, and probably aseptic canning was the first to be utilised with cream.

Now the more familiar plastic, paper and foil laminate cartons are most widely used, but plastic form, fill and seal packages are also used, these being particularly useful for individual portion cups of coffee cream. Figure 16 shows several different packages for UHT creams.

Other forms of integrated sterilising and packaging systems are being developed, and it is likely that these will become commercial realities. Whatever the outcome of such developments, the storage of creams will require special consideration to processing in order that natural destabilising influences are negated.

Freezing

Freezing of cream will inhibit bacterial spoilage, but will also, however, lead to destabilisation, and gross separation of fat and serum results on thawing. Such cream is suitable for reprocessing, and bulk frozen cream has been exported from New Zealand for a number of years (Anon., 1965). The cream may be used as an additive for 'cream soups', where flavour is the prime consideration, or in recombined milk and ice cream where homogenisation is an essential part of the process. Cooper (1978) has described a suitable formulation for the preparation of a functional whipping cream from thawed frozen cream.

Fig. 16. Photograph of aseptic packs for UHT cream.

Freezing of cream is practised in Europe for storage and subsequent utilisation in butter manufacture, which alleviates fluctuations in seasonal production in terms of both quantity and quality of final product. Frozen cream is also used to boost the fat content of market milk when the natural fat content drops below the legal minimum.

In New Zealand, cream is bulk packaged in containers of 20–25 litres capacity. Plastic containers were previously used, but a cheaper bag-in-box system has now been adopted. The containers are placed in racks which pass through a blast-freezing chamber.

In Europe, plate or rotary-drum freezers are commonly used, and Rabich (1969, 1971) has described the problems of freezing cream by these methods. In plate-freezing, the plates contain circulating refrigerant and are arranged vertically, in parallel, with bottom and end seals to form a series of moulds with hydraulic pressure maintaining the plates in place. The cream is poured into the gaps between the plates, and surface freezing is instantaneous at the pre-cooled plate surface; this is essential to prevent adhesion of the cream to the plate. The cream freezes progressively toward the centre with the refrigerant in the plates absorbing the heat, so that finally slabs of frozen cream are formed. The slabs are removed for packaging and subsequent storing by separating the plates.

In drum freezing, a rotating drum containing recirculating refrigerant is immersed in a vat of cream to form a frozen film. The frozen cream is then removed from the drum with a knife, and a flaked product ensues. Such a process gives somewhat more rapid freezing than plate-freezing, is less damaging to the cream and is continuous. The major disadvantage is that the flakes have a lower density than the slabs when packaged in bulk, and more freezer space is required for storage. Figure 17 illustrates the principle of drum freezing.

It is essential that cream is adequately pasteurised to destroy

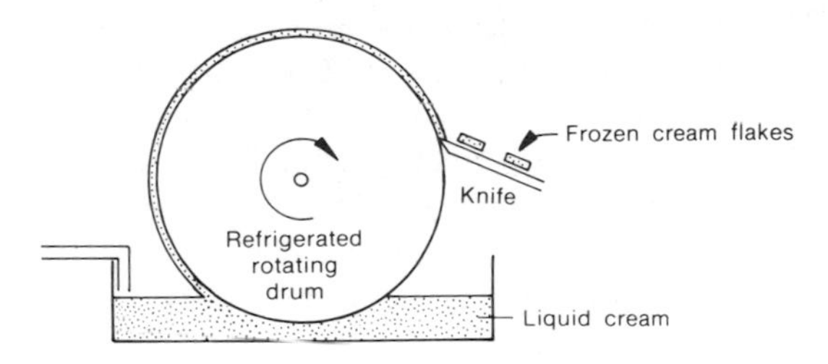

Fig. 17. Diagram of a drum freezer.

enzymes, as many are still active at the low temperatures used in frozen storage. Temperatures less than −18 °C are also recommended for long-term storage of frozen cream, as the rate of deterioration is inversely related to temperature of storage.

Without the use of additives, the stability of cream through a freeze–thaw cycle can only be ensured by the use of rapid freezing, for then only small ice crystals form, and the rupture of fat globule membranes is minimised. Londahl and Johansson (1974) first reported the development of the Pellofreeze machine (Frigoscandia, Sweden), which provides rapid freezing of cream contained as a thin sheet between two continuous stainless steel belts sprayed with low temperature glycol. The resultant frozen sheet is then broken up into flakes. The resultant frozen cream will thaw to form a fluid product which can be used as a normal consumer cream. Whipping cream and double cream are also marketed in such forms, the cream flakes being packaged in heat-sealed plastic bags as used for frozen vegetables.

A frozen novelty machine can be used to freeze cream at a rate which will give a reasonable consumer product, and Chase (1981) has described the marketing of such a cream.

Rapid freezing can also be achieved through a cryogenic process using the latent heat of low temperature boiling liquids to remove heat. In the Cryodrop process (L'Air Liquide, France), liquid carbon dioxide is sprayed into a vessel containing an agitator, and liquid cream is introduced at the same time. The carbon dioxide partially volatilises, and the rest is converted into small particles of solid carbon dioxide (dry ice), which is the more stable form at atmospheric pressure. The liquid cream freezes quickly on contact with the dry ice. An agitator rotates continuously in the vessel to produce a granular end product (Jozon, 1983); Fig. 18 illustrates the principle of the Cryodrop process. Some destabilisation does ensue in this process, and this may be due to the acidity of the carbon dioxide, and the mechanical churning of the agitator and the effervescing gas.

Liquid nitrogen, which has a boiling point of −196 °C, will freeze cream very rapidly. Work with a freezing tunnel has shown that a satisfactory frozen cream for direct consumer use can be produced by such a system; Fig. 19 illustrates the principle of the freezing technique. Cream is poured into suitable containers on a continuously rotating belt which, in turn, feeds them into an insulated tunnel. Liquid nitrogen is introduced at a point towards the other end of the tunnel, and a series of fans distributes the cold nitrogen gas. Freezing is thus accomplished by

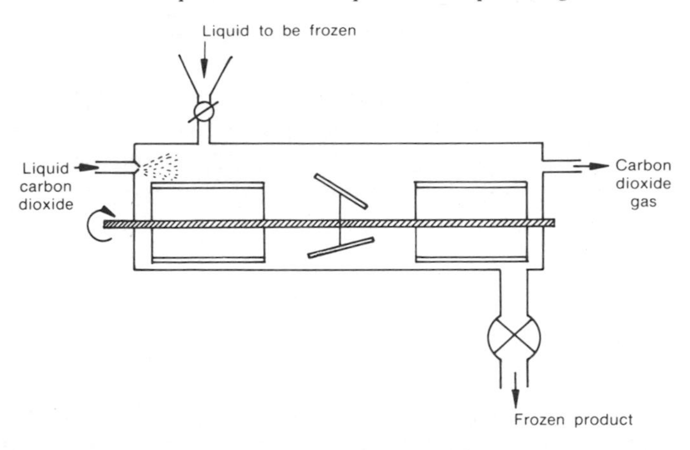

Fig. 18. Equipment used for cryogenic freezing using liquid carbon dioxide.

an essentially countercurrent flow of cold nitrogen gas, and the frozen cream is removed on exiting the tunnel.

Experiments on the cryogenic freezing of cream have shown that the state of the cream is equally as important as the rapidity of the freezing process in the prevention of destabilisation. The preservation of the natural milkfat globule membrane is very important in getting freeze–thaw stability. Homogenisation adversely affects freeze–thaw stability, as does partial churning or separation at high temperature (Towler, unpublished). Additives, such as emulsifiers and stabilisers, assist in the freeze–thaw stability of cream if rapid freezing is not possible. The storage of frozen cream for consumer purposes demands a constant, as well as low, temperature, since temperature cycling may lead to the formation of larger ice crystals with resultant damage to the fat globules.

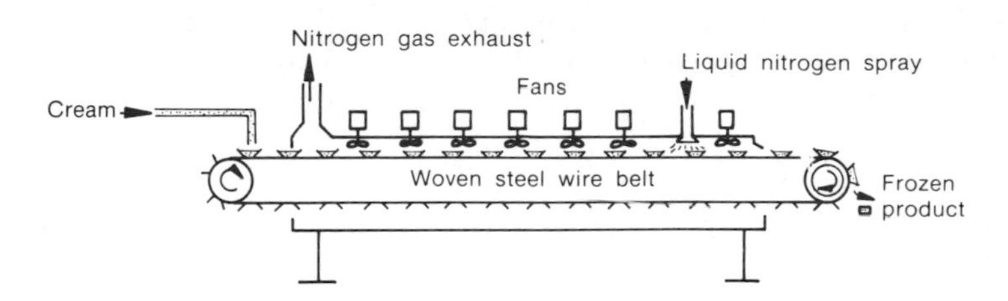

Fig. 19. Equipment used for cryogenic freezing using liquid nitrogen.

Other Methods of Preservation

The addition of low molecular weight carbohydrates (e.g. glucose and fructose) to depress water activity, along with low temperature storage, will preserve cream in a fluid state without the consequential damage caused by freezing. This is the Freeze-Flo® process (Rich Products, USA) as patented by Kahn and Eapin (1979).

Recombined cream

A cream can be reformed from milkfat concentrates and recombined skim-milk or buttermilk by the application of heat followed by homogenisation. For a good recombined whipping cream, it is better to incorporate phospholipids through the use of buttermilk or alpha-serum powder. Alternatively emulsifiers may be used. The homogenisation conditions are critical in obtaining a cream with the desired functional characteristics.

Whipped toppings and coffee whiteners are recombined emulsions designed to be cream substitutes. The use of vegetable fat sources results in much cheaper products, and with a selection of different softening points, functional attributes can be much more easily tailored to meet the product's needs.

CONCLUSION

The word cream has long been associated with being a premium product. Although milkfat may not be regarded as the ideal food by nutritionists, its unique flavour and properties make it a favoured ingredient in many foods. In cream, the milkfat is protected and preserves a special flavour.

New methods of packaging and preservation are going to lead to newer forms of cream for consumer use, and although improved forms of fabricated and simulated products may ensue, natural cream is bound to retain a special place in the eye of the consumer.

ACKNOWLEDGEMENTS

The author would like to thank his colleagues at the New Zealand Dairy Research Institute for assistance in preparing this chapter, particularly Dr Euan Cant for his contribution to Vacreation. The valuable assistance of Alfa-Laval, Lund, Sweden and Westfalia, Oelde, West Germany (through their agents McEwans) is also gratefully acknowledged.

REFERENCES

Alfa-Laval (1980). *Dairy Handbook,* Alfa-Laval AB, Dairy and Food Engineering Division, Lund, Sweden.

Anderson, M., Brooker, B. E., Cawston, T. E. and Cheeseman, G. C. (1977). *Journal of Dairy Research,* **44**, 111–24.

Anon. (1965). *New Zealand Dairy Exporter,* **40** (9), 33–4.

Bechtolsheim, C. von (1888). German Patent 48 615.

Bergman, T. V. and Svedberg, H. A. (1934). US Patent 1, 944, 541.

Chase, D. (1981). *Milk Industry,* **83** (6), 17–20.

Cheeseman, G. C., Anderson, M. and Wiles, R. (1978). British Patent 1, 526, 862.

Cooper, H. R. (1978). *New Zealand Journal of Dairy Science and Technology,* **13**, 202–8.

Cooper, H. R. and Peacock, I. C. (1979). *New Zealand Journal of Dairy Science and Technology,* **14**, 291–8.

Dolby, R. M. (1953). *Journal of Dairy Research,* **20**, 201–11.

Dolby, R. M. (1957). *Journal of Dairy Research,* **24**, 372–80.

Graf, E. and Müller, H. R. (1965). *Milchwissenschaft,* **20**, 302–8.

Griffin, A. T., Amundson, C. H. and Richardson, T. (1970). *Food Product Development,* **4** (7), 49, 52–3, 56.

Hedrick, T. I. (1967). *American Dairy Review,* **29** (9), 36, 120, 122, 124.

Jozon, P. (1983). Preprints of 16th International Congress of Refrigeration, Paris, 576–81.

Kahn, M. L. and Eapin, K. E. (1979). US Patent 4, 154, 863.

Kammerlehner, J. (1973). *Deutsche Molkerei-zeitung,* **94**, 1516, 1518, 1520, 1521, 1637–40, 1742–3.

Kammerlehner, J. (1974). *Deutsche Molkerei-zeitung,* **95**, 1758–61, 1789–92, 1820–5.

Kieseker, F. G. and Zadow, J. G. (1973a). *Australian Journal of Dairy Technology,* **28**, 108–13.

Kieseker, F. G. and Zadow, J. G. (1973b). *Australian Journal of Dairy Technology,* **28**, 165–9.

Kieseker, F. G., Zadow, J. G. and Aitken, B. (1979). *Australian Journal of Dairy Technology,* **34**, 21–4, 112–13.

King, D. W., Russell, R. W., McDowell, A. K. R. and Dolby, R. M. (1972). *New Zealand Journal of Dairy Science and Technology,* **7**, 4–6.

Lang, F. and Thiel, C. C. (1955). *Dairy Science Abstracts,* **17**, 85–107.

Lipatov, N. N. (1976). *Separieren in der Milchindustrie* [*Separation in the Dairy Industry*], VEB Fachbuchverlag, Leipzig, German Democratic Republic.

Londahl, G. and Johansson, S. (1974). *XIX International Dairy Congress,* Brief Communications IE, 649–50.

Mulder, H. and Walstra, P. (1974). *The Milkfat Globule. Emulsion Science as Applied to Milk Products and Comparable Foods,* Commonwealth Agricultural Bureaux, Farnham Royal, Bucks, England.

Pato, T. (1978). *XX International Dairy Congress,* Volume E, 624–5.

Phipps, L. W. (1969). *Journal of Dairy Research,* **36**, 417–26.
Rabich, A. (1969). *Deutsche Milchwirtschaft,* **20**, 645–50.
Rabich, A. (1971). *Dairy Industries,* **36**, 207–11.
Rothwell, J. (1975). (Ed.) *Cream Processing Manual,* The Society of Dairy Technology, Wembley, Middlesex, England.
Snow, N. S., Buchanan, R. A., Freeman, N. H. and Bready, P. J. (1967). *Australian Journal of Dairy Technology,* **22**, 122–5.
Te Whaiti, I. E. and Fryer, T. F. (1975). *New Zealand Journal of Dairy Science and Technology,* **10**, 2–7.
Thomé, K. E. and Eriksson, G. (1973). *Milchwissenschaft,* **28**, 502–6.
Towler, C. (1982a). *New Zealand Journal of Dairy Science and Technology,* **17**, 191–202.
Towler, C. (1982b). *XXI International Dairy Federation,* Brief Communications Volume 1, Book 2, 114.
United Nations Food and Agricultural Organization/World Health Organization (1977). *Milchwissenschaft,* **32**, 278–9.

Chapter 3

Production of Butter and Dairy-based Spreads

R. A. Wilbey
Department of Food Science, University of Reading, UK

The evolution of batch buttermaking practices with the gradual replacement of wooden churns by stainless steel was overtaken in the 1950s by the development of continuous processes. A standard reference for batch buttermaking is that by McDowall (1953).

The various methods of continuous buttermaking developed at that time were described by Wiechers and De Goede (1950). The Fritz process has since become the most common in commercial use, while the principles of the reseparation and phase reversal methods are applied in the manufacture of anhydrous milk fat.

The structure of butter has been reviewed by King (1964), Mulder and Walstra (1974) and more recently by Mortensen (1983). Accounts of the composition and variability of the constituents of the milk fat may be found in either the latter publication, or that by Walstra and Jenness (1984).

For the purpose of this discussion, the term 'Dairy-Based Spreads' is limited to those with more than half their ingredients derived from milk, and with a continuous lipid phase. These products are water in oil (W/O) emulsions, unlike milk and cream which are oil in water (O/W) emulsions.

PRINCIPLES OF BUTTERMAKING

The buttermaking process, whether by batch or continuous methods, consists of the following steps:

(i) preparation of the cream, with or without ripening;
(ii) destabilisation and breakdown of the O/W emulsion;
(iii) aggregation and concentration of the fat particles;
(iv) formation of a stable W/O emulsion;
(v) packaging;
(vi) storage and distribution of the product.

Preparation of Sweet Cream

The separation of sweet (unripened) cream for buttermaking follows the general principles described in the previous chapter. The fat content of the cream should be standardised to the optimum level for the buttermaking equipment to be used. A typical level for continuous buttermakers would be 40 per cent fat.

High temperature–short time (HTST) heat treatment of the cream should be at least at 74 °C, with a minimum hold of 15 s. In the UK, the legal minimum is 72 °C with a range up to 76·7 °C for at least 15 s. (UK Intervention Board, 1982.) A higher temperature (over 79·4 °C) with shorter holding time may be used to give an equivalent heat treatment. More severe heat treatments should be avoided, as the higher the temperature, the greater the migration of copper from the milk serum into the milk fat. Increasing the level of copper in the milk fat makes it more susceptible to the development of oxidative rancidity, and can reduce the shelf-life of the butter. Excessive heat treatment will also lead to the formation of cooked flavours.

After pasteurisation, the cream should be cooled to 4–5 °C, using regeneration and chilled water cooling. (For butter to go into intervention, the cream should be cooled to below 7·2 °C.) The plate heat exchanger design should avoid high pressure differentials, as the high shear rate associated with this could result in premature damage to the milk fat globules. This damaged cream would be unstable and liable to cause problems on subsequent handling prior to buttermaking. Cream may also be damaged by incorrect sizing of pumps, particularly centrifugal pumps. It is recommended that the cream for buttermaking be handled by positive displacement pumps.

Any rubber components in the plant for cream handling or buttermaking must be of a fat resistant type, e.g. nitrile rubber. Rubber compounds that are not fat resistant break down after several days contact, and will contaminate the product.

As an alternative to closed system cooling, the cream may be heated to a higher temperature, e.g. 90 °C, then passed into a vessel at low pressure *c.* 20 kPa/0·2 bar. Evaporation of water from the cream both cools the cream and removes some of the volatiles present in the cream. When cream of indifferent quality must be processed, then this flavour stripping action can be advantageous, and compensate for the additional processing cost. The evaporative cooling process does create additional shear forces on the fat globules, resulting in smaller fat globules. Some of the fat globules appear to survive in the continuous lipid phase of the butter, where the smaller globule size may contribute to an improved texture.

Care is needed in this form of cream treatment to avoid excessive shear forces which would create too many small fat globules. This would result in higher losses of fat in the buttermilk. The more severe vacreation processes, employing direct steam injection heating in addition to evaporative cooling, should only be used where there are taint problems with the cream.

During pasteurisation, all of the fat becomes liquid; on cooling crystallisation will commence. Most of the latent heat of crystallisation of the milk fat is released before the cream leaves the cooler, but there is a significant release of heat in the 2 h following pasteurisation. The temperature rise as a result of this release of heat will depend on the design of the cooler, and on the fat content of the cream. Allowance for the rise of maybe 2 °C, can be made either in the initial cooling of the cream, or by further cooling in the storage vessel.

After pasteurisation and cooling, the cream should be held for a minimum of 4 h, preferably overnight to permit sufficient crystallisation of the milk fat. The fat usually crystallises in mixed crystals of the α and β' forms, varying with the cooling conditions used (Mulder and Walstra, 1974). Some equilibration may be expected on storage. This storage is often referred to as ageing the cream.

For smaller scale operations, cream ageing tanks of 10–20 000 litres capacity may be used, as illustrated in Fig. 1. The tanks are usually vertical with cooling panels or coils built into the walls. Agitation is by slow moving gate-type agitators which may also contain refrigerant. These ageing tanks are suitable for cultured and sweet creams.

At the larger scale of operations however, this size of vessel would require frequent changes during a product run. The feed to the larger continuous buttermakers is commonly 10 000 litres h^{-1}, and may be in excess of 20 000 litres h^{-1}. Any change in the supply would require compensating adjustments to be made to the running of the buttermaker.

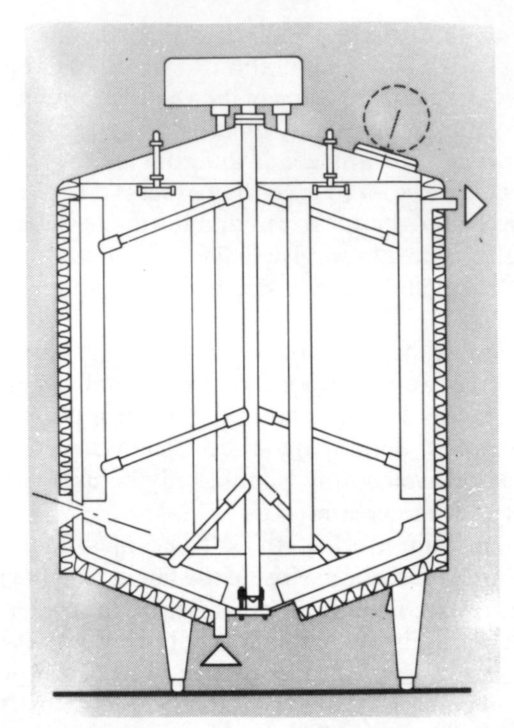

Fig. 1. Sectional drawing of cream processing tank (courtesy Silkeborg Ltd, UK).

Thus for large scale buttermaking, variation in cream supply may be minimised by ageing the cream for a day's production in a single silo. The silo should be equipped for intermittent agitation to avoid separation and stratification. Cooling of cream in the silo may be achieved either by refrigeration through the silo wall, or by recirculation through an external heat exchanger. Where smaller tanks have been used for cooling and ageing the cream, then the creams should be blended in the silo at least 2 h before buttermaking commences.

The salvaging of homogenised cream by blending into cream for butter production should be avoided, as this leads to higher fat losses in the buttermilk.

Ripened Cream

Originally ripened cream was the inevitable consequence of the activity of the microbial contaminants during the lengthy gravity separation, aided by the lack of refrigeration. The introduction of rapid separation, pasteurisation and cooling required the re-introduction of suitable bacteria to produce the ripened cream.

Pasteurisation of cream for ripening is commonly carried out at up to 100–110 °C with no hold. The cream may also be subjected to vacuum treatment during cooling. The fat content of cream for ripening is normally 36–40 per cent, lower than that for sweet cream butter.

In the simplest situation, the cream may be cooled to the ripening temperature and the culture of starter organisms added, e.g. 1–2 per cent starter addition for a fermentation time of 12–18 h at 20 °C. For cream ripening, a mixed starter system should normally be used, including *Streptococcus lactis, Str. cremoris, Str. cremoris* subsp. *diactylactis* and *Leuconostoc cremoris* to produce both acidity and the characteristic flavour.

Diacetyl is a major flavour component in the ripened cream, but a large number of other compounds also contribute at levels both above and below the individual sensory thresholds where synergistic interaction may occur (Mick *et al.,* 1982). Biosynthesis of diacetyl is not significant above pH 5·2. Thus, stopping the fermentation by cooling the cream at pH 5·1–5·3 results in a mild flavour; whilst continuing the fermentation to pH 4·5–4·7 results in a pronounced flavour in terms of both diacetyl and lactic acid.

Cooling the cream inhibits the fermentation, and may be achieved by refrigeration through the walls of the culture vessel, or by passing the cream through an external heat exchanger. Allowance must be made for the continuation of the fermentation during cooling, and the commencement of cooling offset accordingly.

Starter cultures vary considerably in their sensitivity to low temperatures. Cooling to less than 10 °C is usually adequate for short term retardation of starter metabolism, but for longer term storage of ripened cream, then a lower temperature of 3–4 °C is desirable. As with the sweet cream, the low temperature storage will permit further crystallisation of the milk fat.

Modifications to cream treatment

The slow cooling of the cream after it is ripened leads to the formation of larger fat crystals than when the cream is cooled to 5 °C immediately after pasteurisation. This results in ripened creams producing firmer butters than the equivalent sweet cream. Control of the final texture of the butter by varying the fermentation and cooling conditions has been developed to counteract this effect, and to produce a more consistent product throughout the year. The first method was published by Samuelsson and Pettersson, and is often referred to as the Alnarp process.

To produce a softer butter from winter cream with higher melting milk fats, the cream should first be cooled to 8 °C and held for 2 h to promote rapid crystallisation with the formation of fine crystals. The starter culture should then be added, and the cream gently warmed to 19 °C for a further 2 h, after which the cream should be cooled to 16 °C to complete the fermentation. Fermentation under these conditions usually takes 14–20 h, at the end of which the cream should be cooled to 12 °C before churning.

A firmer butter may be produced from summer cream (with relatively lower melting fat) by initially cooling the cream to 19 °C. Starter culture should be added, and after 2 h the cream should be cooled to 16 °C, and held for a further 3 h before cooling to 8 °C and holding overnight. This method requires a higher level of starter addition to compensate for the shorter fermentation time.

These process conditions have been modified extensively to suit individual requirements. A series of processes has been proposed based on increases in the iodine value of the milk fat. A more fundamental approach by Frede *et al.* (1983) used the melting and solidification curves of the milk fat to establish the process conditions. (The transition between solid and liquid milk fat is not regular as fractions of the milk fat change phase at differing temperatures, illustrated in Fig. 2.) For instance in summer, the cream should first be cooled to a fermentation temperature above the upper solidification point of the milk fat, starter culture added and the cream held at that temperature until the desired pH is attained. The ripened cream should then be cooled to 6 °C and held for 3 h, before bringing the temperature up to the buttermaking temperature, which should be below that of the low melting peak. Conversely, a softer butter would be prepared, e.g. from winter cream, by first cooling the freshly pasteurised cream to 6 °C and holding for 3 h. The cream should then be warmed to 2–3 °C above the melting point of the lower main fraction (typically 16–21 °C), the starter added, and the cream held for at least 2 h until the desired pH has been attained. The cream is then cooled to a temperature between the solidification points of the two main fractions before buttermaking, this being a compromise between spreadability of the butter and losses of fat in the buttermilk. As with the original Alnarp process, the starter culture addition would vary between 1 and 7 per cent depending on starter activity, temperature and the fermentation time available.

These processes can be applied to sweet cream as well as cultured cream, but add to the cost of processing. A modified Alnarp treatment using a plate heat exchanger to simplify and shorten the processing was reported by Dixon (1970).

Transfer of Cream to the Buttermaker

Optimum temperatures for ageing the cream are normally lower than that needed for efficient churning. The aged cream is also susceptible to damage if mishandled, which may result in blocked pipework and excessive fat losses in the buttermilk.

Transfer should be carried out using a variable speed positive pump, conservatively rated so that its speed does not normally exceed half the maximum. Feed pipes should also be sized to prevent starvation of the pump, allowing flow rates of 0·2–0·4 m s^{-1} for sweet cream but lower rates for the more viscous ripened creams.

The most economic churning temperature for the cream may be achieved by passing the aged cream through a heat exchanger. Plate heat exchangers are normally used for this duty. There should be a low pressure drop, typically less than 50 kPa/0·5 bar, across the heat exchanger, which should also be operated with a small temperature differential (1–2 °C) between the cream and the water. The cream outlet

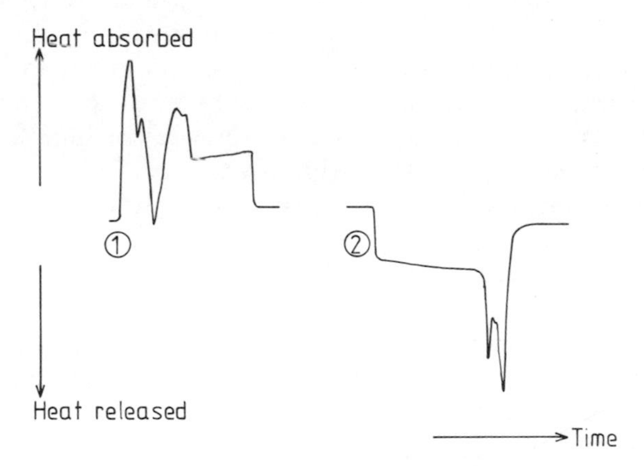

Fig. 2. Example of (1) heating and (2) cooling curves for milk fat — obtained by a differential scanning calorimetry.

temperature must be controlled to within $\pm 0{\cdot}25\,°C$ of the target. Minimising the temperature differential avoids localised overheating, and thus disturbance of the fat crystallisation during the reheating. It also reduces the need for recycling cream from the heat exchanger at start-up and during stoppages.

A consistent feed rate is important in maintaining steady butter-making conditions. One method is to interpose a balance tank between the ageing silo and the pump. This is necessary when a cream recycling line is installed, or when smaller silos are used giving frequent changes of cream during the day. An alternative method, when the reheating of the cream uses small temperature differentials, is to use an in-line flowmeter with a control loop back to the variable speed pump. Pipework, valves and fittings should be designed and installed to transfer the cream with the minimum of stress.

Thus, the supply of cream to the buttermaker should be consistent in terms of:

(i) chemical composition (fat, pH)
(ii) physical characteristics (viscosity, fat crystallisation)
(iii) temperature (max $\pm 0{\cdot}25\,°C$)
(iv) feed rate (max $\pm 0{\cdot}5\%$)

Though it is a general rule to avoid damage to the fat globule before it gets to the buttermaker, some controlled destabilisation can be carried out. Simon Freres produces a turbo-cream feed unit which injects oil-less, filtered, compressed air into the cream line at a rate of approximately $5\ N\ m^{-3}$ air per 1000 litres of cream. The air–cream mixture is passed through a static mixer to produce a foam, thus destabilising the cream emulsion immediately before it enters the buttermaker. The principal claims for this pretreatment are reduced power consumption and fat losses at buttermaking, together with greater flexibility in the types of cream and tolerance of suboptimal process conditions.

Conversion of the Cream to Butter

Once the cream is in the butterchurn, then the fat globules must be disrupted under controlled conditions to destabilise the emulsion and bring about agglomeration of the milk fat.

The aged cream contains approximately equal proportions of solid and liquid fat in the globules. The solid fat is in the form of crystals, possibly lining the globule membrane, with liquid fat occupying the core of the globule. Larger fat crystals, limited by the diameter of the globule, may also be formed. The crystals make the globules less elastic, so that the membrane will be disrupted when the globule is subjected to mechanical stress.

In batch churns, the mechanical stress is created by rotating the partly filled churn so that the cream is lifted up the ascending wall of the churn and then cascades to the base (Fig. 3).

Air bubbles also become entrained in the cream to form an unstable foam. Liquid fat escaping from the disrupted fat globules spreads over the water–air interface in the cream, together with whole and disrupted fat globules. Coalescence or desorption of the air bubbles leads to a reduction of the surface:volume ratio, with concentration of the liquid and globular fat at the surface. This results in an association of the globules into granules bound together by the hydrophobic liquid fat fraction. (Walstra and Jenness (1984) give a more detailed account.) The continuing agitation of the destabilised emulsion leads to a build up of the aggregates to form visible butter grains. Membrane material is lost from the grains into the aqueous portion to form the buttermilk.

Whereas the batch process carries out the destabilisation process relatively slowly with a large mass of cream, the continuous buttermakers operating on the Fritz principle destabilise small quantities at a fast rate.

Cream is fed into the continuous buttermaker (Fig. 4) at the top of the

Fig. 3. Batch churning of butter (courtesy Silkeborg Ltd, UK).

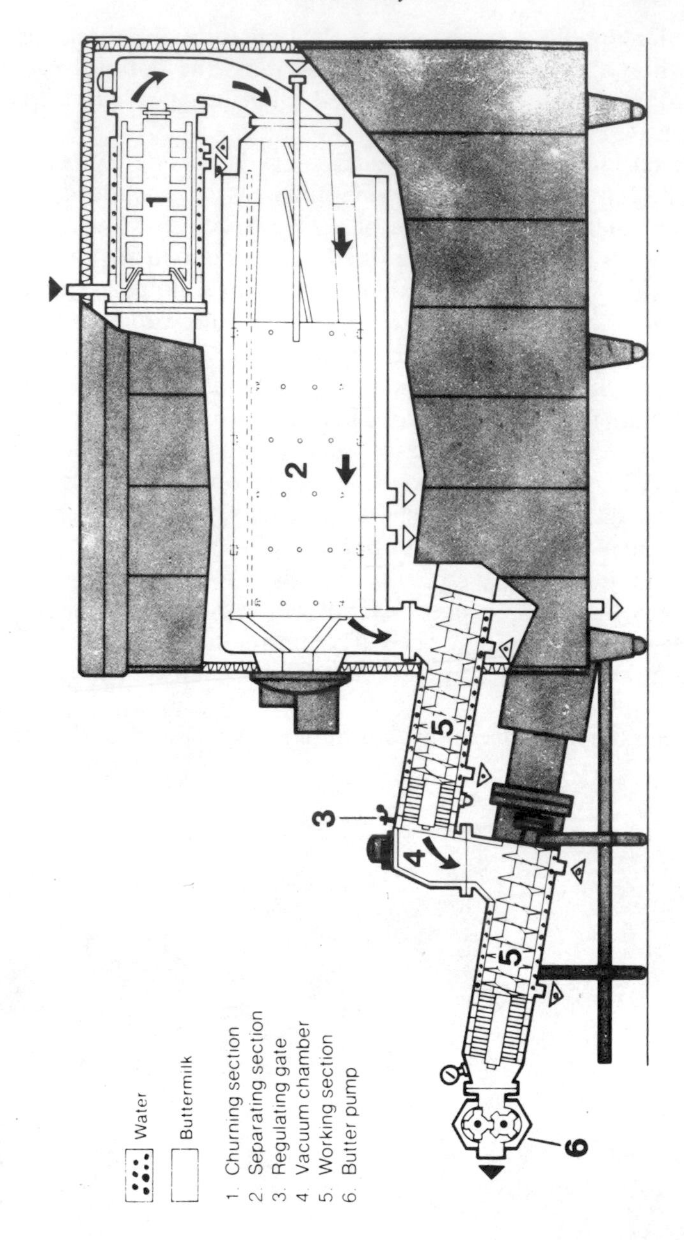

Fig. 4. Section through modern continuous buttermaker (courtesy Silkeborg Ltd, UK).

first churning cylinder. This cylinder is cooled by a chilled water jacket, and breakdown of the cream emulsion is carried out by a multibladed dasher. This dasher is driven through a variable speed gearbox and is normally operated at *c.* 1000 rpm. The action of the dasher both aerates the cream and damages the globules in a period of 1–2 s. The speed of the dasher blades can be adjusted to obtain the desired buttergrain — a faster speed will give larger buttergrains. Excessive speed results in large grains with too much buttermilk retention and a greater formation of colloidal liquid fat particles, giving higher fat losses in the buttermilk.

The mixture of buttergrains and buttermilk drops from the first cylinder into the back of the second cylinder (see Fig. 5) where the buttergrains are consolidated. This second cylinder is built as a perforated rotating drum, larger than the first cylinder and rotating slowly (e.g. up to 35 rpm), so that the buttergrains pass along this inclined rotary screen with a tumbling action. The relatively gentle tumbling action brings about further consolidation and aggregation of the buttergrains, while most of the buttermilk drains away through the rotating screen. The buttermilk is pumped away, but some may be cooled and then recycled to the second cylinder to help cool the buttergrains. Washing of the grains in the later stage of the cylinder may also be carried out, with separate collection of the washwater. This additional process is not desirable under normal conditions, as the disposal of the washwater adds to costs, and the replacement of non-fat milk solids by water results in an increase of the more expensive fat to keep within the maximum moisture content (typically 16 per cent).

From the second cylinder, the moist grains of butter fall into the working section of the buttermaker. This section uses contra-rotating augers to consolidate the grains of butter into a heterogeneous mass, expelling more buttermilk from the grains as they are squeezed together and carried up to a series of perforated plates and mixing vanes. The expelled buttermilk, together with buttermilk coming off the end of the second cylinder, forms a pool at the base of the working section. The level of this pool should be held constant as this will affect the moisture content of the butter. Fat granules in the buttermilk will float to the surface and be reincorporated into the butter; a spinning disc clarifier may be built into the buttermaker to aid fat recovery. Buttergrains may also be recovered by running the buttermilk over a fine sieve before pumping away for further treatment. Better recovery can be achieved by using a vibrating sieve, which will not block so easily and will agglomerate the recovered butter fines. Some fat may also be recovered

Fig. 5. Views of churning (top) and separating (bottom) sections (courtesy Silkeborg Ltd, UK).

from the buttermilk using a milk separator at approximately half its normal capacity. However, recovery of the fat will never be complete, as some phospholipids (from the globule membrane) and colloidal fat droplets will always remain in the buttermilk.

Compacted buttergrains are fed from the auger through the series of alternating perforated plates and impellor blades, examples of which are shown in Fig. 6. The amount of work done on the butter may be increased by reducing the size of the holes in the plates, by increasing the number of plates and impellors, and by varying the angle of some of the impellor blades. Blades with no pitch angle generate the highest shear conditions on the butter, but rely on the other blades and the auger to move the butter through the orifice plates. The shear forces generated not only further the consolidation of the buttergrains, but break up the droplets of buttermilk remaining in the matrix of fat to form a dispersed aqueous phase of the now water in oil emulsion. The droplets of aqueous phase should ideally have a diameter of less than 10 μm.

Fig. 6. View of perforated plate and mixing vanes in working section (courtesy Silkeborg Ltd, UK).

Partially worked butter may contain more than 5 per cent entrained air. This can be reduced by passing the butter through a vacuum section to obtain a denser, finer textured product with less than 1 per cent air incorporation. A second set of augers removes the butter from the vacuum section and forces it through a further set of orifice plates and blades to complete the emulsification before discharge from the buttermaker. This second working section may be driven from the same drive as the first section, or may have an independent drive. The latter method can enable a larger vacuum stage to be incorporated, and give greater flexibility in setting up optimum working conditions for the butter, e.g. the second stage being run at 2–3 times the speed of the first stage. The greater flexibility should be balanced against the need to monitor and control a further set of variables however.

The Simon Freres Contimab machines apply the Fritz principle in a slightly different manner. De-emulsification and granule formation are carried out in a single, larger cylinder. The cylinder may have a standard sand-blasted finish, or also have an expanded metal screen within the cylinder for use with creams needing a more rigorous churning effect, e.g. sweet cream with less than 42 per cent fat, or cultured cream with less than 35 per cent fat. The beater blades operate relatively slower (600–800 rpm), and the blades vary in size to increase the clearance from 3–4 mm up to 6–7 mm at the discharge. The mixture of buttergrains and buttermilk falls into the second, separation section which uses contra-rotating augers to convey and compact the buttergrains. Cooled buttermilk may also be recycled to cool the buttergrains prior to compaction. The action of the augers squeezes buttermilk out of the grains as they are compacted, the unworked butter being then extruded through an orifice plate to drop into the working sections, where surplus buttermilk may also be drained off.

Salted Butter

Batch produced butter may be salted by adding salt to the buttermaker after washing and draining. The salt must be of high quality, e.g. BS 998:1969, with low levels of lead (max. 1 ppm), iron (max. 10 ppm) and copper (max. 2 ppm). The grains of salt should be fine, all passing through a 1·4 mm test size with less than 0·2 per cent retention on an 850 μm sieve.

The salt sets up an osmotic gradient which draws water from the buttergrains, and this effect can lead to free moisture defects if the butter is not adequately worked. Coarse grains of salt which do not completely dissolve will give a gritty product. Fine milling of the salt enables the salt to be more evenly dispersed, and to dissolve faster in the butter.

With continuous buttermaking, the need for a fine salt is increased, and the grain size should not exceed 50 μm. Salt must be dosed as a continuous stream into the buttermaker, preferably into the first working section. The solubility of salt is not very temperature sensitive, and approximates to 26 per cent w/w at ambient temperatures. This restricts the use of a saturated brine to low salt addition rates, less than 1 per cent salt in the final product. Allowance must be made in running the buttermaker to produce buttergrains with a lower moisture content to compensate for the water in the brine. With 16 per cent maximum moisture and 1 per cent salt content, an addition of saturated brine at 3·8 per cent requires a moisture content of less than 13·2 per cent in the unworked butter prior to salt addition.

In the UK, a salt level of 2 per cent is common, which would require an excessive addition rate of saturated brine. The brine should, therefore, be made up as a saturated suspension, normally containing 50 per cent salt, but possibly up to 70 per cent salt for the manufacture of extra salted butter containing 3–4 per cent salt.

The brine suspension should be prepared by dissolving finely milled salt (particle size less than 50 μm) in potable water. During and after the initial dispersion, the mix must be vigorously agitated to ensure the suspension is kept homogeneous. Dispersion of the salt should be well before the brine is used, preferably at least 2 h before use, to ensure that the brine is saturated, and the remaining salt crystals are very small.

A rise in the temperature of the brine during agitation has little effect on the solubility of the salt, but will aid incorporation of the brine into the butter.

The brine slurry should be metered to the buttermaker by positive displacement pumps. The metering accuracy of the pumps may be improved by minimising the level variation in the brine tank. The presence of coarse grains of salt in the brine will obstruct the valves, and result in inaccurate dosing as well as increased wear. The brine must not be allowed to stand in the pump or pipework, as the salt will come out of suspension and cause blockages. At any stoppage, the brine slurry must be recycled till either the buttermaker is restarted, or the dosing equipment washed out.

A 2 per cent salt addition raises the average salt level in the aqueous phase of the butter to over 11 per cent, apparently sufficient to inhibit most spoilage micro-organisms. However, as not all the aqueous phase will be salted, the keeping quality of the butter will be largely determined by the overall handling and process hygiene.

The dosing system may also be used for adding water to unsalted or slightly salted butters to maximise yields. Good bacteriological quality of the water is essential to avoid contamination, especially with psychotrophs which are frequently present in the mains water supply. The water should be treated by filtration, ultra-violet light or pasteurisation.

Alternative Processes for Ripened Butter

In the production of ripened butter, the ripening process adds to production costs, and the migration of copper to the fat increases the susceptibility of the butter to oxidative rancidity. Addition of salt to the ripened butter causes further destabilisation. Lactic buttermilk also presents a disposal problem, as it can seldom be converted economically into other dairy products, and so must often be sold at nominal cost for animal feed.

Several methods have been proposed for overcoming the problems with lactic buttermilk, the most important being that developed at the Netherlands Institute for Dairy Research, and subsequently referred to as the NIZO process (1976).

This process uses sweet cream which is churned to yield sweet buttermilk, suitable for further processing, and unripened buttergrains with a low moisture content (13–13·5 per cent). Starter culture and concentrated lactic acid preparation are added to the unworked butter, which may then be worked and packed in the normal way. The characteristic flavour and aroma develop in the butter on storage to give a similar product to that produced traditionally.

The starter cultures should include both *Streptococcus lactis* subsp. *diacetylactis* (e.g. strain 4/25) and *Leuconostoc cremoris* (e.g. strain Fr 19). The leuconostocs are able to metabolise acetaldehyde produced by the streptococci, and hence avoid the development of a 'green' flavour. Initially the starter organisms were mixed together before injection into the butter, but this was altered to permit separate aeration of the streptococcal culture to increase the diacetyl level in the culture to *c.* 40 ppm, so avoiding problems of reduced diacetyl production on storage below 6°C. The addition of starter, at up to 2 per cent of the butter, is not sufficient to reduce the pH of the butter to less than pH 5·3 as required by Dutch regulations; this problem was overcome by adding a 'culture concentrate' together with the leuconostoc culture.

Culture concentrate is a lactic acid preparation, produced by fermentation of a lactose-depleted whey by *Lactobacillus helveticus.* The fermented whey is then passed through an ultrafiltration plant, and the permeate subsequently concentrated to at least 11 per cent lactic acid; excessive concentration can lead to lactose crystallisation on storage. The addition of 0·7 per cent of this concentrate to the butter, together with the 2 per cent of starters, is sufficient to reduce the pH to less than pH 5·3. Higher levels of concentrate may be used to produce a more acid butter, e.g. pH 4·6. A flow diagram for the NIZO process is given in Fig. 7.

An improved process was claimed by Wiles (1978), avoiding the reliance on the starter organisms for flavour development. The characteristic flavour is achieved by using a flavouring based on diacetyl, preferably a distillate from a specially cultured skim-milk (e.g. Hansen's 15X Starter Distillate). Acidity is increased by the addition of lactic acid. Salt, which would normally inhibit the starter culture, may be added to the mixture of distillate and lactic acid to help flavour the butter to UK market requirements. Cultures may be added separately to bring the level of starter organisms to a similar level to the traditional product (*c.* 5000 cfu g^{-1}). The exact level of starter addition and its metabolic activity are not critical in this application — enabling longer storage of the starter if necessary.

CONTROL OF BUTTERMAKING

Traditional chemical analysis of butter is unacceptably slow for use with modern buttermakers, where the production rate may exceed 80 kg min^{-1}.

Economically the most important parameters to be controlled are the moisture and salt levels. Process parameters on the buttermaker must be set to maximise the moisture and salt levels within the specifications, while producing an acceptable product in terms of moisture distribution, flavour and texture. Microcomputers are now increasingly used to aid the optimisation of processes.

The level of moisture in the butter may be sensed by its effect on the dielectric constant of the butter, and this information fed back to the control system of the buttermaker. Checks are required to maintain the

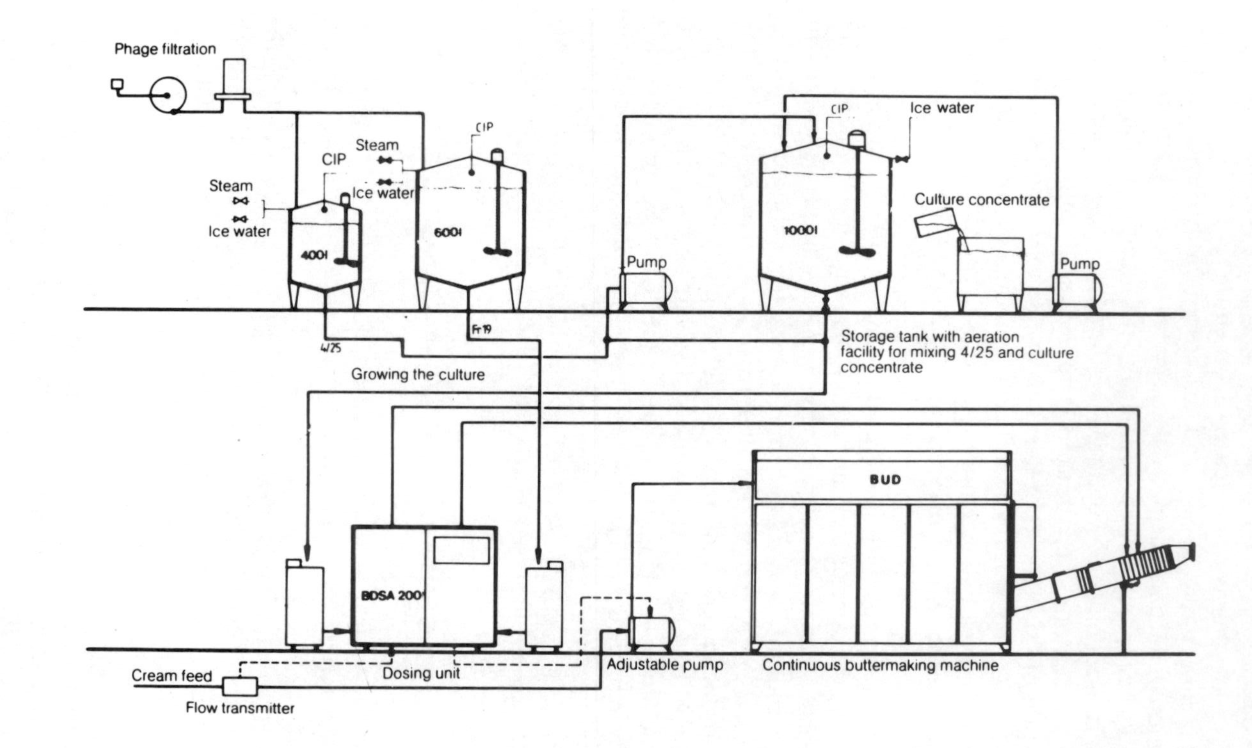

Fig. 7. Process flow diagram for NIZO process (courtesy Westfalia Separator, UK).

calibration of the system. With salted butter, a correction must be made for the effect of the increased electrolyte on the dielectric constant. For systems based on dielectric measurements, an accuracy of ± 0·1 per cent is claimed.

A more sophisticated approach has been developed and is now being marketed as the Alconix system. This uses an array of sensors in the discharge from the buttermaker to measure the levels of moisture, density, temperature, solids-not-fat and salt. The major innovation is the measurement of salt by backscatter of gamma radiation emitted from an Americium-241 source. Accuracy better than ± 0·03 per cent in measuring the salt content is claimed. In creamery operation, Williams (1984) has claimed an increased buttermaking yield. The mean moisture content of the butter increased by 0·16 per cent to 15·84 per cent, and the mean salt content by 0·12 per cent to 1·77 per cent against respective maxima of 16 per cent and 2 per cent to meet UK Intervention Board requirements.

BUTTER HANDLING AND PACKING

Butter from the batch churn may be dropped onto butter trolleys, wheeled to the packer, and then either tipped or shovelled into the feed hopper. This method may still be acceptable for small-scale production, but is labour intensive and potentially unhygienic.

Continuous buttermakers may be set up to discharge directly into a bulk handling system. The simplest approach is to discharge direct into the feed hopper of a bulk packer, but any holdup on the bulk packer will then result in the shutdown of the buttermaker with consequent waste, both in terms of buttermaking capacity and fat in the buttermilk. Shutdown of the buttermaker may be avoided by the introduction of a bulk butter silo. A butter pump should be incorporated into the buttermaker where the silo is pressurised, or cannot be located at the discharge from the buttermaker. These butter pumps use large, interlocking rotors operating at low revolutions and discharging into wide bore pipes (typically 100 mm diameter) with large radius bends (> 500 mm radius).

Two types of butter silo are used. The sealed system is constructed as a vertical cylinder with a telescopic feed pipe mounted onto a piston. The butter in the silo is kept at constant pressure by the action of air cylinders on the piston. Butter is discharged from the base of the silo into butter pumps. This silo system has the advantage of minimising the risks of contamination, but has an operating capacity of only 900 kg.

Larger butter silos are of a more open construction, with capacities up to 10 tonnes, as illustrated in Fig. 8. The sides of the silo are steeply inclined or vertical to avoid bridging. A pair of large, slowly contra-rotating augers in the base of the silo discharge the butter to the butter pumps.

Care is needed in the handling of the butter to control the level of shear forces acting on the product. During the handling procedures, droplets of aqueous phase can come into contact with one another and coalesce, giving rise to free moisture in the butter. Handling should either be very gentle, or else designed to promote re-emulsification. A high shear buttermixer may, for instance, be inserted into the line to ensure a uniform product.

The use of butter production to absorb surplus milk fat results in large seasonal variations of butter supplies in Western Europe. The majority of the butter is initially packed in 25 kg cases, and within the EEC, the packing is normally done to the standards of the national Intervention Boards.

While manual bulk packing systems are still suitable for the smaller

Fig. 8. t h^{-1} continuous buttermaker discharging into 10 t butter silo (courtesy Alfa-Laval).

buttermaking lines, automated bulk packaging systems (Fig. 9) are now available to handle the output of the larger buttermakers. These lines erect and weigh the empty cartons, which may be lined with either parchment or plastic film fed from reels. The filling operation is in two stages — the first fill is to give approximately 100 g shortweight. The carton is then reweighed, the tare for that carton deducted, and a second fill made to bring the weight to within 10 g of the target. The case may then be sealed, labelled and palletised. Integration of the control systems with those of the buttermaker and butter silo will minimise the manning levels required. For short periods, the process could be supervised by a single operator.

The most efficient method of packing consumer portions, typically 250 g size, is by direct feed from the buttermaker via a silo to the fillers. A closed system using a pressure compensator at the filler is the most hygienic system, but feeding to an open hopper is an acceptable alternative. Butter fed direct to the fillers is relatively soft during the filling operation, so support must be provided to the packaging during this process.

The rigidity of cold-stored butter and the high cost of plastic packaging, coupled to the very cost sensitive market conditions, has largely precluded the use of preformed, plastic containers for butter. Most consumer butter packs use a film wrap — either vegetable parchment, or a parchment-lined, aluminium foil. Vegetable parchment is the cheapest of the materials used, and is found extensively in the UK.

Fig. 9. Automated bulk packing line (courtesy Alfa-Laval).

The major disadvantage of parchment is its permeability, and allowance must be made for loss of water from the packs on storage. Ultra-violet light can penetrate the packaging and promote the development of oxidative rancidity, though this may be reduced by the incorporation of pigments such as titanium dioxide. Under high temperature storage and similar abuse the packs may also become greasy. Laminates containing aluminium foil have much lower weight-loss factors and are impermeable both to fat and ultra-violet light, so that the added cost may be balanced against the potential for a longer shelf-life and increased product quality when it reaches the consumer.

A small proportion of butter is packed in transparent films, e.g. cellophane. The packaging is attractive but, as it has good UV transmission properties, the butter is susceptible to off-flavours due to oxidative rancidity. The fluorescent lighting in refrigerated display cabinets is a common source of UV radiation.

The only significant market for butter that uses plastic packaging is the individual portions for catering and institutional use. In this market, parchment packs are in the minority, most packs being either miniature foil bricks or plastic containers. These plastic containers are normally produced from a PVC sheet using form-fill-seal techniques, the containers being sealed by foil or aluminised films. These small portions, typically 9–14 g net weight, warm very quickly on removal from refrigeration, and the use of a rigid plastic container helps to maintain the quality of presentation of the product at ambient temperatures.

Repackaging of Bulk Butter

Bulk-packed butter is a relatively stable commodity at low temperatures, and Murphy (1981) has reviewed the bacteriological aspects of butter production. Storage for six months at −10 °C has no significant effect on the quality of salted, fresh cream butter; the current Intervention Board standards require storage at temperatures colder than −15 °C. Many public cold stores, however, operate at lower temperatures, down to −30 °C, where other more perishable commodities may be stored in the same chamber as the butter. Under these conditions, the butter will remain in satisfactory condition for more than one year. Cream may also be stored deep-frozen for subsequent processing into butter (or anhydrous milk fat).

Repackaging of bulk-butter into consumer portions reconciles the discrepancies between consumer demand and the variations in butter production. Before repackaging, the frozen butter must be allowed to reach an optimum of 6–8 °C, and during this tempering stage, the humidity of the chambers should be controlled to avoid excessive condensation on the butter. Heat for the tempering process may be provided economically by carrying out the first stage of the tempering in the packet butter store, or by re-using low grade heat recovered from other processes. During the tempering, the batches of butter should be checked for compositional and organoleptic quality. Before repacking, the butter must be comminuted and reblended to break down the matrix of fat crystals and reintroduce plasticity. This reblending stage provides an opportunity to increase the salt and moisture levels to the maxima permitted (by legal or other specifications), thus improving the conversion yield and the economies of the operation. Any problems with free moisture in the butter may also be overcome at this stage.

For small-scale repacking operations, the blocks of tempered butter are comminuted by dropping through a shiver. The comminuted butter is then blended into a plastic mass using a batch blender with heavy duty contra-rotating 'Z' shaped blades. These blenders should preferably be driven through a variable speed gearbox, though a two-speed drive is acceptable. Passage of the bulk-butter through a metal detector before shiving is desirable to minimise the risks of both product contamination and damage to the process plant.

Continuous butter blenders are now available for efficient handling

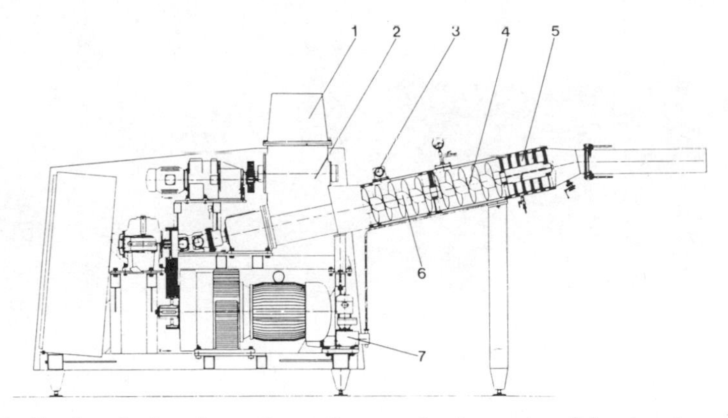

Fig. 10. Sectional drawing of continuous butter mixer/blender (courtesy Westfalia Separator, UK): 1, Feed hopper; 2, chopping device; 3, dry-salt dosing connection; 4, vacuum chamber; 5, blending section; 6, transporting augers; 7, metering pump.

of large quantities of butter, as illustrated in Fig. 10. The continuous blender includes a chopper for the bulk butter, and provision for in-line addition of salt, water and cultures. Construction of blending and working sections is similar to that of continuous buttermakers. The capacity of the blender depends on the nature and temperature of the butter, e.g. 10 t h^{-1} as an additional working stage for freshly produced butter, dropping to 7 t h^{-1} with stored butter at 7–8 °C, and less than 5 t h^{-1} with thawed butter at 0·5–2 °C. The reblended butter may be discharged into a butter silo and pumped to packaging machines (Fig. 11) in a way similar to that for freshly produced butter.

SPREADABLE BUTTERS

Physical methods to control the variation in butter hardness, and to impart a softer texture have had limited success. The softening achieved may be largely confounded by temperature variations in distribution and the home.

Most butters have too high a level of solid fat for easy spreading when cold, and although normally sufficiently plastic to spread at 15 °C, when

Fig. 11. Pumped distribution of butter to packing machines (courtesy Alfa-Laval).

less than 40 per cent of the fat is solid, easy spreading characteristics are only achieved at solid fat levels of 20–30 per cent. The obvious solution in temperate climates is for the consumer to leave sufficient butter for immediate use at room temperature, as was done before refrigerators became a normal household item. However, this approach is inconvenient, and is not acceptable in hot weather or warmer climates when the butter tends to melt and deteriorate rapidly. To achieve a spreadable butter at domestic refrigeration temperatures (i.e. 5–10 °C), a substantial modification of the properties of the milk fat is needed. Methods of achieving this modification without the inclusion of additional ingredients have been investigated. The principal methods used have been modification of the diet of the cow, and the recombination of fractionated milk fats to form butter.

The diet of cows has been modified by feeding protected supplements. These protected supplements were produced by encapsulation of the feedstuff by a protein–aldehyde condensation product. This modification of the protein, typically by reaction with formaldehyde, makes the proteinaceous outer layer of the supplement resistant to the rumen microflora. The particles of the protected supplement should then pass unaffected through the rumen to be broken down and digested in the intestine. By-passing the rumen microflora increases the availability of the supplement to the cow, and enables feeding a higher level of energy, without reducing the appetite, to give an increased milk yield. The protected fat by-passes the natural degradation and hydrogenation systems of the rumen microflora, and this results in the fat being absorbed as fatty acids and glycerides typical of the feedstuff, rather than as short-chain, saturated acids.

Change in the feed is reflected in the nature of the milk fat produced (Storry *et al.,* 1974), as well as the quantity, and the degree of change depends on the type of feed, and efficiency of the protection of the fat. In work with protected rapeseed oil (Sporns *et al.,* 1984), there was a significant change in the butter composition, though not as large as was expected.

Production of a softer butter by this means would require a major change in the feeding practices in a large area. A less saturated milk fat, suitable for the manufacture of a soft butter, would also cause problems if used in other dairy products, e.g. whipping cream. At this stage, there are still substantial technological problems to be overcome in producing an effective, economic and acceptable, protected feed supplement.

The alternative to changing the nature of the milk fat at source is to select fractions of the milk fat for use in a spread. A double fractionation of the milk fat is needed to obtain a high and a low melting fraction; the fraction with an intermediate melting range would require a separate market.

A spreadable butter product was developed in New Zealand (New Zealand Dairy Research Institute, 1977) using a low melting, milk fat fraction which was substantially liquid at 0 °C. This was blended with a high melting fraction, water, salt and milk solids-not-fat to produce a spread with acceptable consistency at 5–22 °C. The technology required for such products is similar to that for margarine, with the use of batch or continuous mixing followed by processing through swept-surface heat exchangers and a worker unit. The capital cost for this plant, in addition to the anhydrous milk fat production and fractionation plants, would require a premium price for the spread over that of butter.

DAIRY SPREADS

The alternative to modifications of the milk fat itself is to blend the milk fat with other fats that are liquid at 0–5 °C. This method can help reduce the cost of the product so that it may compete more readily with other spreads in the market. The mixing of milk and other fats, however, is not permitted in some countries, e.g. Germany, Netherlands and Denmark.

In some countries, up to 10 per cent butter may be included in margarines, e.g. UK and Switzerland, whilst blends of 25 per cent butter and 75 per cent margarine have been marketed in the USA. These products may be regarded as upgraded margarines rather than true dairy spreads.

A good example of a dairy spread is 'Bregott', developed and marketed by the Swedish Dairies Board. 'Bregott' was originally developed using ripened cream (*c.* 35 per cent fat, pH 4·6–4·7) to which refined, deodorised soy bean oil was added in the batch butterchurn. The soy bean oil was added at approximately 20 per cent of the total fat, giving a soy bean oil content of 16 per cent in the final product. The level of soy bean oil may be varied seasonally to compensate for variations in the milk fat, and hence standardise the spreadability of the product.

The mixture of cream and soy bean oil should be churned at a lower temperature than for ripened cream alone. The buttermilk is then drawn off, and the grains washed with cold water. A small quantity of salt may be added prior to working the mixture in the churn. Fat losses during churning were found to be higher for 'Bregott' (average 0·86 per cent) than for normal buttermaking (average 0·59 per cent) (Joost, 1977). A batch of 'Bregott' (1500–3000 kg) may be produced by this process in 2 h. The product is then packed into tubs, as the traditional butterfoil would not provide sufficient support for such a soft product.

The introduction of 'Bregott' was intended to defend the decreasing level of butter sales in Sweden by increasing the proportion of milk fat in the total yellow fats market. The success of this is demonstrated in Fig. 12, for not only was the decline in milk fat sales halted, but there has been some recovery in the sales of milk fat.

Manufacture of Bregott-type products is not limited to batch churning. A continuous method was subsequently developed (Johansson 1980) using a PSM buttermaker. Ripened cream, with a fat content of 36–38 per cent and pH 4·6, is cooled to 3–4 °C by pumping it through a plate heat exchanger, and held for a minimum of 3 h, during which time the temperature may rise to 5 °C as a result of latent heat released during further crystallisation of the milk fat. The aged cream is then pumped from the ageing tank. Soy bean oil (also at 5 °C) is continuously injected into the stream of cream by a dosing pump to give a total fat content of over 40 per cent, preferably 43–45 per cent, and the two are blended together, using an in-line static mixer, before entering the buttermaker. Though thc oil must be well dispersed in the cream, it is essential that very fine droplets are not formed, since these would result in higher fat

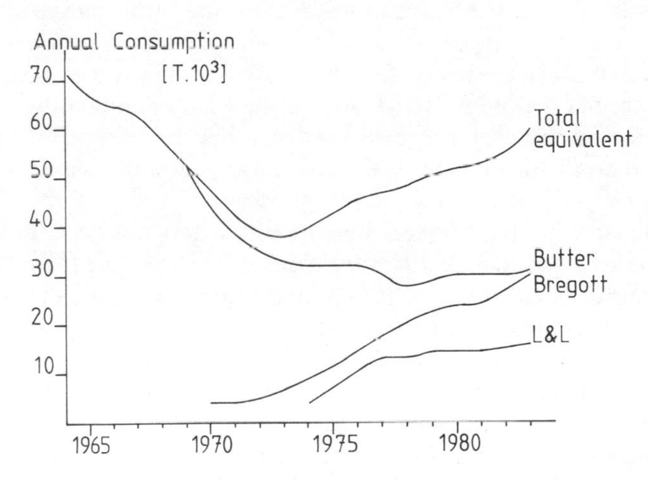

Fig. 12. Consumption of dairy-based yellow fats in the Swedish market (after IDF, 1984).

losses during churning. The cream–soy bean oil blend may be churned at 5–7 °C, and the finished product is extruded from the buttermaker at approx. 10 °C; the lower temperature partially compensates for the lower solid fat content in the product.

Vegetable oils have been added at other stages of the process, for instance to the milk prior to separation, to the cream before pasteurisation and to the butterstream, both in the continuous buttermaker and on reblending. In all cases, satisfactory products could be made, but with injection of the vegetable oil into the butter, higher shear rates were needed to ensure adequate mixing (Dixon *et al.*, 1980). However, product flow in the dairy is eased, and the risk of accidental contamination of other dairy products is reduced, if the vegetable oil addition is not made any earlier than during the feed to the buttermaker.

Similar products to 'Bregott' have now been produced in other countries (IDF, 1984). In the UK the concept has been taken a stage further by Dairy Crest (the manufacturing subsidiary of the MMB of England and Wales) with their product 'Clover'. The presence of a subsidy on packet butter in the UK market reduced the price differential between butter and margarine to aid butter sales. The subsidy was not available for hybrid products, so the degree of substitution had to be increased to compensate for the higher cost of the milk fat. Simply increasing the level of a vegetable oil, such as soy bean oil, would make the product too soft, so a third, more saturated, fat component was introduced. Under UK regulations, both butter and margarine should contain at least 80 per cent fat and have a maximum moisture content of 16 per cent, and the milk fat content of margarine is limited to a maximum of 8 per cent. Reduction of the fat content avoided possible conflicts with the margarine regulations, but the addition of a monoglyceride emulsifier was needed to ensure adequate dispersion of the now larger proportion of aqueous phase.

Processing of the 'Clover' has been carried out with continuous buttermakers, and also using margarine technology; the former method being found more appropriate. The method adopted for the consumer trials and the subsequent launch of 'Clover' in 1982 was to blend the non-dairy fats (soy bean oil, more saturated fat and a monoglyceride emulsifier such as Dimodan S) with the pasteurised sweet cream. The mixture may then be allowed to age overnight. The non-dairy fat should have a solid fat content of 15–35 per cent at 5 °C and 7·5–25 per cent at 20 °C, e.g. a blend of soy bean oil and partially hydrogenated soy bean oil. To maintain the dairy claim, fat addition levels were at less than

50 per cent of the total fat. A typical blend of dairy and vegetable fats would have a solid fat content of 42 per cent at 5 °C and 12 per cent at 20 °C (Wiles and Lane, 1983). Churning of the aged mixture at 7 °C may be preceded by air injection. The product extruded from the buttermaker may then be pumped to the filling machines, and filled into rectangular plastic tubs with a 250 g fill. The packed product would be cooled before despatch.

Protein-enriched Spreads

Following the development and launch of 'Bregott', a reduced fat product Latt och Lagom (L & L) was developed by the Swedish dairy cooperative, Mjolkcentralen Arla. The objective was to produce a spreadable product with half the fat of butter or margarine, while using dairy products as the principal ingredients. Swedish regulations require a fat content of 39–41 per cent for low-fat spreads. In several countries, a market for low-fat spreads had already been identified by the margarine industry, and a large number of patents had been filed. Products on the market at the time were relatively simple emulsions of a saline, aqueous phase in a blend of fats and emulsifier. These products have relatively poor organoleptic qualities.

The approach used for L & L was to prepare an aqueous phase derived from concentrated buttermilk and containing more than 13 per cent milk protein. This aqueous phase was then dispersed in a lipid phase blended from milk fat, soy bean oil and monoglycerides. Using buttermilk in the L & L, together with milk fat, linked the product closely to butter and anhydrous milk fat production. In Sweden, most of the butter produced has a pH of 4·6–4·7, and its production creates equivalent volumes of lactic buttermilk. Retail sales of cultured buttermilk products were relatively low, so the main outlet was disposal for animal feed. Recovery of the buttermilk protein, so that only the whey went for animal feed, would give a better return.

The method of production was patented (Strinning and Thurell, 1973) and demonstrated a combination of the technologies associated with the dairy and margarine industries. Preparation of L & L commences with pasteurisation of the lactic buttermilk (pH 4·6) to kill the starter bacteria; inactivation of enzymes and some denaturation of whey proteins would also take place. The pasteurised buttermilk is then passed through a nozzle bowl separator, where most of the protein is removed from the whey and ejected from the separator as a viscous, white paste of buttermilk protein concentrate. This concentrate, with a protein content of 13–20 per cent, may then be cooled and stored till needed. The concentrate contains relatively high levels of fat globule membrane, but must be converted into a suitable form for the aqueous phase by changing the casein into a soluble form. The addition of sodium citrate and sodium phosphate achieves this change by raising the pH and sequestering calcium from the casein, so that the micelles dissociate and the casein hydrates further. This is accompanied by a visual change from a white paste to a creamy viscous liquid. A typical protein level would be 13 per cent for a product of 7·5–8 per cent protein content. Sodium chloride may be added at this stage, but at a lower level (e.g. 1 per cent) than for an equivalent salt-flavour in butter; the presence of the other salts and the protein contribute to the saltiness of the product. Standardisation of the pH to 6·4 is achieved by adding alkali, e.g. sodium hydroxide. Preservatives, such as potassium sorbate, may also be added at this stage. The temperature of the completed aqueous phase should be at a standard level, e.g. 45 °C.

The lipid phase is made by blending the vegetable oil, typically soy bean oil, with anhydrous milk fat and minor quantities of monoglyceride emulsifier (e.g. 0·3 per cent), β-carotene for colouring, and vitamins A and D. These minor ingredients may be pre-dissolved in a part of the milk fat. Milk fat may be used as the principal fat (Strinning and Thurell, 1980), but the product, though softer than butter, is less easily used direct from the refrigerator. The level of monoglyceride is kept low as it plays only a secondary role in the formation and stabilisation of the emulsion, with the buttermilk protein concentrate as the principal emulsification agent. Unlike the earlier protein-free, low fat spreads, a more saturated monoglyceride may be used (Madsen, 1976). Higher levels of monoglyceride cause destabilisation of the water in oil emulsion. The emulsion should be formed by adding the aqueous phase to the lipid phase, the latter normally being at a higher temperature, e.g. 50 °C. Addition of the lipid phase to the aqueous phase would result in the formation of a stable O/W emulsion that could not readily be reversed to the desired W/O emulsion on subsequent processing. This emulsification is carried out as a batch operation as illustrated in Fig. 13, with careful control of both the addition rate and the level of agitation being needed to minimise the risk of phase reversal. The presence of both the cultured buttermilk concentrate and the milk fat in the emulsion will give an acceptable flavour to the product, but this may be enhanced by addition of aroma compounds to the emulsion.

Fig. 13. Batch emulsification tank using load cells for accurate control of ingredients (courtesy St Ivel Ltd, UK).

The completed emulsion may then be pasteurised and cooled using scraped-surface heat exchangers as illustrated in Fig. 14. During the cooling of the emulsion, the viscosity remains relatively low until the temperature is reduced to below 35°C, when the viscosity increases significantly. The viscosity increase is due to crystallisation of the higher melting, saturated fats and monoglycerides. Within the scraped-surface heat exchanger, the crystals formed on the surface of the barrel are scraped off the surface by rotating knives (as shown in Fig. 15), and mixed back into the bulk of the product. This acts as a seed to promote rapid crystallisation of the fat in the emulsion.

Cooling to 35–40°C may be achieved using water as the refrigerant, and further cooling using a direct expansion refrigerant, such as liquid ammonia. On a commercial scale, at least two barrels would be needed for the latter stages of cooling the emulsion. Shear forces developed within the scraped-surface heat exchanger continue the emulsification process, creating a wide range of particle sizes for the dispersed aqueous phase. The emulsion would be stabilised at this stage by the gelation of the aqueous phase, and by the reduced mobility of the now plastic, continuous lipid phase. The texture of the product may be further

Fig. 14. Scraped-surface pasteuriser and cooler (courtesy St Ivel Ltd, UK).

modified and improved by incorporating a worker unit — a tubular structure with both static and rotating pins through which the product passes.

After cooling to 8–12 °C, the pasteurised product should be piped direct to the filler for dosing into plastic tubs. The product is sealed under an aluminium foil, with the foil then covered by a plastic reclosable lid. The closed system with direct feed to the filler is desirable, not only to produce a consistent product (as crystallisation of the fat continues through the filling process), but to minimise the risks of post-processing contamination.

In low-fat spreads, the aqueous phase makes up a much larger proportion of the product than in butter (approx. 60 per cent compared to less than 20 per cent). The droplets of disperse phase may also be larger than in a well worked butter, and the level of salt in the aqueous phase considerably less than in a salted butter (2 per cent as compared to 10 per cent). In the case of the protein-enriched spreads, there are also ample nutrients, including protein and lactose, present in the aqueous phase. Pasteurisation is sufficient to reduce the level of microflora to a safe level, provided that the subsequent storage and

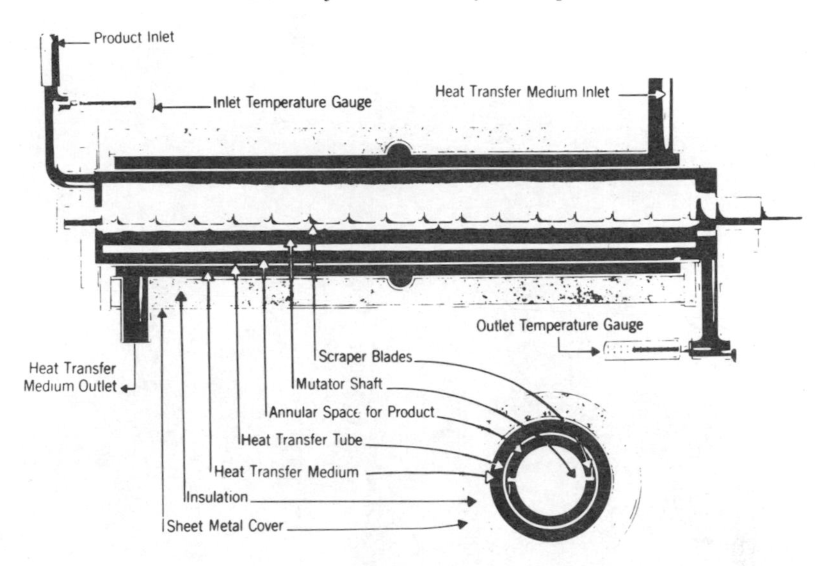

Fig. 15. Section through a scraped-surface heat exchanger (courtesy Chemtech Ltd).

distribution is under chilled conditions as for most other dairy products.

The major commercial spoilage risk with low-fat spreads is that of mould spoilage, particularly by those species of *Penicillium* and *Cladosporium,* which can grow under cold storage conditions. For this reason care is needed not only in the hygiene of the packaging operation (illustrated in Fig. 16) but also in the procurement and handling of the packaging materials. The addition of potassium sorbate to the aqueous phase, though less effective at pH 6·4 than at pH values below 6, provides a significant measure of protection to the product.

As with 'Bregott', considerable interest was shown in the L & L product. The first licence taken for the process was by Unigate in the UK, where different technical and marketing constituents applied. Since most butter in the UK is from sweet cream, the large-scale fermention of buttermilk was required. The use of potassium sorbate in low-fat spreads was not permitted at that time either, and the presence of potentially conflicting patents provided some formulation constraints. The UK product was launched as 'St Ivel Gold' in 1977, and as with 'Clover', the operation of the consumer butter subsidy was a major

Fig. 16. Multi-lane filler with immediate sealing to minimise contamination risks (courtesy St Ivel Ltd, UK).

commercial factor. Since the subsidy would not be extended to 'St Ivel Gold', it was necessary to reduce the ingredient cost by phasing out the milk fat addition to the product. This was a significant change from the initial L & L concept, where defence of milk fat sales was a major consideration. The milk fat was replaced in stages by blends of vegetable oils and hydrogenated vegetable oils, developed so that the original characteristics of the product would be retained. The reformulation was based on established principles of margarine manufacture (e.g. Wiederman, 1978; Erickson *et al.*, 1980) using blends where crystallisation would be directed towards the β' state. The preservation of the flavour characteristics may be attributed to the buttermilk protein concentrate, which still formed the major ingredient of the product.

The use of buttermilk as a base for low-fat spreads is a constraint on the location and extent of their manufacture, particularly where butter production is seasonal in character. This constraint has been overcome in a number of ways. In Japan, a variant of L & L has been produced using casein as the protein source. Processes employing skim-milk as an alternative to butter have also been developed (Wallgren and Nilsson, 1978).

To use skim-milk as an alternative starting material, it should first be pasteurised, then incubated with starter organisms and preferably with rennet action at *c.* 0·001 per cent. The starter organisms should be homofermentive, since the production of carbon dioxide leads to foaming and processing problems. Fermentation temperatures depend on the starter organisms selected, and would be in the range 20–30 °C for a cottage cheese starter culture. As with quarg, the fermentation should continue to the isoelectric point of the milk; deviation from this would lead to less efficient separation of the protein.

The coagulum must then be broken up, and the suspension heated to 52–55 °C by passing it through a heat exchanger. Small temperature differentials within the heat exchanger are essential to avoid damage to the protein, as grains of overheated protein may easily block the heat exchanger, and separator, and would not be readily redissolved at later stages in the process.

Once heated, the milk is held for 15–60 min, and during this holding time, calcium diffuses from the micelle into the serum. Without the hold, the protein would not be stable to subsequent processing, and would also have a less efficient emulsifying action. (It is during this hold that any carbon dioxide present in the fermented milk comes out of solution, forming a foam in open tanks, causing operational problems and some loss of protein.)

Following the hold, the skim-milk should be subjected to a further heat treatment, bringing the temperature up to 65–70 °C — depending on the stability of the fermented skim-milk. This heating must be carried out rapidly, e.g. using steam injection, with the hot skim-milk fed directly into the nozzle bowl separator. The function of the second heating is to aid separation by reducing the viscosity of the skim-milk, to further denature enzymes, and inactivate any surviving starter bacteria. Rapid cooling of the hot concentrate is essential to prevent the development of granularity, and a reduction in the emulsifying properties of the concentrate. Similarities should be noted between the production of protein concentrates from skimmed milk and the production of quarg.

The use of skim-milk under these conditions produces a blander product than that based on buttermilk, thus placing greater emphasis on the role of either milk fat or flavourings to achieve a satisfactory flavour.

Reduced-fat Dairy Spreads

This discussion has concentrated on a small number of examples which have achieved commercial success. Mann (1981, 1984) has reviewed a range of products and claims.

In general, the lower the fat content, the lower the inherent stability of the W/O emulsion. The relatively simple emulsion of a saline, aqueous phase in milk fat may easily be produced down to *c.* 40 per cent fat, but has poor organoleptic qualities. The addition of milk protein, preferably the products of fermentation processes, gives a much better product. The presence of protein at higher fat levels, e.g. *c.* 60 per cent, is beneficial, but at lower fat levels (*c.* 40 per cent), the tendency for milk proteins to promote O/W emulsions is a serious problem. This may be overcome by using high levels of milk proteins (an increasingly expensive commodity) together with heat treatments to modify their characteristics. Satisfactory organoleptic properties can be achieved at lower protein addition rates, but careful selection of both emulsifiers and hydrocolloid additives is needed to maintain the emulsion stability. This area is the subject of many patent claims.

Production of butter is currently in excess of its commercial market needs. Competition from less expensive spreads of marine, animal and vegetable origin (together with the exploitation of real and speculative health risks) is unlikely to diminish. Butter must be expected to continue as a premium product in the foreseeable future. A large sector of the total yellow fat market is likely, however, to be taken by spreads combining the superior organoleptic qualities of the dairy ingredients with the convenience and economic advantages obtained by blending with other ingredients, e.g. vegetable oils. Further improvements in flavour technology will also intensify the competition between dairy and non-dairy products. The greater interest in diet, especially fat and energy intake, increases the opportunity for reduced-fat, hybrid products. With the development of these hybrid products, one can expect a greater transfer of technologies across what were traditional divisions in the food industry.

REFERENCES

BS 998 (1969). British Standard Specification for Vacuum Salt for Butter and Cheese making and other food uses. British Standards Institution, London.

Dixon, B. D. (1970). *Australian J. of Dairy Technology,* **25** (2), 82–4.

Dixon, B. D., Cracknell, R. H. and Tomlinson, N. (1980). *Australian J. of Dairy Technology,* **35** (2), 43–7.

Erickson, D. R., Pryde, E. H., Brekke, O. L., Mounts, T. L. and Falb, R. A. (Eds) (1980). *Handbook of Soy Oil Processing and Utilisation,* American Soy Assoc. & AOCS.

Frede, E., Precht, D. and Peters, K. H. (1983). *Milchwissenschaft,* **38** (12), 711–14.
IDF (1984). Doc. 170 The World Market for Butter.
Johansson, M. S. J. (1980). UK Patent 1 582 806; US Patent 4 209 546.
Joost, K. (1977). *Svenska Mejeritindningen,* **69** (3), 13–14.
King, N. (1964). *Dairy Science Abstracts,* **26** (4), 151–62.
Madsen, J. (1976). *Food Product Development* (April), 72–80.
Mann, E. J. (1981). *Dairy Industries International,* **46** (12), 18–19.
Mann, E. J. (1984). *Dairy Industries International,* **49** (10), 13–14.
McDowall, F. H. (1953). *The Buttermakers Manual,* New Zealand University Press.
Mick, S., Mick. W. and Schreier, P. (1982). *Milchwissenschaft,* **37** (11), 661–5.
Mortensen, B. K. (1983). In: *Developments in Dairy Chemistry,* Vol. 2 (Ed. P. F. Fox), Applied Science Publishers, London.
Mulder, H. and Walstra P. (1974). *The Milk Fat Globule,* C.A.B. & Pudoc.
Murphy, M. F. (1981). In: *Dairy Microbiology, Vol. 2* (Ed. R. K. Robinson), Applied Science Publishers, London.
New Zealand Dairy Research Institute (1977). UK Patent 1 478 707.
Sporns, P., Rebolledo, J., Cadden, A. M. and Jelen, P. (1984). *Milchwissenschaft,* **39** (6), 330–2.
Strinning, O. B. S. and Thurell, K. E. (1973). UK Patent 1, 455, 146; US Patent 3 922 376.
Strinning, O. B. S. and Thurell, K. E. (1980). UK Patent Application 2 066 837 A
Storry, J. C., Brumby, P. E., Hill, A. J. and Johnston, V. W. (1974). *Journal of Dairy Research,* **41**, 165–73.
UK Intervention Board for Agricultural Produce (1982). Pamphlets MS/BUT/17.
Wallgren, K. and Nilsson, T. (1978). UK Patent Application 2011 942 A.
Walstra, P. and Jenness, R. (1984). *Dairy Chemistry and Physics,* Wiley, New York.
Wiechers, S. C. and DeGoede, B. (1950). *Continuous Butter Making,* North-Holland, Amsterdam.
Wiederman, L. H. (1978). *J.A.O.C.S.,* **55** (11), 823–9.
Wiles, R. (1978). UK Patent 1 597 068.
Wiles, R. and Lane, R. (1983). European Patent Application 0 106 620.
Williams, G. R. (1984). *J. Society of Dairy Technology,* **37** (2), 66–8.

Chapter 4

Drying of Milk and Milk Products

M. E. Knipschildt
APV Anhydro A/S, Søborg-Copenhagen, Denmark

HISTORY AND DEVELOPMENT

Milk is extremely perishable, and yet for a number of reasons, it is desirable to preserve it for later consumption. Today, drying is the most important method of preservation. The advantage is that, with modern techniques, it is possible to convert the milk to powder without any loss in nutritive value, i.e. milk made from powder has the same food value as fresh milk. The disadvantage is that energy consumption is high, in fact, no other process in the dairy industry has such a high energy demand per ton of finished product as drying. This is due to the fact that approximately 90 per cent of the milk is water, and all the water has to be removed by evaporation by the application of heat. Modern membrane techniques makes it possible to remove some of the water mechanically without the use of heat, but this method is not widely used in practice due to its limitations. Nevertheless, the sharp increase in the cost of energy has caused significant developments in process and equipment, making it possible today to convert milk into powder with an energy consumption per ton of finished product of approximately half of that required before.

Butter and cheese were for a long time the only dairy products known with extended keeping qualities, but in 1856, a new dairy product with long shelf-life was introduced. It was sweetened, condensed milk, followed some 30 years later by evaporated milk. The manufacture of condensed milk grew steadily until approximately 1950 (Hall and Hedrick, 1966), but since then a constant decline has occurred.

From 1860, private and cooperative dairies started to be built in USA, and from 1880 in Europe. In these dairies, butter and cheese were manufactured on a large scale. The skim-milk was sent back to the farmers for animal feed or dumped, and the whey was disposed of in rivers and lakes. This was the usual pattern for many years, in fact, right up till about 1930, when utilisation of the by-products slowly commenced.

Drying of milk was introduced at the beginning of this century on a modest scale. It took many years before equipment and processes were developed for commercial use. From approximately 1930, the spray drying process started to gain importance, but the great developments have taken place since the Second World War.

Dairy Research Institutes throughout the world have been somewhat slow in recognising dry milk as a product deserving the same attention as the classical products, cheese, butter and liquid milk. Substantial developments have been made by equipment manufacturers and dairy factories. However, during the last 20–30 years, useful work has been carried out at several Dairy Research Institutes with regard to the manufacture of dry milk and whey, and the importance of these products is now generally recognised.

Milk drying has become an essential part of the long chain between the farm producer and the final domestic consumer, and dried milk manufacture has grown into a large industry of considerable international importance. The diversification of the dry milk industry has been described by Robinson (1982) outlining the potential use in the food industry of the non-fat milk solids and the whey proteins.

Due to the dry milk industry, it is possible to make milk available in regions that are unfavourable for dairying, or where the milk production is insufficient. Milk produced in Australia, New Zealand or in Europe is consumed in the Far East, in the Middle East, or other countries where milk powder is converted into liquid milk or UHT-milk in recombining dairies. In most countries where milk production has become highly organised, the industry has to face great seasonal variations in the milk supply. The most important way of dealing with the surplus milk in the flush season is to dry it. During the lean season, it is often necessary 'to stretch' the milk by adding powder (to tone the milk). In this way, it is, for instance, possible to keep cheese production at a constant level all the year round to meet the demand for cheese, and to obtain full utilisation of equipment and manpower.

The drying process plays a decisive part in achieving two objectives which today are topics in the developed part of the world. One is to meet the increasing shortage of proteins in the developing countries, as milk powder is one of the food products most widely used in relief programmes. In this context it should be mentioned that UNICEF several years ago started to fight starvation by means of milk proteins in dry form, and UNICEF have contributed to the erection of milk powder plants in several countries all over the world. The other objective is to fight pollution, and in achieving this, the drying process is of great importance. It would not be possible to build the large cheese factories we have today without a drying plant to take care of the large quantities of whey. A drying plant is the most efficient disposal plant for the whey.

The removal of water from the milk takes place in two stages. The first stage is concentration by vacuum evaporation, and the second stage is drying; 90 per cent of the water in the milk is removed in the evaporator and only 10 per cent in the spray dryer. However, the energy required per kg water evaporated in the spray dryer is 16–20 times the energy required per kg water removed in the evaporator.

In absolute terms, the energy consumption of the dryer is approximately twice that of the evaporator if a six-effect evaporator and a two-stage spray dryer is used. If a four-effect evaporator is used, the energy consumption of the evaporator and the dryer will be approximately the same. The above illustrates that although only 10 per cent of the water is removed in the dryer; this should not lead to the assumption that the efficiency of the dryer is of minor importance.

EVAPORATION

Concentration of milk at low temperature by vacuum evaporation is based on the physical law that the boiling point of a liquid is lowered when the liquid is exposed to a pressure below atmospheric pressure. The milk is brought to boil at low temperature in an apparatus with negative pressure. The maximum boiling temperature accepted in a modern evaporator is approx. 70 °C corresponding to an absolute pressure of 230 mm (9·05 in.) Hg.

The first to use vacuum evaporation for the concentration of milk was Gail Borden, who started a condensing factory in Connecticut, USA in 1856. Concentration alone does not improve the keeping quality of milk and, therefore, sugar (sucrose) was added as the sugar inhibits bacterial growth in the concentrate; that was the start of sweetened, condensed milk.

As often happens when a new product is introduced, the consumers were slow in accepting the product, and the factory had to close a few years later. However, he believed in his product, and started a new factory a few years later. Gail Borden died in the town of Borden, Texas in 1874, but his company 'The Borden Condensed Milk Company', grew steadily, and is today one of the largest in the industry.

The next important development was the introduction of concentrated milk with a long shelf-life, but without the use of sugar as a preservative. It took place in Switzerland, where The Anglo-Swiss Condensed Milk Company started to manufacture sweetened, condensed milk in 1866 in the first condensed milk factory in Europe built in Cham, canton Zug. The operator of this factory, John Meyenberg, conceived the idea of giving the concentrated milk a heat-treatment severe enough to destroy the enzymes and to reduce the bacterial count. The revolutionary idea in his process was that of avoiding recontamination after heat-treatment; he suggested putting the evaporated milk in cans, sealing the cans and then sterilising the sealed cans. He designed a revolving steriliser in which the cans were heated to 115 °C, and this innovation became the basis of a new industry, the manufacture of unsweetened, condensed milk to be known as evaporated milk. Mr Meyenberg was, in 1884, granted patent on his process of preserving milk without addition of a preservative, and in the following years he built several factories in the USA to produce evaporated milk. In 1902, The Anglo-Swiss Condensed Milk Company sold its entire American interests to Borden's Condensed Milk Company, and the factory in Switzerland merged with Henry Nestlé of Vevey, Switzerland.

Since the Second World War, the manufacture of sweetened, condensed milk and evaporated milk has been in steady decline in favour of milk powder. However, there are still some traditional markets in the Far East and in South America for these products. As the countries where condensed milks are favoured have insufficient milk production, milk powder is often used for the local manufacture of sweetened, condensed milk and evaporated milk; this calls for powder especially manufactured for the purpose.

Principle of Evaporation

It was several years before vacuum evaporators were used to concentrate milk prior to drying. The development mainly took place after the Second World War, and the rise in energy costs since 1973 has led to a new generation of evaporators. The first evaporators to be used in conjunction with dryers were installed around 1930, and the material used was copper or aluminium. Since the Second World War only stainless steel has been used, and this material offers a great advantage as it permits chemical cleaning. Whereas all parts of an evaporator made of copper or aluminium had to be accessible for manual cleaning, this is no longer necessary when chemical cleaning can be used, and it presents new possibilities in design.

For the last 30 years, the tubular falling film evaporator has been universal in the dairy industry, and its only competitor of importance has been the plate evaporator introduced in 1957. Richard Seligman, the founder of the APV Company in England, invented, in 1923, the plate heat exchanger which has been of paramount importance to the dairy industry ever since. It was, therefore, natural for APV to design an evaporator where the heating bodies consist of plate heat exchangers. This evaporator, known as the plate evaporator, has been very successful, and offers the advantage of low building height making it possible to install it in existing buildings. A further advantage is that the heating surface can be varied easily by adding or removing some of the plates; this facility makes the plant very versatile and capable of handling a wide range of products. The plate evaporator held a strong position in the dairy industry for approximately 10 years, but since then it has been used only in other industries, as the capacities called for in the dairy industry have been outside the range of the plate evaporator.

To fully comprehend the advantages of the tubular, falling film evaporator, the principle of the tubular, natural circulation evaporator used in the dairy industry for many years shall first be explained.

Natural Circulation Evaporator

The natural circulation evaporator is also known as the climbing or rising film evaporator (Fig. 1). The milk is brought to the boil in tubular heat exchangers (calandrias) consisting of a large number of tubes through which the product rises. The tubes are heated externally by a steam chest. The milk enters the bottom of the calandria and boiling takes place near the base of the tubes. The vapour bubbles expand rapidly drawing a thin film of product at high velocity up the walls of the tubes. From the top of the calandria, the product is discharged into a separator vessel where the vapour is removed and used as the heating medium in the following effect. The vapour from the last effect is led to

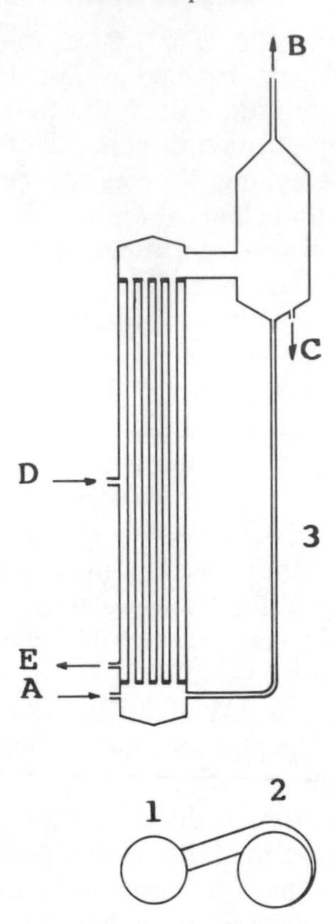

Fig. 1. Circulation evaporator. A: product; B: vapour; C: concentrate; D: boiler steam; E: condensate; 1: calandria; 2: separator; 3: recirculation tube.

a condenser. The product is piped back to the base of the calandria (recirculated) until the desired concentration is reached. Natural circulation evaporators are usually built as double effect plants; the tube length would normally be 4 m, and the diameter 25 mm.

Falling Film Evaporator

In the years after the Second World War, tank collection of milk was rapidly adopted in countries with high levels of milk production. The consequence was much larger dairy factories, a development which has continued to take place as the fabrication costs per ton of finished product are reduced with increased throughput.

The milk powder factories demanded evaporators with much larger capacity than known before, and, at the same time, higher concentration, better steam economy and improved product quality were called for. These factors led to the design of the falling film evaporator first introduced in 1953 by Wiegand in Germany. The falling film principle offers essential advantages, and was soon adopted by all the leading evaporator manufacturers in Europe.

The tubular falling film evaporator has for many years been the only evaporator installed in the dairy industry. High steam economy is obtained by recompression of the vapour created by the boiling of the milk. This is done by adding energy to the vapour, either by a steam jet, or by a mechanical compressor. The two methods of recompression have given the names to the two types of evaporators used today; TVR stands for thermal vapour recompression, and MVR for mechanical vapour recompression.

TVR Evaporators

Calandrias are used as heat exchangers as in natural circulation evaporators, but the tubes are up to 15 m long and 50 mm (2 in.) in diameter (see Fig. 2). The milk is introduced at the top of the calandria, and moves by gravity down the inside surface of the tubes as a thin film aided by the vapour which also moves downwards. The vapour is separated from the milk in a separator placed at the base of the calandria. The vapour is used as the heating medium in the following effect, and from the last effect, the vapour goes to the condenser.

A part of the vapour from the first effect's separator is used, together with boiler steam, as the heating medium in the first effect's calandria. The pressure — and consequently the temperature — of the vapour from the separator is increased by use of a thermocompressor (Fig. 3). Boiler steam (usually at a pressure of 6–10 bar) is introduced through a nozzle creating a steam jet in the mixing chamber whereby vapour from the separator is sucked into the mixing chamber. The speed of the mixture of boiler steam and vapour is reduced in the diffusor and, consequently, the pressure and the temperature are increased, making the mixture suitable as a heating medium in the first effect's calandria.

The amount of vapour sucked into the recompression unit will often be equal to the amount of boiler steam. The design of a thermal vapour recompressor is critical, but the unit is inexpensive, and the saving it provides in the steam consumption is very considerable. It has no

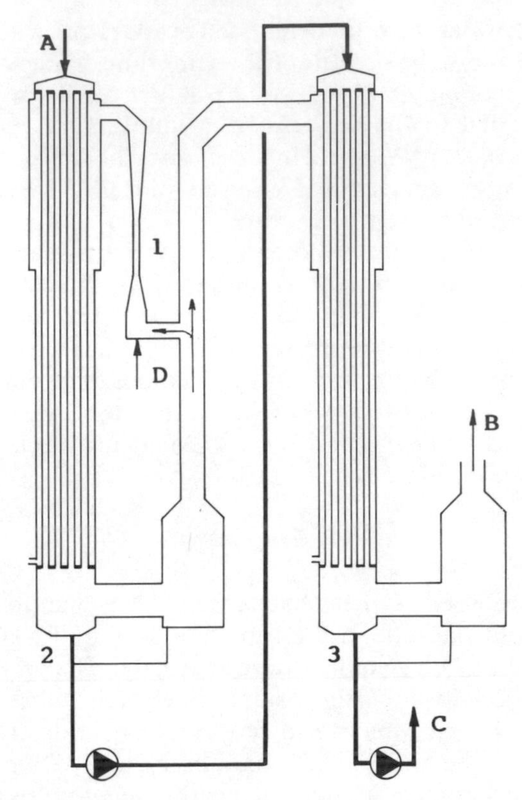

Fig. 2. TVR falling film evaporator. A: product-inlet; B: vapour; C: concentrate; D: boiler steam; 1: thermocompressor; 2: first effect; 3: second effect.

moving parts and requires no maintenance. The thermocompressor must be effectively sound insulated. The performance of a thermocompressor depends on the pressure of the boiler steam, and on the temperature difference (Δt) between the jacket temperature in the first effect (temperature of the mixture of boiler steam and vapour) and the temperature of the vapour from the separator. The suction ratio (amount of vapour to amount of steam) increases with the steam pressure, and it also increases with decreasing Δt:

1 kg steam of 9 bar will suck:

1·0 kg vapour at Δt 13°C
1·4 kg vapour at Δt 10°C
2·3 kg vapour at Δt 6°C

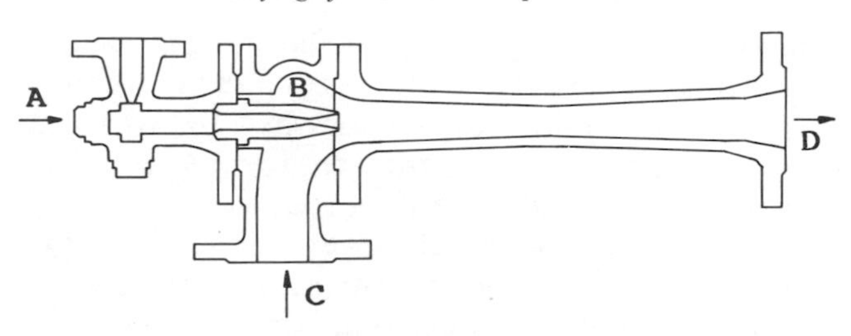

Fig. 3. Thermocompressor. A: boiler steam; B: steam nozzle; C: vapour from separator; D: recompressed vapour.

Uniform distribution of the milk over all the tubes is essential, as burning-on, i.e. deposits or scale in the tubes, will occur quickly if the tubes are 'starved'. A distribution device is placed at the top of the calandrias, and the various manufacturers use different designs of the distribution head. The milk is supplied to the top of each calandria at a temperature slightly higher than the boiling temperature in the calandria, so that the liquid flashes on entry. The flash vapour will assist in uniform distribution of the liquid over the tubes.

The purity of the condensate is governed by the efficiency of the separators. To keep entrainment loss (i.e. loss of milk solids in the vapour) at a minimum, it is essential to ensure high vapour velocity in the separators with tangential inlet to give centrifugal motion. The adoption of the cyclone principle in the design of the separators has removed the need for the non-sanitary deflecting plates used earlier in the separators.

The most important advantages of falling film evaporators over circulation evaporators are as follows:

(1) A condition for the operation of a climbing film evaporator is a fairly high temperature difference (Δt) between jacket temperature and boiling temperature. Below a certain Δt, the evaporator will not operate at all, and a temperature difference of 15–20 °C is not unusual. The large temperature difference between the tube wall and the product promotes scaling of the tubes.

In a falling film evaporator, the energy needed to move the product through the tubes is no longer provided (as in climbing film evaporators) by the difference in temperature outside and inside the tube. This means the falling film evaporator can operate at a much lower Δt, which gives a number of advantages. The lower the Δt, the smaller the risk of scale formation in the tubes, i.e. the operating time before cleaning is required can be extended. As deposits in the tubes are usually small even after 22 h operation, cleaning can be done by CIP, whereas the tubes in a circulation evaporator often require manual cleaning.

The low Δt opens the possibility of using six or seven effects. The temperature drop available over the effects is the difference between the boiling temperature in the first effect and the boiling temperature in the last effect (condenser temperature). To avoid protein denaturation, the boiling temperature in the first effect should not exceed 68–70 °C, and it is hardly feasible to have a lower temperature in the last effect than 43 °C. This gives 25 °C to be divided over six effects, i.e. a temperature difference of only 5 °C between the effects, or between heating medium and product. A specific steam consumption (kg steam per kg water evaporated) as low as 0·09 is obtained with a multiple-effect evaporator, whereas a double-effect plant would use more than four times as much. In addition, there is, normally, practically no scaling in the tubes even after long operation times due to the small Δt. Of course, the investment in a multiple-effect evaporator is substantially higher, but with today's high energy costs, the payback time for the additional effects is only a few years.

(2) The falling film evaporator offers gentle heat treatment due to single-pass operation and short holding-time. Undesirable bacteriological, physical and chemical changes in the milk resulting from excessive heat treatment are thus avoided. Whereas the circulation evaporator holds a large milk volume, the milk quantity in a single-pass evaporator is small. From the start of a circulation evaporator, it can take up to 1 h or more before the desired concentration is reached, but in a falling film evaporator, the concentrated milk is drawn off within a few minutes from the start of the plant.

(3) The low Δt, which it is possible to use in falling film evaporators, is favourable to vapour recompression and, consequently, the energy consumption is low. In a natural circulation evaporator, it is necessary to operate with a high Δt in order to obtain a high concentration, due to the increased viscosity of the product. The low Δt used in a falling film evaporator makes it possible to reach high concentrations without scaling of the tubes.

(4) The tendency to foam formation in the separators of circulation

evaporators, which has an adverse effect on the entrainment loss, is eliminated in single-pass evaporators.

Preheaters

Extensive utilisation of the heat in the vapour for preheating the feed to the evaporator is essential in order to obtain a high thermal efficiency. The milk is normally drawn from milk silos at 4 °C, and heated in spiral heaters placed in the condenser and in the calandrias (Figs 4 and 5). The preheater in the condenser also serves the purpose of condensing some of the vapour, thereby saving cooling water, but provision must be made in the design for adequate condensing facilities, in case an elevated feed temperature is used. Whey is normally stored at 65 °C, and supplied to the evaporator at that temperature. Therefore, the first preheaters will be bypassed when the whey is evaporated, and the condenser must be designed for the full load without contribution from the milk preheater. The spiral preheaters have the disadvantage of not being accessible for mechanical cleaning and inspection, but cost less than external, straight-tube preheaters placed adjacent to the calandrias. The spiral preheaters should not be used for high temperatures, and special preheaters are required for the manufacture of heat classified powders.

Care should also be taken in using the spiral preheaters for partly concentrated milk. In the case of 'backwards' product flow, to avoid a high concentration in the effect where the temperature is lowest, the product will, for instance, go from the last effect to the second last effect. The product temperature is thus below the boiling temperature in the calandria to which it is pumped, and the milk with high total solids must, therefore, be heated. For this purpose, an external, straight-tube preheater is best. In some cases, where a spiral preheater has been used, it has become blocked-up as a consequence of power failure or other disturbance. If that happens, it is necessary to bypass the preheater and seal it off, or try to force out the congealed concentrate with a high-pressure pump (homogeniser); assuming the preheater is designed for the high pressure.

Condenser

A shell and tube condenser should be used where the vapour moves through the tubes and cooling water is passed over the tubes. This type of condenser is called a surface condenser, and it ensures that no carry-

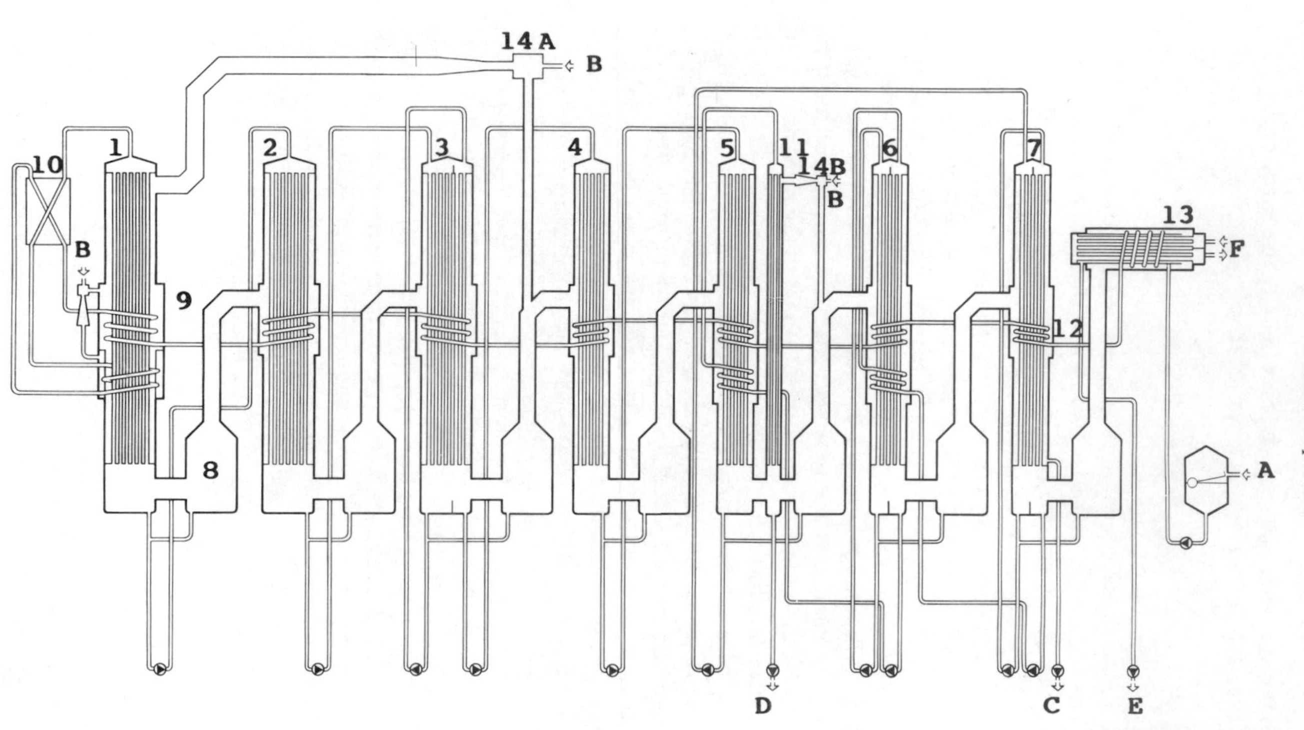

Fig. 4. Seven effect TVR evaporator with finisher. A: feed product; B: boiler steam; C: condensate outlet; D: product outlet; E: vacuum; F: cooling water; 1: first effect; 2: second effect; 3: third effect; 4: fourth effect; 5: fifth effect; 6: sixth effect; 7: seventh effect; 8: vapour separator; 9: pasteurising unit; 10: heat exchanger; 11: finisher; 12: preheater; 13: condenser; 14a and 14b: thermocompressor. (Courtesy APV ANHYDRO A/S, Denmark.)

Fig. 5. Calandria (courtesy APV ANHYDRO A/S, Denmark).

over from the one medium to the other takes place. In a jet condenser, the cooling water is sprayed into the vapour so that the cooling water may be contaminated, and in case of incorrect operation, some cooling water may be sucked into last effect contaminating the concentrate.

Vacuum Producing Equipment

The purpose of the vacuum equipment is to produce and maintain a vacuum in the plant by removing the air and non-condensable gas. Normally, a water ring pump is used, but the vacuum can also be produced by steam jet ejectors, which have low maintenance costs and, therefore, may be preferred if the cost of the steam for the ejectors is lower than the cost of the power required for the pump.

Finisher

Where high total solids in the concentrate is required, it is usual to provide the evaporator with a finisher, specially designed to handle viscous products. It is advantageous that the finisher is a small, separate evaporator where the boiling temperature can be chosen to suit the product. Further, where automatic control of the total solids in the concentrate leaving the evaporator is used (automatic density control), a quick response is obtained when the density controller governs the steam supply to the finisher, rather than the steam supply to the main evaporator or the milk intake. Variations in these parameters will give a slow response in a multiple-effect evaporator.

Condensate

The condensate from the first effect is usually led back to the boiler. To protect the boiler against contaminated condensate, the purity of the condensate can be measured by a conductivity meter which activates a bypass valve in case the conductivity exceeds a certain, predetermined value. The condensate from the other effects may be used for preheating the air to the spray dryer and, after that, for cleaning.

To ensure maximum heat transfer in the calandrias from the steam (vapour) side to the product, provision must be made for efficient drainage of condensate, and for venting of the steam side to remove all air.

Pumps

A large evaporator requires many circulation pumps. Cavitation in the pumps causing objectionable noise can be eliminated by placing adjustment valves on the pressure side of all the pumps.

Wetting Rate

The wetting rate is the proportion between product and heating surface, and is a key design figure for a falling film evaporator. It is equally critical whether the heating surface is too large or too small for a specified evaporation duty. Every tube requires a certain minimum quantity of product to cover the entire surface. Therefore, in a single-pass evaporator, the number of tubes per calandria will decrease from effect to effect, because the milk volume is constantly reduced as the concentration is increased. The first effect is always the largest, and in large evaporators, the calandria will be divided into two stages in order to overcome the problem of distribution of the product over a very large number of tubes, and to obtain the correct wetting rate. When the first effect is in two stages, the milk flow will be in series, but the steam flow will be parallel. Partitions are sometimes placed in the calandrias, and the product is pumped from one section to the other, so that the correct wetting rate is maintained over a large heating surface.

In comparing different proposals for an evaporator, one should compare the heating surface offered. Some suppliers specify the inside surface of the tubes, others the outside surface without making it clear which area they refer to. The difference between the inside and outside tube areas can be approximately 5 per cent. Figure 6 shows a six effect TVR evaporator.

Fig. 6. Six effect TVR evaporator (courtesy APV ANHYDRO A/S, Denmark).

Matching Evaporator and Dryer

If the duty is only one product, for instance skim-milk, it is obviously no problem, but if different products should be evaporated and spray dried, it becomes necessary to use a feed rate other than the design figure. The spray dryer will operate equally well at a reduced feed rate, as it is simply a question of reducing the air inlet temperature. This, of course, will reduce the thermal efficiency of the dryer, but otherwise, it will not cause any problem.

As far as the evaporator is concerned, it is possible to reduce the evaporation capacity by reducing the supply of steam to the evaporator. However, the evaporator will not operate satisfactorily when the milk intake is reduced so much that the wetting rate becomes too low. The range within which the capacity of the evaporator can be varied with no adverse effect is referred to as the turn-down ratio, and the desired turn-down ratio should be specified initially as it is a design parameter. The following example illustrates the problem. Let us assume an evaporator and a spray dryer are designed to handle 10 000 kg h^{-1} skim-milk of 9 per cent solids, and as a secondary duty, whole milk (full cream milk) of 12 per cent solids is to be dried. The performance data are as in Table I. If the intake of whole milk exceeds 7382 kg h^{-1}, the dryer will not be able to handle all the concentrate.

Turn-down ratio of evaporator:

in terms of product: $$\frac{7382 \times 100}{10\,000} = 74 \text{ per cent}$$

in terms of evaporation: $$\frac{5537 \times 100}{8125} = 68 \text{ per cent}$$

Normally, an evaporator can be made with the above turn-down ratio, if it is specified at the design stage. Two thermocompressors will be

TABLE I

Product/total solids	skim-milk (9%)	whole milk (12%)
Intake, kg h^{-1}	10 000	7 382
Evaporation in evaporator, kg h^{-1}	8 125	5 537
Concentrate, 48% solids, kg h^{-1}	1 875	1 845
Evaporation in spray dryer, kg h^{-1}	937	937
Powder, kg h^{-1}, moisture content	938/4%	908/2·5%

supplied — one for the full duty and one for the reduced duty — as a thermocompressor has a narrow operation range.

Of course, it is possible to operate the evaporator at its full capacity on whole milk, but then the excess of concentrate has to be stored. That is seldom the right solution, because the provision of storage facilities, and the energy for cooling and reheating the concentrate, bring nothing but extra costs.

The total solids of skim-milk may vary and, therefore, sometimes a powder manufacturer specifies that a plant is to be designed for only 8 per cent solids in the milk 'to be on the safe side'. When, in actual fact, the milk contains 9 per cent solids, the powder manufacturer finds to his surprise that the dryer cannot cope with the rated intake of milk. This is hard for him to understand as the milk contains less water than assumed. However, at the rated milk intake, the dryer has to have 12·5 per cent higher capacity if the milk contains 9 per cent solids instead of 8 per cent; the evaporator has, of course, spare capacity. Exactly the opposite is the case if the milk has a lower total solids than the plant is designed for, so that if the solids in the milk vary, the evaporator must be designed for the minimum solids, the dryer for the maximum solids in the intake milk.

In recent years, the size of the tubular, falling film evaporator has increased, and it is now available with a capacity up to 60 tonnes water evaporation per hour.

MVR Evaporators

Mechanical vapour recompression has been known for many years, and was used in milk evaporators in Switzerland during the Second World War due to lack of fuel for raising steam. Electric driven turbocompressors were installed for recompression of the steam, but due to the high cost of this machine, MVR evaporators were not further developed in the dairy industry for many years.

Nuclear power and the steep rise in fuel costs have changed the relative costs of steam and power, and during the last 5–6 years, MVR evaporators have found increasing use in the dairy industry. At the same time, advances have been made in the recompression techniques making an MVR evaporator as reliable as a TVR evaporator. Today, the choice between the two systems is an economic exercise. Previously, the position was that where the running costs of MVR were lower, the annual utilisation of the plant had to be high to justify the higher investment, but today there is not a great difference in the costs of the different evaporators. Therefore, the difference in direct running costs will normally be the decisive factor. It is not always a question of comparing the costs of steam and electric power. With natural gas being available in many countries, it can be economical to drive the compressor by gas. Either the compressor can be driven by a gas motor or electric power to drive the compressor can be generated by a gas motor.

If a dairy factory has insufficient steam to supply a TVR evaporator, there may be a good case for an MVR plant, eliminating the need for a new boiler. However, the available power supply should also be looked into as an MVR evaporator requires an electric motor of a size not normally used in dairies; as an example, an MVR evaporator with capacity 30 tonnes water evaporation/h requires a 400 HP motor for the compressor.

Whereas only a part of the vapour is recompressed in a TVR evaporator, all the vapour is recompressed in an MVR evaporator. Normally, the evaporator consists of only one effect (Fig. 7), and the boiling temperature can be chosen as the most suitable for the product. The calandria is divided into sections, and the product is fed by pumps through the sections in series. The number of tubes in the sections is reduced ensuring the correct wetting rate as the product volume is reduced with increasing concentration. For capacities exceeding approximately 15 tonne h^{-1} water evaporation, two or more calandrias may be used in parallel, i.e. the boiling temperature is the same in the two calandrias.

Apart from the steam used for start-up, an MVR evaporator requires no steam and no cooling water. However, if high total solids is to be achieved, it is advantageous to use a steam heated finisher for the last concentration. In this way it will be easier to control the evaporator. If skim-milk has to be concentrated to 50 per cent solids, it can, for instance, be concentrated to 42 per cent solids in the MVR effect, and further up to 50 per cent in the steam-heated finisher. In this case, 96 per cent of the evaporation will take place in the MVR effect, and only 4 per cent in the finisher. It will thus be seen that the amount of steam and cooling water required by the finisher is insignificant. Extensive preheating of the feed takes place in plate heat exchangers utilising the heat in the condensate.

Suitable mechanical vapour compressors in stainless steel are available from various manufacturers. It is an expensive item, and it is, therefore, costly to keep a spare compressor. It is often more favourable to design the evaporator in such a way that it also can be used as a TVR evaporator, i.e. a thermal vapour recompressor is installed as 'spare' instead of a spare mechanical compressor. If the mechanical compressor breaks down, the evaporator can be operated on steam during the hours when steam will be available. This seems to be the right solution if prices for fuel and electric power are likely to change later on, so making a steam heated plant more favourable than an electric driven plant. In addition, having a dual system is a good insurance against changing costs of fuel and power, which could lead to an incorrect investment in the initially chosen system. Likewise, if initially a TVR evaporator is chosen, it would be wise to design the evaporator in such a way that it can be converted later into an MVR, and to reserve space in the layout for a mechanical compressor.

With a temperature rise of 10 °C of the vapour from the suction side of the compressor to the pressure side, the power consumption of the compressor will be 0·042 kW per kg water evaporated. The power consumption will be reduced with decreasing Δt, but, of course, the heating surface has to be increased at unchanged capacity. To optimise an evaporator is a complex task, and it is tedious to do without the aid of a computer.

An important development has taken place during the last few years which makes MVR evaporators more attractive. Instead of using a compressor to raise the pressure and temperature of the vapour, it is now possible to use a high pressure, heavy duty fan specially designed for the purpose. The fan is a far simpler machine than a compressor, and significantly less costly. Several evaporators have been fan operated for a number of years with very good results. The fan operated evaporator represents a new generation, and it will, no doubt, contribute essentially to make mechanical vapour recompression more widely used than before. The fan must be designed to operate under vacuum with a casing to be vacuum-tight at a pressure below 0·1 bar. Special stuffing boxes and special impellers are used. To provide continuous variation of the speed of the fan, the motor should either be a DC-motor or an AC-motor with frequency converter; the frequency converter is expensive for large effects. The DC-motors require frequent attention, but operate well when they are carefully serviced.

Some entrainment from the separators cannot be avoided and, therefore, the recompression unit must be cleaned at intervals. It is not easy to clean a compressor, whereas a fan can be cleaned by water supplied from an injection system permanently placed in the fan. This

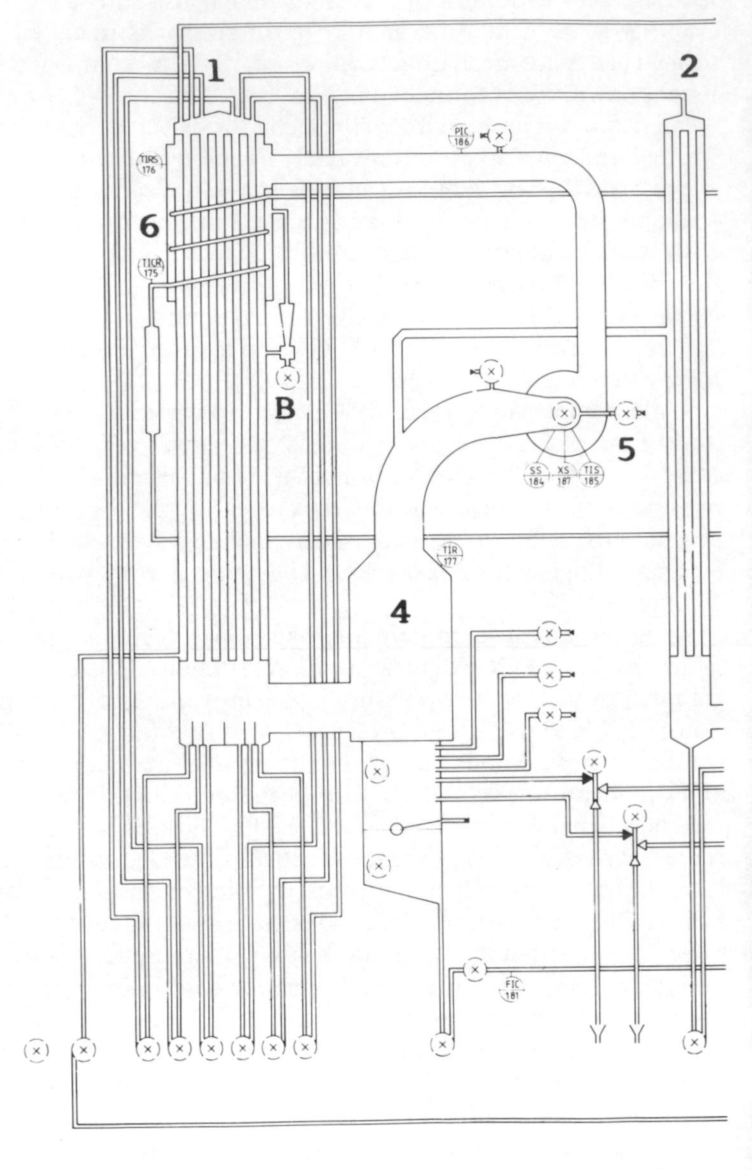

Fig. 7. MVR evaporator. A: feed product; B: steam for preheating; C: condensate outlet; D: product outlet; E: vacuum; F: cooling water; 1: MVR effect; 2 and 3: finisher; 4: vapour separator; 5: mechanical compressor/high-pressure

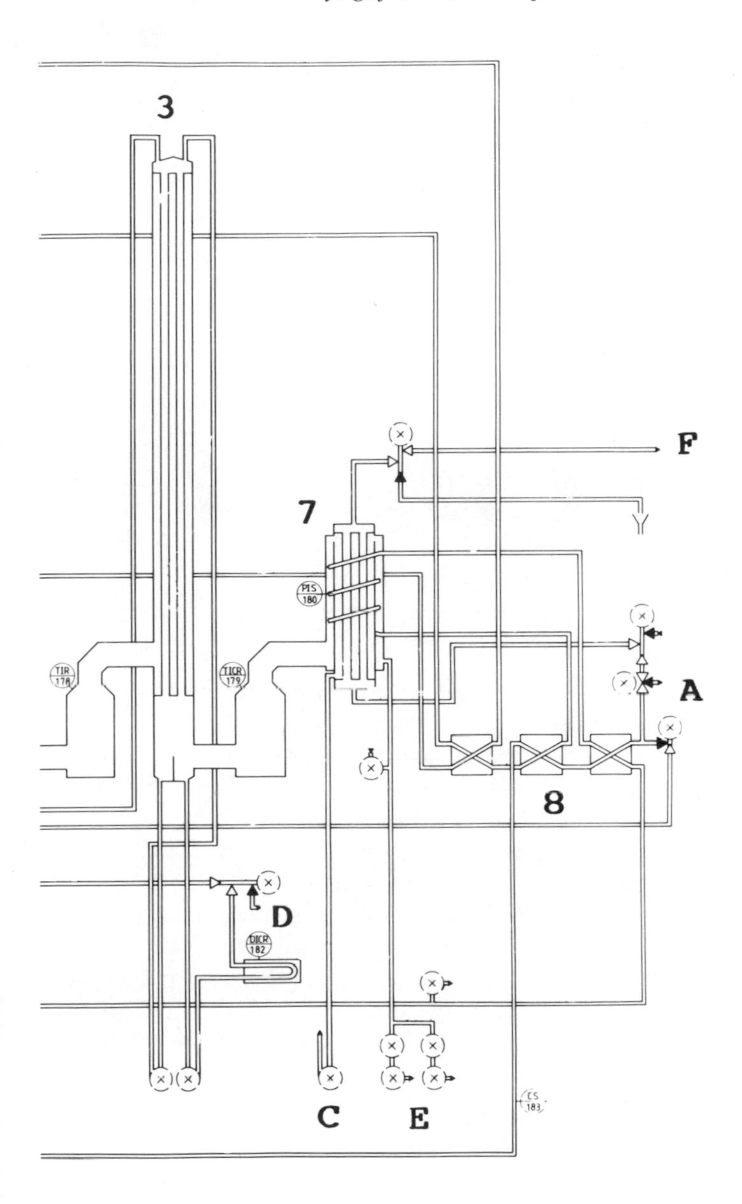

fan; 6: pasteurising unit; 7: condenser; 8: preheater. (Courtesy APV ANHYDRO A/S, Denmark.)

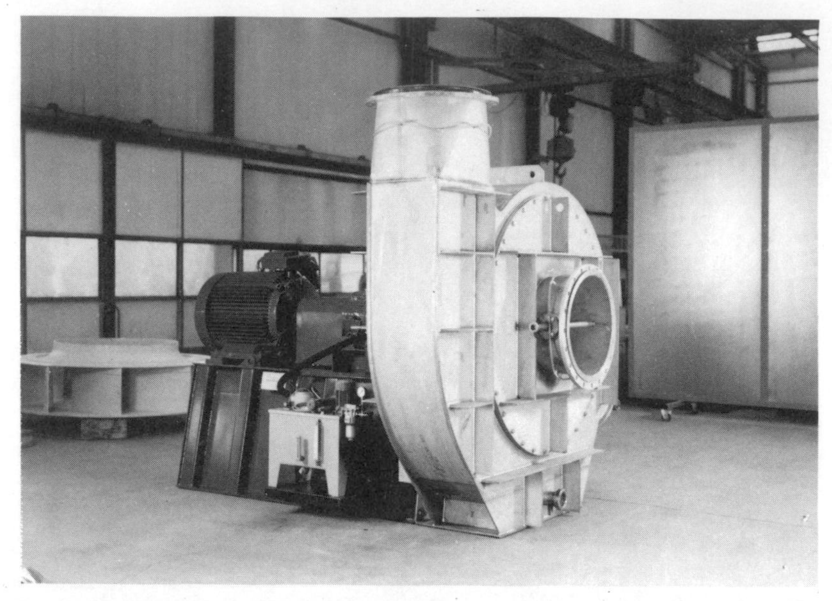

Fig. 8. Heavy duty radial fan for MVR evaporating plants (courtesy Piller, Germany).

means that the fan can be included in the daily CIP-cleaning of the evaporator. The correct operation of the fan can be checked from the control panel by electric temperature sensors measuring the temperature of the bearings as well as a device for vibration measurements. The fan should be provided with sound-proof insulation, and should be connected to the ducts by compensators preventing vibrations being transmitted to the rest of the plant, and allowing the ducts to expand when heated. An impeller can be kept as spare, but is normally not required.

A temperature rise of up to 5 °C from the suction side of the fan to the pressure side can be obtained, and a higher Δt can be achieved by using two fans in series. A fan operated evaporator is often designed for a Δt of only 4 °C, which ensures long runs without scaling of the tubes. The corresponding power consumption of the fan is 0·012 kW per kg water evaporated. Figure 8 shows a heavy duty fan used in milk evaporators.

Reverse Osmosis

Instead of evaporating the water, it is possible to remove some of the water from the milk mechanically by using the membrane techniques.

Whereas ultra-filtration is a separation process, reverse osmosis (RO) or hyper-filtration is a concentration process removing essentially only the water. The fact that the concentration takes place without exposing the product to heat is of no importance in this context, because only low levels of concentration can be achieved and, therefore, RO must be followed by thermal evaporation. Only for products that are heat-sensitive to such an extent that thermal evaporation cannot be used — as for example egg-white — is it a decisive factor that the water removal takes place in an RO plant without the use of heat.

Milk and whey can be concentrated by RO in the proportion of 1:2, i.e. skim-milk from 9 per cent solids to 18 per cent and whey from 6 per cent solids to 12 per cent. To illustrate the proportion between thermal evaporation alone and RO plus thermal evaporation, we assume, as an example, that 10 000 kg skim-milk at 9 per cent solids should be concentrated to 50 per cent.

Thermal evaporation alone:

10 000 kg h^{-1} skim-milk, 9 per cent solids
8200 kg h^{-1} water removed in evaporator
1800 kg h^{-1} concentrate, 50 per cent solids.

RO plus thermal evaporation:

10 000 kg h^{-1} skim-milk, 9 per cent solids
5000 kg h^{-1} water removed by RO
5000 kg h^{-1} concentrate, 18 per cent solids
3200 kg h^{-1} water removed in evaporator
1800 kg h^{-1} concentrate, 50 per cent solids.

Approximately 60 per cent of the total water removal takes place in the RO plant.

The difference in the costs of an evaporator to take the full load and an evaporator to take only 40 per cent of the evaporation will not cover the cost of the RO plant. Therefore, the justification for RO must be a saving in the running costs. The driving force in RO is electric power for the circulation pump. The justification for the removal of 60 per cent of the water by RO would be that the cost of the power required by the RO plant would be less than the cost of the steam required by a TVR evaporator to remove the same amount of water. However, in the cases where electricity should be used for concentration rather than steam, it is more advantageous to install a MVR evaporator than an RO plant, and do the whole evaporation in one machine. The power consumption of a modern MVR evaporator per kg water evaporated is about the same as for an RO plant and, therefore, MVR is the answer rather than RO when electric power is used for concentration of milk. With the development of MVR evaporators, the interest in using RO in the dairy industry has decreased.

The running cost of an RO plant is not only the power consumption. Maintenance in terms of replacement of membranes entails a significant cost, and the cost of enzymes for the cleaning of cellulose acetate membranes should also be included. Another problem, which seldom is appreciated at the time when an RO plant is purchased, is that the loss of milk solids into the water penetrating the membranes (the permeate) is at a level where disposal of the permeate presents a problem. Experience has shown that the biological oxygen demand (BOD) of the permeate is too high to permit disposal of the permeate into rivers and lakes. It must be discharged into an effluent disposal plant, whereby costs are incurred. The condensate from an evaporator will normally be sufficiently pure to allow direct disposal, but the condensate is often used for cleaning instead of hot water. However, the permeate from an RO plant is cold and has, therefore, no value for cleaning purposes.

In special cases, if for example whey from cheese or casein plants has to be transported to a drying station in a different location, it may be considered economic to use RO at the location where the whey is produced, and thereby reduce the volume to half of the original, but normally RO will not be able to compete with thermal evaporation in the dairy industry.

DRYING

Milk and milk products are dried by either the roller process or the spray process, the latter being the dominating one today.

Roller Drying

Roller drying was the first method used for the drying of milk. It originates from the USA, where it was invented in 1902 by John A. Just. In the USA, the process is also referred to as drum drying. James Hatmaker of London, England, purchased the Just patent and improved the process. The resulting modification is known as the Just–Hatmaker process, and was widely used for many years all over the world. In several countries, a roller dryer was referred to as 'a Hatmaker'.

The principle of the roller drying process is that the milk is applied in a thin film upon the smooth surface of a continuously rotating, steam-heated, metal drum. The film of dried milk is continuously scraped off by a stationary knife located opposite the point of application of the milk. The dryer may consist of a single drum or a pair of drums. The milk may be preconcentrated before it is applied to the drum, and the degree of concentration varies with the design of the dryer. The rule is that a single-drum dryer can handle milk of higher concentration than a double-drum dryer. The dried milk film is milled to break up the film. The particles of a roller-dried powder are solid, flat, irregular flakes containing no air. The absence of occluded air retards oxidation, resulting in good keeping quality.

The first Just–Hatmaker roller dryer was installed in 1903 in the creamery at Sherborne, Dorset, England by the company Cow & Gate. This machine was so successful that the company installed a further two roller dryers a year later in Somerset and Dorset to produce 'dried full cream milk, dried half-cream milk and dried separated milk from pure English milk' as the advertisement read. The new product was marketed as 'germ-free and water-free milk'. The claim for a 'germ-free' product was no exaggeration, because it is characteristic of the roller process that the milk is exposed to a high temperature heat treatment which safe-guards the bacteriological quality of the dried product. For this reason, roller drying became a standard process for manufacture of infant foods for many years.

However, the excessive heat treatment results in a high degree of protein denaturation. The solubility of the dried product is poor, and the reconstituted milk has a cooked flavour. These features render it unacceptable as a substitute for liquid milk. The chocolate industry prefers roller-dried milk, which is claimed to give a desired flavour to the chocolate. When manufacturers of spray powder are requested by the chocolate industry to produce a powder with properties closely resembling roller powder, it can be done by giving the milk an intense heat treatment prior to drying.

The maintenance costs of roller dryers are high, as resurfacing of the drum must be done at frequent intervals to keep a smooth surface, and the knife to remove the product from the drum requires frequent sharpening. The capacity per unit is low, and with the centralisation of dairy factories, it became impossible to handle the large milk quantities by the roller process. Another factor, which today is a decided dis-advantage, is the pollution from roller dryers. Above the drum is installed a hood which collects and carries the vapour that arises from the milk drying. The hood exhausts to the outside through a stack, and as the entrainment of fine milk flakes into the vapour can be considerable, the exhaust from roller dryers presents a pollution problem unless the vapour is led to a wet scrubber.

Spray drying was without practical importance in the dairy industry until the First World War. Between the First and Second World Wars, roller drying and spray drying developed in parallel, but after the Second World War, the dryers installed in the dairy industry have been mainly spray dryers. The roller dryer does not meet today's requirements with regard to powder quality, high capacity, high preconcentration, and low operating and maintenance costs.

Manufacturers of milk powder within the European Economic Community (EEC) have the advantage of a guaranteed outlet with minimum price from the so-called Intervention Board, but this arrangement only applies to spray powder. The fact that roller powder is not covered by the intervention system has been a further incitement for powder manufacturers within the EEC to shift from roller drying to spray drying.

Spray Drying

Spray drying can be described as instantaneous removal of moisture from a liquid. To achieve this, the liquid is converted into a fog-like mist ('atomised'), whereby it is given a large surface. The atomised liquid is exposed to a flow of hot air in a drying chamber. The air has the function of supplying heat for the evaporation and, in addition, it acts as carrier for the vapour and the powder. When the atomised product is in contact with the hot air, the moisture evaporates quickly and the solids are recovered as a powder consisting of fine, hollow, spherical particles with some occluded air.

Spray drying is a gentle drying method. The material to be dried is suspended in the air, and the drying time is very short. Air inlet temperatures up to 215 °C are used for the drying of milk, but due to the evaporation, the temperature drops immediately in the drying chamber, and the milk solids will not reach a temperature approaching the air inlet temperature. The temperature in the drying chamber is equal to the air outlet temperature, which is approximately 95 °C when single-stage drying is used, and lower with multi-stage drying. The product temperature will be 20–30 °C below the air outlet temperature provided a co-current dryer is used, i.e. a dryer where product and air move in the same direction.

The evaporating capacity of the dryer is determined by the difference between the air inlet temperature and the air outlet temperature. Also the heat loss from the dryer affects the capacity, and efficient insulation of the hot air ducts to the dryer and the drying chamber is essential. Further, the humidity of the intake (ambient) air affects the capacity. The humidity of the ambient air varies within a wide range, and can have a considerable adverse effect on the capacity in a warm and humid climate. The low air temperature in the winter gives a low humidity in the air and, consequently, an increased capacity of the dryer. The powder is cooled immediately as it comes out of the dryer, either in a pneumatic, powder cooling system, or in a fluid-bed; Fig. 9 shows two different types of single-stage dryer. In (A), the drying chamber has a flat bottom from which the powder is removed pneumatically by means of a rotating powder collector, which sucks the powder from the chamber. The powder is separated from the air in the cyclones. Under each cyclone is placed a rotary valve (air lock) through which the powder is discharged into the pneumatic, powder cooling system. In (B), the drying chamber has a conical bottom from which the greater portion of the powder is removed by gravity, and discharged into the fluid bed, powder cooler; only the fine powder particles are carried with the air to the cyclones.

In the following section, a description of the various components of a spray dryer is given.

Components of a Spray Dryer

Air filtration

Efficient air filters for the drying air and the cooling air are essential to avoid contamination from the air of the powder, only dry filters are used today — either cleanable or throw-away filters. The use of filters where the pressure drop over the filter is automatically kept constant ensures a constant amount of air in the dryer and this prevents the capacity of the dryer being reduced as a result of reduced air intake due to dirty filters. The filters are designed in such a way that clean filtering material is supplied from a roll activated by the pressure drop over the filter. The used filter material is simultaneously collected on another roll. The saving in labour for supervision and cleaning of air filters contributes to the cost of new filter rolls.

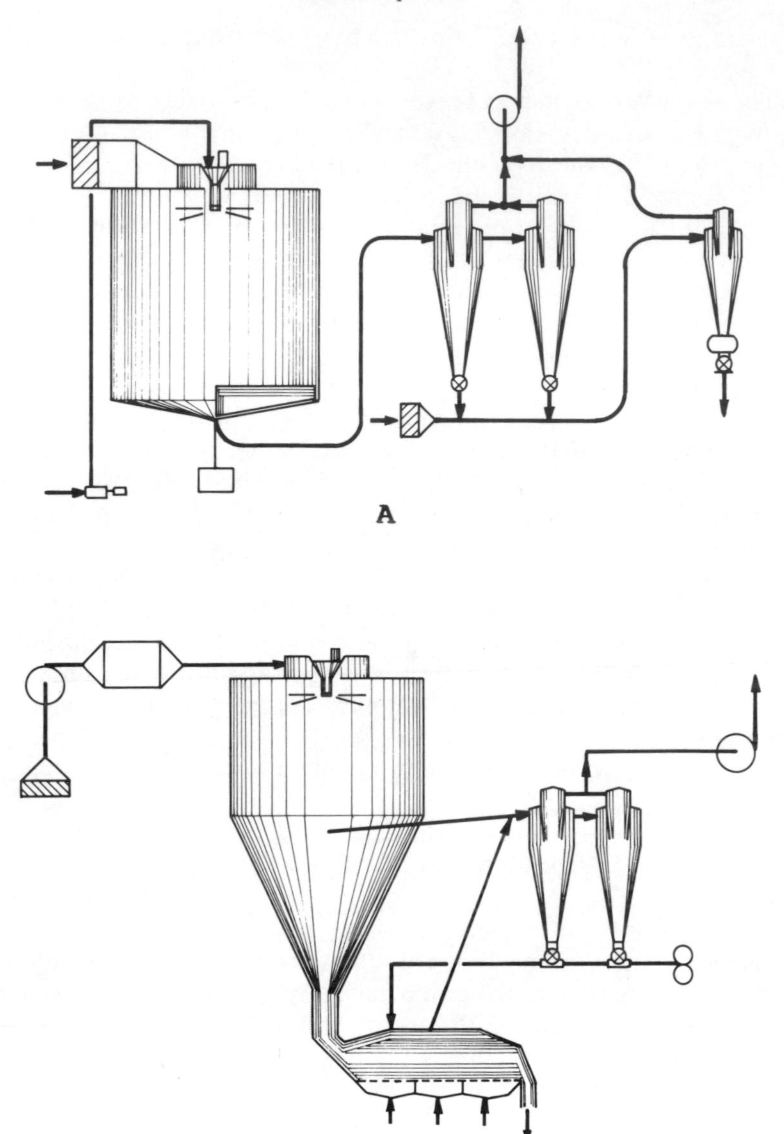

Fig. 9. Single-stage spray dryer. A: drying chamber with flat bottom; B: drying chamber with conical bottom and fluid-bed. (Courtesy APV ANHYDRO A/S, Denmark.)

The filter for the drying air is not required to destroy possible airborne bacteria, as it is unlikely that bacteria will survive when the air is heated. However, as the cooling air is not subjected to this 'pasteurisation' it would be recommendable to use bacteria-absorbing filters for the cooling air in places where airborne bacteria are likely to be found. If an air filter inadvertently gets wet, it must be thoroughly cleaned or replaced immediately, as there will be milk solids collected in the filter which provide an excellent growth medium for bacteria when moisture is present.

Today, it is quite common (as explained later) to use fabric filters at the end of the dryer to collect the fine material escaping from the cyclones. If the fabric filter is designed in such a way that the powder collected in the filter is suitable for mixing into the powder from the dryer, special attention must be paid to the filter for the intake air. If the filter for the intake air is less efficient than the fabric filter for the exhaust air, there will, obviously, be an accumulation of impurities in the fabric filter. Therefore, the filter for the intake air must be more efficient, or at least as efficient, as the fabric filter for the discharged air.

Feed pump

The feed pump for a rotating atomiser must be a sanitary, volumetric pump with variable speed drive. A nozzle atomiser requires a feed pump to give a pressure up to 300–400 bar, and it must be a three-piston pump to avoid pulsations.

Atomiser

Atomisation of the concentrated milk is the principle of the spray drying process. By atomisation, the concentrate is converted into droplets varying in size between 10 and 200 μm, with the greatest portion in the range of 40–80 μm. Two systems are used, stationary pressure nozzles or rotating atomisers. The use of pressure nozzles originates from USA where they are still used to a great extent, whereas the rotating atomiser was developed in Europe, where it is still the preferred system.

The pressure nozzle is a simple and inexpensive item, but it requires a high pressure pump which is costly. The pump must have a variable speed drive, unless the feed rate is controlled by a by-pass valve after the pump. This latter system is not desirable as it involves recirculation of the concentrate at a temperature where bacterial growth can take place and, further, it can cause formation of foam.

The principle of the rotating atomiser involves feeding the concentrate onto a disc rotating at high speed. The product is fed into a chamber at the centre, and moves to the periphery of the disc through a number of vanes; as the product leaves the disc it is broken up into droplets. The rotating speed of the disc is usually 10 000–12 000 rpm, and the diameter of the disc between 220 and 400 mm; discs of various configuration are used (see Fig. 10).

The rotating atomiser is a high-speed precision piece of equipment and is, therefore, expensive, but the feed pump is considerably less costly than the high-pressure pump required for nozzle atomisation. The high rotating speed of the disc can be obtained by using a direct coupled high-speed electromotor with a frequency converter, or by using a standard motor with gear or belt transmission. The belt transmission is simple and reliable provided the right kind of belt is used,

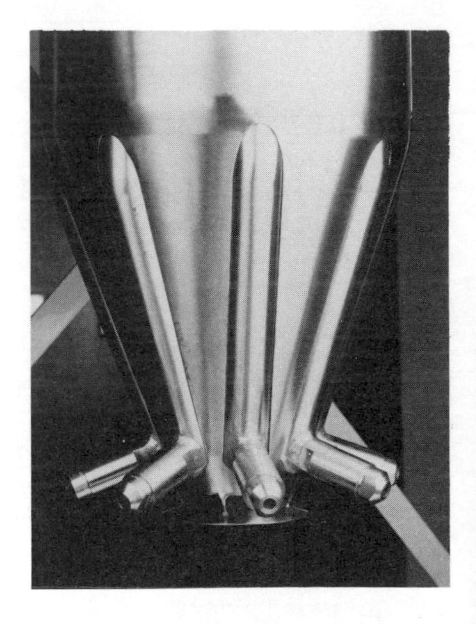

Fig. 10. Rotating atomiser (left) and nozzle atomiser (right) (courtesy APV ANHYDRO A/S, Denmark).

and it is easy to change the rotating speed of the disc by fitting another size pulley to the spindle. The centrifugal force available in a rotating atomiser to break-up the liquid is much higher than the force from the pressure in a nozzle atomiser and, therefore, the rotating atomiser can handle more viscous products (i.e. concentrate with high total solids) than a nozzle atomiser.

A nozzle is exposed to considerable wear and, therefore, the orifice must be of a hard metal. The control of a dryer with a nozzle atomiser is more difficult than a dryer with a rotating atomiser, because the feed-rate cannot be varied without changing the pressure, unless another orifice is fitted; this change cannot be done during operation. When the pressure is changed, the atomisation is also changed which is undesirable. When a rotating atomiser is used, a variation in the feed-rate will not affect the atomisation, as it is governed only by the rotating speed of the disc.

To take up variations in the total solids of the feed to the atomiser, it is necessary to vary the feed-rate to the atomiser so as to maintain a constant temperature of the outlet air from the dryer, and thereby, a constant moisture content of the powder. This is done automatically by means of a temperature sensor in the duct carrying the outlet air. The temperature sensor governs the speed of the feed pump, which will supply more or less concentrate to the atomiser if the outlet temperature is above or below the set point. Great variations in the total solids of the feed cannot be tolerated when nozzle atomisation is used, as variations in the feed-rate give a significant change in the atomisation.

The question of whether the one atomisation system is better than the other cannot be answered in a simple way. For some products the one system is better than thc other and, therefore, if the aim is to have a versatile dryer, the type of spray dryer should be installed where both atomisation systems can be used. It is not necessary to have two feed pumps because the high-pressure pump can be used without pressure with the rotating atomiser.

Nozzle powder has less occluded air than powder from a rotating atomiser and, therefore, a higher bulk density. A bulk density of 0·7 g cm^{-3} is normal for skim-milk powder produced by nozzle atomisation, whereas approx. 0·6 is normally achieved when a rotating atomiser is used.

Whole milk powder and fat-filled powder produced by nozzle atomisation are claimed to contain less free fat than powder produced by means of a rotating atomiser. It is obvious that the high-pressure pump and the passage through the nozzle has a homogenising effect which will reduce the amount of free fat. Practically the same result can be obtained by homogenising the concentrate prior to feeding it to a rotating atomiser. The rotating atomiser can handle concentrate of higher total solids. A nozzle atomiser is not suitable for whey concentrate containing lactose crystals, as the crystals will wear the nozzles.

If powder from different products or a powder with certain characteristics is to be produced, the dryer must have both atomisation systems.

Air heating system

Steam has traditionally been used for heating the drying air, but in a modern milk dryer, the air inlet temperature may be as high as 215–220 °C, and steam is seldom available at the pressure required to produce such high temperatures. When steam is used, the air temperature is usually boosted by another heating system.

Saturated steam of 10–12 bar gives an air temperature of 170–180 °C, as the heating surface is normally rated to give an air temperature of 8–10 °C below the temperature of the steam. Superheated steam is not suitable, as it requires an air to air heat exchanger which is much more costly than a saturated steam to air heat exchanger. If only superheated steam is available, it can be saturated by injection of water before it enters the air heater.

The corrosive nature of steam and condensate makes it necessary to use stainless steel tubes welded into stainless steel inlet and outlet headers. Aluminium fins are tension wound onto the tubes to give good heat transfer characteristics. The condensate stage is connected to the steam stage by means of manifolds incorporating a stainless steel orifice plate. This system has the advantage that no steam traps are required. Care must be taken that the steam/air heater is manufactured according to the code accepted by the authorities. The code may be national or international, such as TUV, Veritas or Lloyd's.

Air cooling has to be used for the powder cooling system when the ambient air temperature is above 10–15 °C. The air cooler is constructed as a steam/air heater with the tubes carrying chilled water instead of steam. The air cooler must be connected to a drain to take away the condensed humidity from the air, and it must be provided with an efficient droplet eliminator. It must be designed to cool the air below the required temperature, as the air has to be reheated a few degrees to bring it above saturation temperature. Condensate or hot water can be used for the reheat section.

Liquid phase heating (hot oil system) is widely used to heat the drying air, either to do all the heating or to boost the temperature from a steam/air heater. Figure 11 shows a diagram of a liquid phase heating system. An oil- or gas-fired heater is used to heat a heat transfer fluid to 330–350 °C. The hot fluid is pumped to an air heater battery similar to that employed when steam is used. A high thermal efficiency is obtained by preheating the return fluid in the stack of the unit used for heating the heat transfer fluid.

The economiser built into the stack is designed to cool the flue gases from an oil burner from about 500 °C to 180 °C by means of the return fluid at approximately 160 °C from the air heater. The stack temperature must, of course, be kept above the condensation temperature for the flue gases. Where a gas burner is used, it is possible to cool the flue gases further by the return oil. The thermal efficiency of a liquid phase heating system with economiser is claimed to be close to 90 per cent. It is a simple system operating in closed circuit at atmospheric pressure, and it is well suited to heat the drying air to the desired high temperature.

Another system widely used for heating the air is indirect oil or gas heating. In this system, the oil or gas is burnt in a cylindrical chamber, and the air is drawn over the external surface thus gaining heat. Before

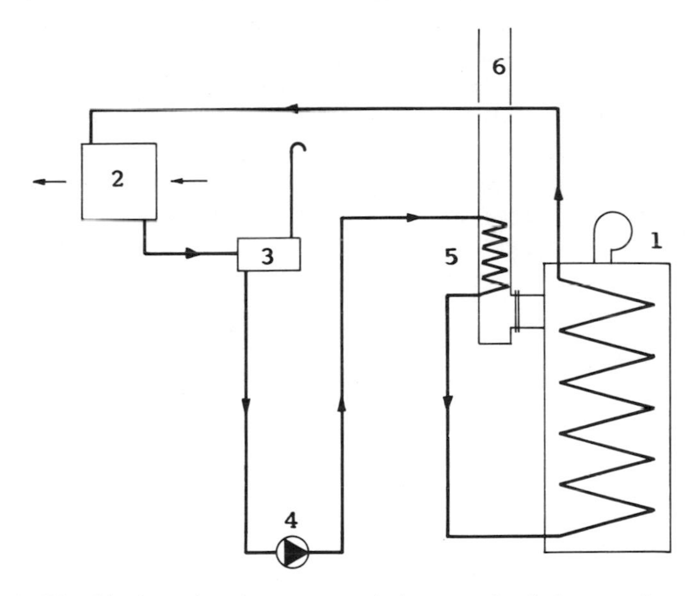

Fig. 11. Liquid phase heating system. 1: burner; 2: air heater; 3: expansion tank; 4: circulation pump; 5: economiser; 6: stack for flue gases.

the combustion gases are exhausted to the atmosphere, they are cooled by the intake air, but the thermal efficiency will not exceed 85 per cent. The material used for a heat exchanger of this kind should be heat-resistant stainless steel. The heater will be damaged if it is not constantly cooled by the air and, therefore, the burner must be electrically interlocked with the fans making it impossible to operate the burner without the fans being in operation. Electric air heaters can be used, but are normally not competitive with regard to the running costs. They are, however, sometimes used as booster systems.

Direct gas-fired air heaters have been in use for several years in milk dryers in North America. The availability of natural gas and the high fuel prices have, in recent years, created a considerable interest in this system; Fig. 12 shows a direct gas-fired air heater. No heat exchanger is required. The mixture of the combustion gases from the burner and ambient, filtered air is led to the drying chamber as the drying medium. The absence of a heat exchanger brings the thermal efficiency to practically 100 per cent, i.e. a saving in fuel consumption of 10–20 per cent is obtained compared with indirect heating. Further, the capital investment in a direct system is substantially less than for an indirect system. Maintenance costs are lower, and the quick response and flexibility of a direct system is also an advantage.

In spite of the great attraction offered by direct gas firing, it is not yet widely used outside North America. The reason is recognition in recent years of the danger of contaminating the milk powder product by direct gas firing. The contamination is produced by the oxides of nitrogen contained in the combustion gases, and it is, therefore, imperative to minimise the quantity of nitrogen oxides in the combustion gases. When nitrogen oxides get in contact with food products, they can react with amines associated with the proteins to produce nitrosamines

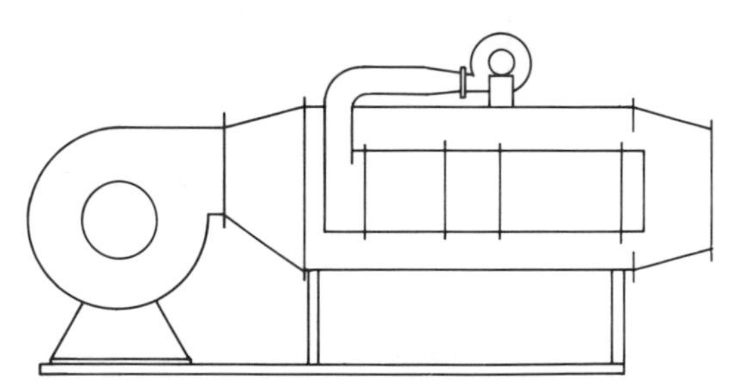

Fig. 12. Direct gas-fired air heater.

(Challis *et al.,* 1982), which are classified as cancer promoting when assimilated. Nitrosamines can also be formed when amines react with nitrite present in food products. Nitrite is sometimes added to food products, or it is formed from nitrate by bacteriological conversion.

Nitrites and nitrates are, therefore, undesirable products in food commodities, as they can lead to formation of nitrosamines which constitute a health hazard. Research work has shown that an increased content of nitrite and nitrate is found in milk powder prepared in direct gas-fired dryers, as much as double of that found when indirect heating was used (Harding and Gregson, 1978). It was found in Australia (Rothery, 1968) that when direct gas firing is used, the nitrite content of the spray dried sodium caseinate is higher than in skim-milk powder. The interesting discovery was made by Rothery that the amount of nitrite could be significantly reduced by an injection of steam into the flame, whereby the flame temperature was reduced, and hence smaller amounts of nitrogen oxides were produced. The nitrite content of spray dried caseinate was, in this way, reduced from about 28 ppm to about 3 ppm, and the nitrite content of skim-milk powder was reduced from 1 ppm to less than 0·1 ppm. The work done by Rothery clearly demonstrates the requirement for minimising the amount of nitrogen oxides created by the combustion, and this recognition has, in recent years, led to the development of a new generation of gas burners to be used in the food industry.

Oxides of nitrogen are formed as part of all combustion processes in which air is used as the oxidant. When nitrogen and oxygen are heated, nitrogen oxides are formed by thermal fixation of the nitrogen. Another source is the conversion of chemically bound nitrogen in the fuel. The nitrogen oxides found in combustion gases are mainly NO and NO_2, but also small quantities of higher oxides are formed. NO_x is conveniently used as common designation for all the nitrogen oxides in the combustion gases, both those created from the air by the thermal process, as well as those originating from nitrogen in the fuel. It is only during the last five to six years that it has been recognised that special burners producing low NO_x in the combustion gases are required in the food industry, in order to prevent the formation of nitrosamines in food products exposed to the combustion gases.

Fuel-bound nitrogen is present in oil and coal, where it contributes significantly to the amount of NO_x in the combustion gases. Fuel-bound nitrogen can account for over 50 per cent of the amount of NO_x in the combustion gases from oil, and approximately 80 per cent when coal firing is used. When gas is burned, the whole amount of NO_x in the combustion gases is generated from the air as a result of the thermal process, as there is practically no fuel-bound nitrogen in gas. The statement has been made that it should be perfectly safe to use methane, propane or butane from refineries in direct fired milk dryers, because these gases contain no impurities. This is a fundamental misunderstanding. Natural gas also contains very few impurities, and the combustion is practically complete, for it undergoes a cleaning process so that possible impurities are stripped-off before the gas is supplied to the consumers. It is not impurities in the gas that give cause for concern — it is the NO_x formed by the combustion. This process takes place, of course, regardless of the kind of gas used.

The rate of thermal formation of NO_x from oxygen and nitrogen is strongly temperature dependent. The amount of combustion generated NO_x is governed by the flame temperature and the duration of the top temperature phase. At flame temperatures in the range of 1500–1600 °C, the formation of NO_x starts to be significant and, therefore, the flame temperature should be kept below this level, but high enough to ensure complete combustion. The duration of the top temperature phase should be short, but combustion must not be inhibited too early by dilution with the air to be heated, as this will produce undesired intermediate combustion products in the air. A low flame temperature can be obtained, as mentioned earlier, by injection of steam into the flame. However, this is not a method suitable in practice in connection with spray drying, as it will increase the amount of moisture in the drying air, which will necessitate a higher outlet temperature of the air to maintain the specified moisture content of the powder. In low NO_x burners, a flame temperature of 1200–1300 °C is obtained by using overstoichiometric combustion with an excess air level of up to 90 per cent. Uniformity of the air/gas mixture is essential in ensuring not only controlled combustion, but also the absence of pockets of gas-rich mixtures, the combustion of which would result in localised high temperatures and high NO_x levels.

Figure 13 shows a low NO_x burner. The venturi gas/air mixer is designed to give a uniform mixture of the gas and air before it enters the flame stabiliser. Increased stability of the flame is achieved by using a refractory lined fire tube. The NO_x level ex-burner (i.e. before dilution with air) is guaranteed to be below 1 ppm by volume. Conventional gas burners will produce an NO_x level ex-burner of between 50 ppm and several hundred ppm (typically about 150 ppm). The NO_x level of the drying medium getting in contact with the atomised milk concentrate will be substantially lower than the NO_x level ex-burner, due to dilution by the air to be heated. At an NO_x level of 1 ppm ex-burner, the NO_x level of the drying medium will be 0·12 if a temperature of 200 °C of the mixture of combustion gases and air has to be achieved, and lower in case of a lower drying temperature. The figure 0·12 ppm should be understood as the increase in NO_x level (i.e. NO_x added from the combustion), as ambient air can contain variable concentrations of NO_x produced by industry, power stations and the exhaust gas from motor cars.

It is only in recent years that it has been recognised that combustion generated NO_x in process air can react with food products whereby nitrite, nitrate and nitrosamines are formed. It is also of recent date that sufficiently sensitive methods for detection of nitrosamines have been developed. Food products are in contact with the combustion gases from bakeries, from the gas ovens used in households and from open grills with the combustion gases from coal. However, as the surface of such food products is small compared with the volume, it has been found that the increase in nitrite, nitrate and nitrosamines is so small that it is not possible to measure it with the methods available today.

It is a different matter when milk is spray dried, because the product is given a very large, contact surface in a spray dryer. It has been proved that when direct, gas-fired spray dryers are used, nitrite, nitrate and nitrosamines are formed in the powder at a rate in direct proportion to the amount of nitrogen oxides in the drying air. Therefore, special burners are required, designed to produce the minimum amount of combustion generated NO_x; no more than 1 ppm of NO_x from the burner mouth can be accepted.

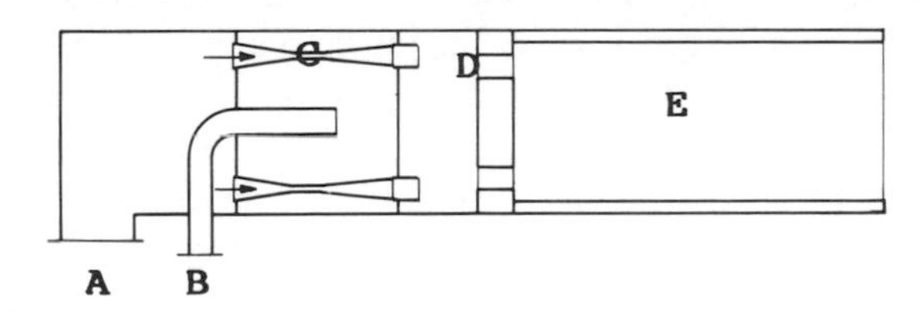

Fig. 13. CAX low NO_x gas burner. A: combustion air entry; B: gas entry; C: gas–air mixer; D: pre-mixer burner stabilising nozzles; E: combustion chamber. (Courtesy Urquhart-Engineering Co. Ltd, UK.)

Direct gas firing was used in milk dryers in North America long before the problems associated with this system were recognised. Practically only skim-milk and whey are spray dried in North America, whereas in Europe the production of whole milk powder and fat-filled powder is significant. It may be that the problems are more pronounced when powder containing fat is produced, but further research work is required to verify that.

The authorities in Europe and in other countries realise that the use of direct gas firing for spray drying of milk presents a complex problem and, therefore, they do not cause any restrictions. It is not allowed and it is not forbidden. In Holland, the answer from health authorities to applications to use direct gas firing in spray dryers for edible milk powder has been that the heating system used is not subject to any legislation, but it is the manufacturers' responsibility to ensure that certain limits for the content of nitrite, nitrate and nitrosamines in the powder are not exceeded. It is specified that the content of nitrosamines must be below 1 ppb, and WHO has issued standards for the maximum content of nitrite allowed in food products. Several milk dryers in Holland are equipped with direct gas firing. The authorities have the right to demand analysis for nitrite, nitrate and nitrosamines in samples taken at any time. In practice, samples are taken every third month to ensure the correct operation of the burners. These analyses are costly and tedious to make.

Another product formed by the combustion of gas is water. This does, of course, not constitute a health hazard, but it affects the capacity of the dryer. From the combustion of one standard m^3 gas is formed approximately 1·7 kg water, which is present as vapour in the drying medium. It has been shown (Muir *et al.,* 1981) that the water of combustion in a direct, gas-fired spray dryer is equivalent to approximately 20 per cent of the water to be evaporated in the dryer. The increased moisture content of the drying medium has the effect that a higher outlet temperature of the dryer must be used than when the dryer is indirectly heated. This means that a higher air quantity is required to obtain the same evaporating capacity of the dryer, i.e. the economic advantage from direct gas firing is partly offset by the loss in the efficiency of the dryer caused by the water formation occurring when gas is burned. In a warm and humid climate, where the ambient air has a high moisture content, the extra moisture in the drying air generated by the combustion can lead to such a high outlet temperature, when single-stage drying is used, that a serious loss in powder solubility can occur.

To underline the significance of the above, the following incident may be relevant. A powder manufacturer decided to change from indirect heating to direct gas firing for economic reasons. He approached the health authorities who advised him to install a low NO_x burner. He did not get in contact with the supplier of the spray dryer, but went to a company that specialised in the supply of gas burners. They made it clear that they knew nothing about spray drying, but they undertook to deliver a burner with guaranteed low NO_x formation. The burner was installed, but to the surprise of the powder manufacturer, it was only possible to obtain the specified moisture content of the powder at an air outlet temperature very much higher than used before. This, of course, reduced drastically the capacity of the dryer, but worse than that, the powder was useless as the solubility was very poor. The powder manufacturer had followed the request from the authorities to install a low NO_x burner. The burner supplier had fulfilled his guarantee, and there was nobody to blame for the poor result of the conversion of the air heating system. A consultant found that a burner using steam injection to obtain the low NO_x content had been installed. This extra moisture together with the water of the combustion of the gas raised the moisture content in the hot air supplied to the dryer by an amount equal to 40 per cent of the water to be evaporated in the dryer. Consequently, a very high air outlet temperature was required to obtain the specified moisture content of the powder. The direct air heating system had to be abandoned, and the indirect system was reinstalled.

Drying chamber

Two different shapes are shown in Fig. 9, the chamber with flat bottom, from which the powder is sucked from the chamber by means of a rotating powder collector, and the chamber with a conical bottom, from which the greater portion of the powder is removed by gravity.

The advantages of the flat bottom chamber are low building height, forced removal of the powder from the chamber (i.e. controlled residence time), and easy access for cleaning. It is also an advantage that hammers or vibrators on the chamber are not required.

The reasons for the use of a conical chamber are:

(1) the incorporation of two- or three-stage drying;
(2) a requirement for the separation of the powder into a coarse fraction from the chamber, and a fine fraction from the cyclone. By recycling the fines to the atomisation zone, a dust-free, partly agglomerated powder is obtained;

(3) the manufacture of a powder with a fat content exceeding 35–40 per cent. Powder with a high fat content has a tendency to stick to the interior surface of the ducts and cyclones, and therefore, the greater portion of the powder must be discharged direct from the chamber, so that only the small quantity contained in the air will pass through the ducts and cyclones.

Only skim-milk powder is comparatively free-flowing, for both whole milk powder and fat-filled powder have poor flow characteristics. To avoid a build-up of powder on the conical bottom, it is necessary to use external hammers or vibrators on the cone, which presents two problems. One is the problem of noise, which is very objectionable and very difficult to overcome. The other is the risk of the development of cracks in the stainless steel sheet, due to metal fatigue caused by the external hammers. The impact of the hammers can normally be adjusted, but it is important to equip the system with a blocking device to prevent the hammers from causing an impact higher than what is regarded as safe to prevent development of cracks. The operator tends to use the maximum possible impact to facilitate operation and cleaning of the dryer.

It is difficult to detect fine cracks in stainless steel sheet, but it can be done by applying a special liquid to the sheet so that the cracks become visible. There have been incidents of bacteriological contamination of the powder originating from an accumulation of moist powder in cracks in the stainless steel sheets. To avoid these problems it is essential to use a steep cone (small included angle), so that the number of external hammers and their impact is reduced to a minimum. The steep cone does, of course, increase the height of the chamber which adds to the cost of the building, but it is a necessary feature if the powder is taken out of the chamber with high moisture content for subsequent drying, as the moisture will reduce the flow characteristics of the powder.

The two parts of a spray dryer requiring most skill and experience to design are the atomiser and the air distributor placed on top of the drying chamber. The hot air is admitted to the chamber through the air distributor. It contains a number of vanes to give the air a rotary movement following the direction of rotation of the atomiser, and so impart a swirl pattern to the particle cloud to move it downwards into a vortex. The intimate mixing of the drying air and the atomised product is essential to obtain the highest utilisation of the heat, i.e. to produce a low air outlet temperature. Further, the air distributor must direct the product cloud downwards to prevent moist powder impinging on the walls of the chamber, particularly at the level of the atomiser disc.

When a nozzle atomiser is employed instead of a rotating atomiser in a chamber designed for a rotating atomiser, four or more nozzles are used. The nozzles are not in a vertical position, but are placed at an angle to the vertical direction. In this way, the pattern of the product cloud is similar to that produced by a rotating atomiser, which is a condition for the use of the same chamber configuration and the same air distributor for both atomisation systems.

The chamber ceiling will be heated to a temperature equal to the air inlet temperature at the point where the hot air enters the chamber. To avoid charring of milk solids at this point, the ceiling is provided with an annular cooling ring around the air distributor. Slightly heated air is blown through the cooling ring to remove enough of the heat to bring the surface temperature down; unheated ambient air should not be used as it may cause condensation.

The chamber volume is sized in ratio to the volume of drying air in such a way that the number of air changes gives an air residence time in the drying chamber of 25–30 s. The drying chamber is the most space demanding part of a spray dryer, and a building to house a large drying chamber represents a considerable investment. Drying chambers can be designed to be erected in the open without a building, and Fig. 14 shows a spray dryer for milk where the drying chamber and the vertical, indirect oil-fired air heater are placed externally; the feed pump, control panel, cyclones, fans, and bagging-off equipment are placed in an adjacent building.

The insulation of a chamber installed in the open is increased from 100 mm to 200 mm. The temperature inside the chamber can, for the purpose of assessing the heat loss, be assumed to be 100 °C. The outside temperature may be approximately 15 °C if the chamber is housed in a building, but could be −20 to −30 °C if the chamber is installed in the open. The temperature difference across the wall is thus increased from 85 °C to say 130 °C, i.e. by approximately 50 per cent. By increasing the insulation by 100 per cent, ample compensation has been made for the larger Δt even when the heat loss due to convection is considered, and therefore, the external installation of the chamber does not cause any loss in the efficiency of the dryer. The insulation is provided by a cladding of plastic-coated, mild steel sheets. The cladding is extended downwards to cover the support for the chamber and upwards to cover a light steel structure (penthouse) on top of the chamber to accommodate the atomiser. In this way, the chamber looks like the milk silos, which long ago were moved to the outside. There is no operation, cleaning or other function to be done externally. The door giving access to the

Fig. 14. Spray dryer with drying chamber and indirect oil-fired air heater in the open (courtesy APV ANHYDRO A/S, Denmark).

chamber is placed in a wall recess in the adjacent building, and the access to the penthouse for the atomiser is also from this building. The space under the chamber is connected to the building by a short corridor, where a fluid bed can be installed.

Milk dryers with external chambers have been built in several countries. The only problem encountered, in some cases, has been to obtain planning permission when a building is deemed to suit the surroundings better.

Powder recovery system

Cyclones have been used for many years to separate the powder from the air. The principle of cyclone separation is based on the centrifugal force exerted on a particle similar to the principle of a centrifuge. The separation efficiency of a cyclone varies inversely with the diameter of the cyclone, and directly with the square of the tangential velocity of the air, and with the mass of the particles to be separated. It can be expressed by the formula:

$$\frac{mV^2}{d}$$

where m is the particle mass, V is the air velocity and d is the diameter of the cyclone.

Of course, the shape of the cyclone, i.e. ratio between cylindrical part and conical part also affects the efficiency.

The loss of solids from a system of cyclones for a milk dryer is often expressed as a percentage of the total powder production, although only a smaller portion of the powder (20–30 per cent) may pass through the cyclones. The greater portion, discharged direct from a conical drying chamber into a fluid bed, does not pass the cyclones, and even if it did, it would contribute little to the loss, as it consists of large particles. The fines suspended in the air are conveyed to the cyclones, and the particles below 5–10 μm constitute the loss to the atmosphere.

For the powder manufacturer, it is logical to express the efficiency of the powder recovery system as a percentage of the total powder production, but to assess the loss with regard to the environment, it is necessary to express it as the amount of solids in relation to the volume of the discharged air. Normally the loss is expressed in mg solids per standard m^3 air. In North America, the loss is expressed in grains per standard ft^3, and the relationship is 1 grain/st.ft^3 = 2275 mg/st.m^3.

One effect of the dryers getting bigger is that the absolute amount of milk solids discharged to the atmosphere from the cyclones has reached such levels that it no longer can be tolerated for environmental reasons, in particular where the dryer is close to a residential area. In most countries, legislation has been introduced in recent years to limit the emission levels by specifying the maximum permitted amount of solids contained in the discharged medium; in the cases where the dust does not constitute a health hazard. In Europe, the permitted emission level from milk dryers varies between 50 and 100 mg/st.m^3 air discharged. In the USA, the permitted emission level from milk dryers varies from 0·01 grain/st.ft^3 in California to 0·04 on the East Coast (23–91 mg/st.m^3). In practice, the cyclone loss varies within wide ranges depending on several technical and technological factors; Table II shows some typical cyclone losses from three dairy products.

It will be seen that the loss from cyclone separators in all cases exceeds the permitted emission level. Therefore, today it is not possible to install a milk dryer where the powder recovery system consists of cyclones alone. Further cleaning of the air is necessary to bring the loss to the atmosphere within the permitted limit. The powder discharged to the atmosphere represents a significant financial loss, and therefore, when the air is cleaned further, it should be done in such a way that the solids recovered can be added to the powder from the dryer to increase the yield. The following three methods can be considered:

(1) The use of a wet collector or wet scrubber after the cyclones to trap the greater portion of the solids by scrubbing the air with water. The water is recirculated and periodically renewed.

The disadvantages are that the milk solids are not recovered, disposal of the scrubbing water presents a problem, and the scrubber itself is hygienically unacceptable in a dairy factory, because the conditions in the scrubber are ideal for rapid growth of micro-organisms. To overcome the first two problems — recovery of the solids and disposal of the scrubbing medium — the so-called milk scrubber was introduced about ten years ago. Unconcentrated milk is used as the scrubbing medium instead of water to absorb the solids from the air. The milk goes from the scrubber to the evaporator, and the solids recovered will be contained in the concentrate to the dryer and will thus increase the yield. However, the third problem — the bacteriological contamination — was not solved. On the contrary, this problem became highly critical, because all the product passes through the scrubber, whereas when water is used, the scrubber is kept separate from the product.

TABLE II
Powder loss from cyclones

Product	*Loss*	
	Percentage of total production	*(mg/st.m^3)*
Skim-milk	0·5	200
Whole milk	0·35	150
Whey	1·0	400

The idea of using the milk as the scrubbing medium originated in the USA where it was used 50 years ago, but abandoned due to the bacteriological hazard. Thus, there is a wet zone in a scrubber, and at the air exit there is a dry zone. The transition between wet and dry represents a moist zone where conditions for the growth of bacteria are ideal. By assessing the bacterial count in the milk before and after the scrubber, it is found that the number of bacteria is increased immensely after a few hours operation. High temperature pasteurisation of the milk is necessary before it enters the evaporator. In a place where a large number of bacteria are present, there is a considerable risk of the generation of toxins. The toxins are not destroyed by the heat treatment commonly used, and it is difficult to analyse the powder for toxins. By employing a milk scrubber, the powder manufacturer runs the risk of producing milk powder contaminated with toxins, which have boundless consequences.

It is incomprehensible that milk scrubbers have been put on the market in recent times. Milk producers are encouraged to produce milk with low bacterial counts, and then an unhygienic piece of equipment is installed in the factory that generates bacteria at a high rate. It is against all logic, and against the continuous trend of raising product quality.

Milk scrubbers were installed around 1976, and during the following years, on dryers in several countries, as they were marketed by well reputed companies alleging that precautions were taken in the design to prevent bacterial build-up. Today they have practically all been taken out, or are only used where powder for animal food is produced. A duplex or triplex installation is required for edible products, as the scrubber must be cleaned after 3–4 h operation. The complicated changeover procedure must be designed to exclude the possibility of the cleaning solution getting into the product. The rinse water from cleaning five times during a 20 h run puts a considerable load on the effluent plant. The introduction of milk scrubbers into a modern dairy factory has been a step backwards, and should not be repeated.

(2) Another attempt to minimise the amount of dust discharged to the atmosphere has been made by employing highly efficient cyclones. These cyclones are used in the chemical industry, and the principle is that secondary air at high pressure is introduced into the cyclone to increase the velocity, and thereby increase the efficiency. However, these cyclones have not been successful in the dairy industry as they are unsuitable for powder containing fat. When the powder is exposed to the high speed of rotation in the cyclone, a high amount of free fat

(surface fat) will be generated, and the powder sticks to the interior surface of the cyclone. As the cyclones are not sanitary, they are not used either for non-fat powder.

(3) The third method of collecting the dust in the air from the cyclones is to insert a textile filter (fabric filter or bag collector) before the air is discharged to the atmosphere. The filter is not used as the main collector, i.e. the cyclones are retained, and the filter is employed as a secondary collector.

A textile filter is not a piece of equipment that really belongs in a dairy factory. However, the fact that a filter is the only known solution to the problem has caused considerable development in the design of filters suitable for milk dryers. If a filter of conventional type is installed, the recovered product is kept separate and sold as animal food. This will make it difficult to justify the filter from an economic point of view. By installing a filter specially designed for milk dryers, a first grade product is recovered, which can be continuously blended back into the main stream of powder from the dryer. In this way, the filter serves the purpose both of protecting the environment as well as increasing the yield.

The loss of powder from the filter is extremely small. Normally less than 10 mg/st.m^3 air is guaranteed, which will meet even the most strict standards for environment protection. If the air from the cyclones contains milk solids in the order of 200 mg/st.m^3 air, the filter will recover 95 per cent. To illustrate what is involved, it is useful to look at an example. We assume a spray dryer produces 4 ton/h of skim-milk powder and is in operation 4000 h per year. With a stack loss from the cyclones of 0·5 per cent and a filter with an efficiency of 95 per cent, the powder recovered from the filter amounts to 76 ton per year. At a value of £900 per ton, it represents a financial contribution of £68 400 per year.

Figure 15 shows a filter designed for milk dryers. The air from the cyclones is introduced into the bottom part of the filter from where it enters the inside of the bags. The solids are thus collected inside the bags contrary to conventional filters where the solids are collected on the outside of the bags. The air goes through the filter bags into the filter housing and then to the atmosphere.

The filter bags are provided with stainless steel rings preventing the bags from collapsing when exposed to negative pressure. The bags are continuously cleaned by a slowly rotating suction arm which is connected to two bags in each position, where the arm, by means of a timer, is stationary for a short time. In this way, the powder accumulated in the

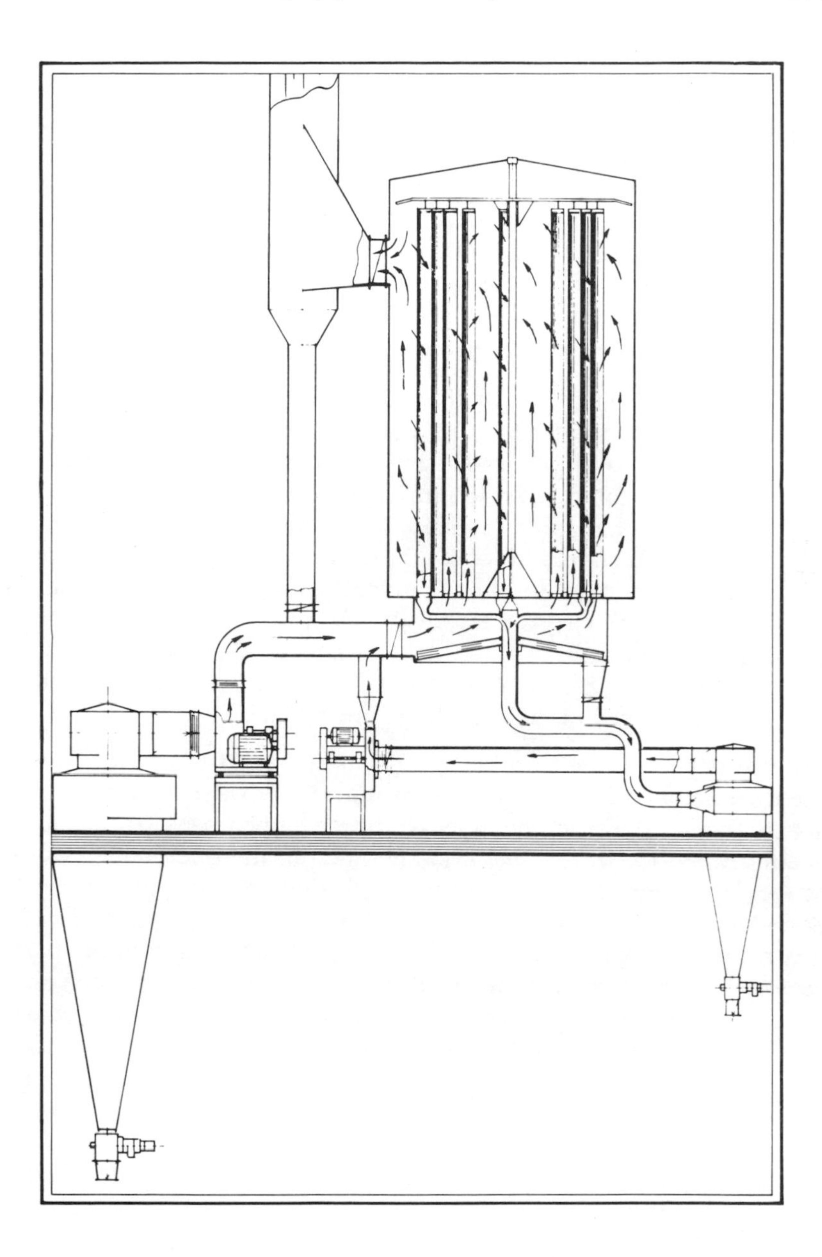

Fig. 15. Textile filter (courtesy APV ANHYDRO A/S, Denmark).

bags is sucked out by a high pressure, stainless steel fan, and separated from the cleaning air in a small cyclone which has a higher efficiency than the large cyclones for the drying air. The air from the small cyclone is passed back into the bottom part of the filter, i.e. the pneumatic cleaning system operates in a closed circuit. Powder that drops from the bags into the bottom part of the filter is discharged into the pneumatic cleaning system by means of a rake attached to the rotating suction arm. Compressed air, normally used for cleaning of filter bags, is not required. The product does not come into contact with the baghouse which only contains the filtered air to be discharged. The absence of product in the baghouse has the following advantages:

(1) A baghouse is not sanitary, and product sticking to the interior surface can become a 'nest' of bacteria — particularly if care is not taken to avoid condensation at start-ups. Dry cleaning of a baghouse for powder would, in practice, not be feasible.
(2) The baghouse can be constructed of mild steel. Only the bottom and the chamber into which the dust laden air is introduced must be made of stainless steel. The chamber is provided with cleaning doors giving easy access for dry cleaning.
(3) Due to the fact that the baghouse only contains the cleaned air, it does not constitute any hazard for dust explosion, and therefore, special precautions are not required.

The powder recovered from the filter is discharged from the cyclone of the pneumatic, bag cleaning system into the powder from the dryer. Where the powder from the dryer is conveyed pneumatically to silos, it is convenient to connect the cyclone outlet to the pneumatic powder conveying system. The powder from the filter consists of very small particles, i.e. the texture is different from that of the powder from the dryer, but this will of course not be apparent by continuous back-mixing. The filter powder does not differ from the powder from the dryer with regard to solubility and bacterial count. Sediment tests made on the filter powder can sometimes show disc B indicating that the filter needs cleaning, but that is without practical importance, because the quantity is too small to have any influence on the sediment of the powder resulting from back-mixing.

It is imperative that a by-pass with air tight dampers is installed to connect the dryer directly to the atmosphere when it is wet cleaned. When the dryer is idle, the filter bags must be protected against moisture pick-up, either by supplying heated air to the filter, or by blowing dehumidified air through the filter. If the main fan for the dryer is placed between the cyclones and the filter, care must be taken that the powder is not contaminated in the fan, i.e. the fan must be provided with an efficient, rust protecting coating, or be made of stainless steel. The filter acts as silencer for the fan reducing the external noise level. The fan can also be placed after the filter, and while it can then be standard, a separate silencer must be provided.

Besides giving environment protection and increased yield, the textile filter has another important function. It provides clean air for heat recovery, i.e. transfer of heat from the outgoing air to the inlet air; for this purpose a heat exchanger is required. If the air is used direct from the cyclones, the milk solids in the air will foul the heat exchanger; this will not happen when the air has been cleaned in the textile filter. By recovery of the heat from the outgoing air, a fuel saving of 20–25 per cent is obtained. The installation of a textile filter is often initiated by environmental requirements, but can be a profitable investment, providing both solids recovery and heat recovery.

Heat recovery

Several methods have, in recent years, been introduced to transfer heat from the outgoing air to the inlet air, either by air to air heat exchangers, or by air to liquid heat exchangers.

Air to air heat exchangers can be of the plate type, or tubular type. The air intake is usually far from the point where the air is discharged, and therefore, an air to air heat exchanger requires considerable ductwork. Fouling of the surface in contact with the outlet air will occur if the heat exchanger is used without a filter after the cyclones. The fouling will reduce the heat transfer, and therefore, the heat exchanger must be cleaned frequently. The fouling does, of course, reduce the amount of powder discharged to the atmosphere, but the milk solids are lost, and the frequent wet cleaning adds to the load on the effluent plant. In case of a blocked cyclone, the heat exchanger will be clogged-up with powder.

A much simpler, heat recovery system can be established when the air from the cyclones is cleaned in a filter. In this case, exhaust air is passed over the surface of a finned, tube heat exchanger transferring heat to the liquid inside the tubes. The heated liquid, which can be water with glycol, is pumped to an air heater battery in the inlet air stream to preheat the inlet air. It does, of course, involve heat transfer twice, and therefore, the resultant preheat temperature is slightly lower than when an air to air heat exchanger is used, but the system is very much simpler and less costly to install. In case an indirect oil- or gas-fired air heater is used, the temperature of the water in the heat recovery system can be raised by running the water through an economiser in the flue gas stack for the air heater. In this way, the water temperature is boosted and a higher preheat temperature is achieved.

It is, of course, most rewarding to recover heat from the outgoing air in a system where the temperature of the the outgoing air is high; that is the case when single-stage drying is used. If we assume an air inlet temperature of 215 °C, ambient temperature 15 °C, and a preheat temperature of 70 °C obtained by heat exchanging the ingoing air with the outgoing, it will be seen that instead of heating the air 200 °C, the air will only have to be heated 145 °C with heat recovery, i.e. a saving of 27·5 per cent is obtained. In a three-stage dryer where the temperature of the outgoing air is lower, the saving by heat recovery will be approximately 20 per cent. The pay-back time for a heat recovery system will be short when the yearly utilisation of the dryer is high.

Thermal efficiency

The thermal efficiency of a spray dryer is the ratio between the heat utilised for evaporation and the total heat consumption. The heat required for secondary drying is normally not included in this context, and therefore, the thermal efficiency in percentage is expressed by

$$\frac{T_i - T_u}{T_i - T_a} \times 100$$

where T_i is air inlet temperature, T_u is air outlet temperature, T_a is ambient air temperature.

It is apparent that the efficiency is increased with increasing T_i and T_a and with decreasing T_u.

The maximum air inlet temperature to be used in a milk dryer is 215–220 °C, and the ambient air temperature can be raised by employing heat recovery.

The aim of increasing the efficiency by reducing the outlet temperature has led to the introduction of two- and three-stage drying. Multi-stage drying is based on the fact that whereas most of the water is readily evaporated from the concentrate, a more severe heat treatment — higher temperature or longer drying time — is required to remove the last few percent of the moisture. Thus the rate of evaporation is fast until a dry particle surface has been formed, but it takes more heat or longer time to evaporate the moisture contained in the particles.

Two- and Three-stage Drying

Figure 16 shows a two-stage dryer with fluid-bed secondary drying. The powder is discharged from the drying chamber at 5–6 per cent moisture into the vibrating, fluid-bed dryer and cooler (Fig. 17). The fluid-bed consists of a stainless steel, perforated plate through which air is blown upwards. The powder is 'fluidised' by the air, and high turbulence occurs which provides excellent conditions for heat transfer between air and powder.

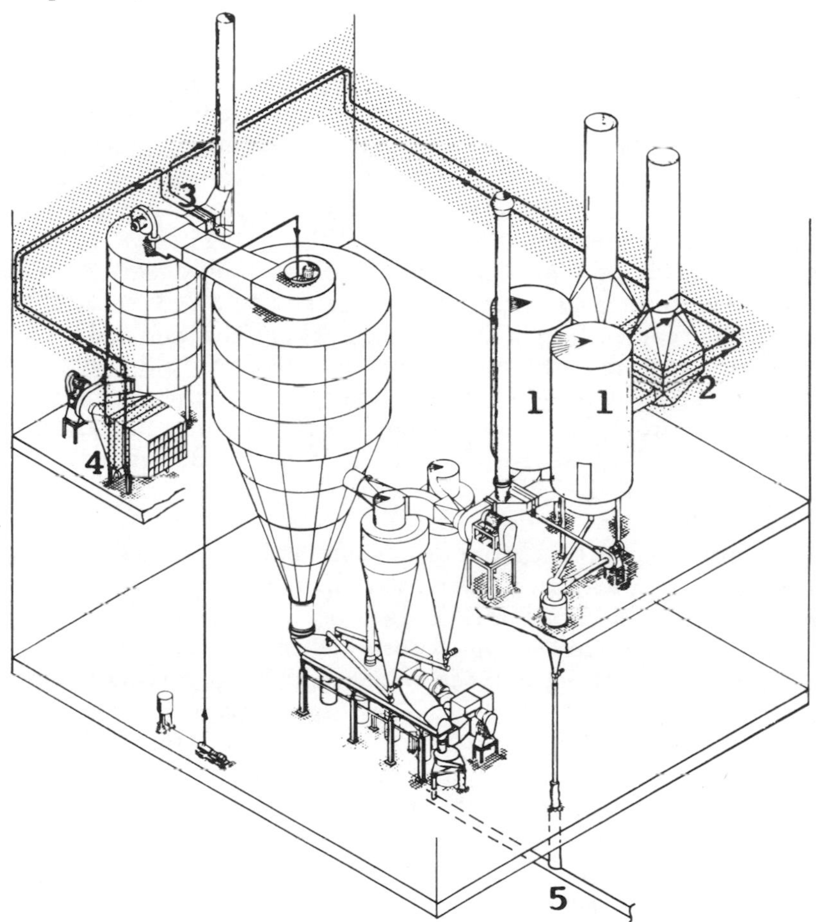

Fig. 16. Two-stage dryer with solids recovery and heat recovery. 1: textile filter; 2: heat exchanger for heat recovery; 3: booster in flue gas stack; 4: air preheater; 5: fluid lift system to powder silo. (Courtesy APV ANHYDRO A/S, Denmark.)

In the first section, the powder is dried to its final moisture content by air at 100–120 °C, and in the second section, the powder is cooled by air at approximately 10 °C. The powder layer is approximately 100 mm high, and the fairly long residence time in the fluid-bed (10–12 min) makes it possible to perform the drying at a low air temperature. The air used for drying and cooling is collected in the hood over the perforated plate, from where it goes to the cyclones for separation of the fines carried with the air. The high moisture in the powder from the chamber is obtained by using a low, outlet temperature. The low, air outlet temperature improves the thermal efficiency of the dryer, and it also improves the powder quality — especially in terms of solubility, bulk density, and free fat. The dryer shown in Fig. 16 has both solids recovery (textile filter) and heat recovery. Liquid is used to transfer the heat from the outgoing air to the intake air, and the temperature of the liquid is boosted in the flue-gas stack for the indirect, oil-fired air heater.

Three-stage drying (Fig. 18) was introduced in recent years in order to improve further the thermal efficiency of the drying process by transferring a greater portion of the evaporation from the first stage to the second and third stages. The second stage is a fluid-bed placed at the outlet of the conical part of the drying chamber. This fluid-bed is static (non-vibrating), and the semi-dry powder drops over an adjustable weir into an external fluid-bed for final drying and cooling. A low outlet temperature from the drying chamber can be used, as the powder is delivered from the first drying stage to the integrated fluid-bed at a high moisture content.

The air leaves the drying chamber through a vertical duct placed at the base of the cone in the centre of the chamber. The off-centre discharge of the powder makes this arrangement possible which is an improvement, as the conventional, horizontal, outlet air duct in the chamber is difficult to clean inside and collects powder deposits on the outside. Further, the horizontal air duct prevents the swirl pattern of the drying air being extended into the cone, as the duct is obstructing the rotary movement of the air. Without the horizontal air duct, it is possible to support the swirl movement by the so-called 'wall sweep arrangement', i.e. the tangential introduction of secondary air through a number of narrow, vertical slits placed in the drying chamber wall at the transition point between the cylinder and the cone. The temperature of the sweeping air is slightly lower than the temperature of the air in the chamber. The high-speed air rotation prevents the moist powder from lodging in the cone, and it also reduces the amount of powder carried with the air to the cyclones. The

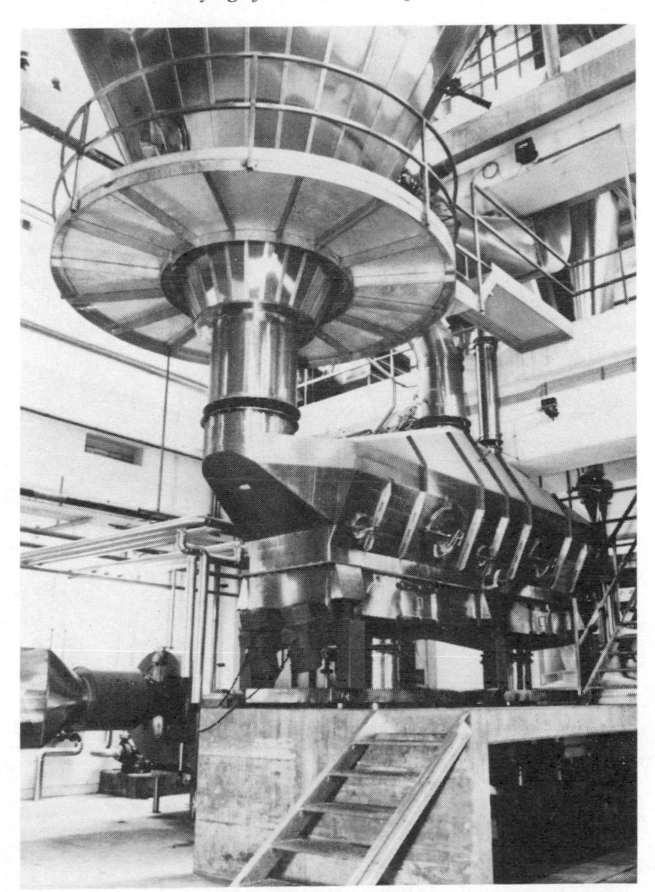

Fig. 17. Fluid-bed (courtesy APV ANHYDRO A/S, Denmark).

combination of wall sweep and a steep cone reduce the requirement for external hammers, i.e. the number of hammers as well as their impact can be reduced.

The external fluid-bed can be substituted by a pneumatic system for drying and cooling the powder, if the height under the chamber is insufficient for the installation of a fluid-bed; this may be the case, when an existing, single-stage dryer is converted to two- or three-stage drying. Both the two-stage dryer as well as the three-stage dryer can operate with either rotating atomiser or pressure nozzle atomiser, and Table III shows the heat required to evaporate 1 kg water in single-, two- and

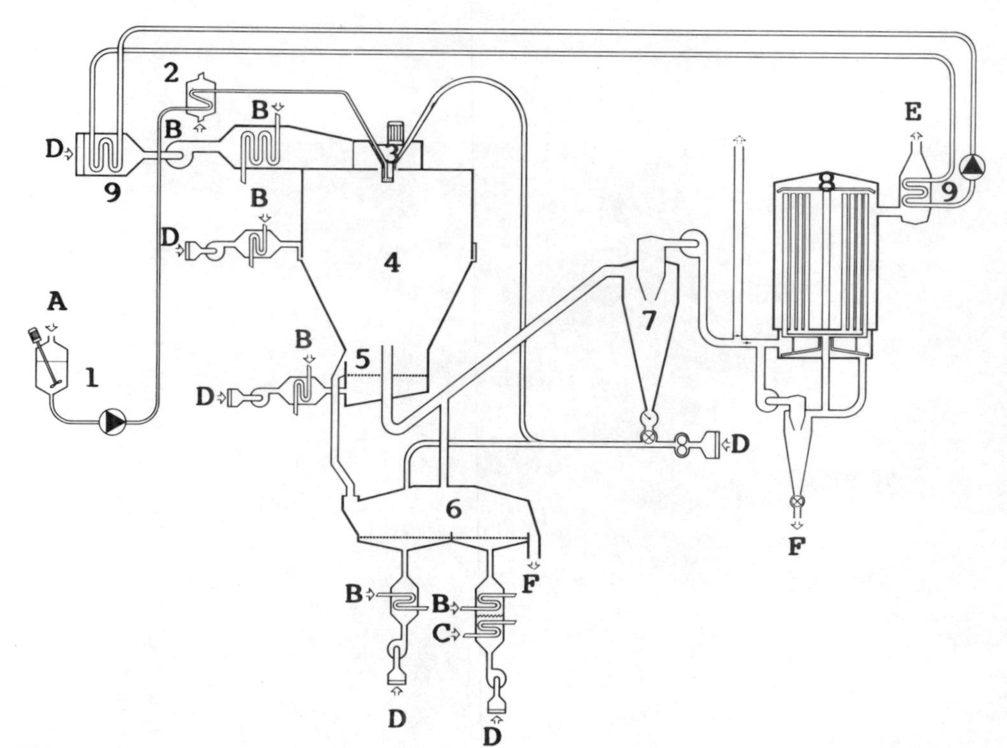

Fig. 18. Three-stage dryer with solids recovery and heat recovery. 1: feed tank; 2: concentrate preheater; 3: atomiser; 4: spray drying chamber; 5: integrated fluid-bed; 6: external fluid-bed; 7: cyclone; 8: bag filter; 9: liquid coupled heat exchanger; A: product inlet; B: steam; C: cooling water; D: air inlet; E: air outlet; F: product outlet. (Courtesy APV ANHYDRO A/S, Denmark.)

TABLE III
Heat required to evaporate 1 kg water. Air inlet temperature 215 °C

Type of dryer	*kcal*
Single-stage	1200
Two-stage	980
Three-stage	850

three-stage dryers. The air inlet temperature, is in all three cases, 215 °C, and the dryers are without heat recovery. As heat recovery reduces the heat consumption by approximately 20 per cent, it will be seen that the heat consumption of a three-stage dryer with heat recovery approaches half of the heat consumption of a single-stage dryer without heat recovery. This clearly illustrates the progress made in energy conservation, and the saving per ton of powder will be even greater when the contribution from a low energy demanding evaporator is included in comparison with a conventional evaporator.

The advantages offered by three-stage drying are reduced heat consumption, reduced space demand, and improved powder quality. The amount of air required to obtain a certain capacity is the smallest possible when three-stage drying is used, and thereby the cost of a textile filter for cleaning the outgoing air is reduced.

CIP-cleaning

Wet cleaning of a dryer should be restricted to a minimum to reduce the risk of bacteriological contamination, as moisture is favourable for growth of bacteria. However, to maintain disc A when the test for scorched particles is made, it is necessary to wet clean a spray dryer at certain intervals. In fact, this test is often used as an indicator of when wet cleaning is required. If the dryer is being used for skim-milk, wet cleaning should only be performed about every three weeks, but often the requirement for cleaning is dictated by a change of product.

For many years, the operators have had to dismantle the ducting and other plant items for manual cleaning, but as the dryers are getting bigger and more complex, automatic cleaning in the shortest possible time is required. Cleaning in place is well established in the dairy industry, but has only in recent years been applied to spray dryers. This is done by fitting spray balls in the plant in such a way that the spray balls cover the entire interior surface. They are permanently placed in ducts, cyclones, and fluid-beds (above and below the perforated plate) as shown in Fig. 19. The spray balls are stationary, and in a large dryer 60–80 spray balls are required. It is possible to leave the spray balls in powder-carrying ducts, cyclones and fluid-beds, because when the plant is in operation, heated, purging air is blown through the entire CIP-system, thereby preventing the product from penetrating into the spray balls.

The spray balls are connected in sections, as it would require too great a quantity of water to clean the whole plant at the same time. The cleaning cycle can be programmed, as experience will tell how long a time is required for each section to be cleaned. Normally the aim is to clean the dryer in 2–3 h including subsequent drying with hot air, but a longer time is required if the dryer has, for instance, been used for fat-filled milk which tends to form hard deposits. It is important that all surfaces in the dryer are completely dry before operation commences, and air at 90–100 °C should be drawn through the plant after wet cleaning to sanitise the dryer.

Only the drying chamber is not cleaned by a built-in cleaning device (the sprinklers shown in Fig. 19 in the chamber ceiling are for fire protection). The chamber is cleaned by removing the atomiser, and lowering a rotating-jet tank cleaner down into the chamber. The rotating-jet cleaner is very effective, as the water hits the surface at high pressure, thereby producing a mechanical action for the removal of wall deposits.

All cleaning water leaves the plant through the fluid-bed and can be collected in tanks. As the cleaning water from the first rinse contains some milk solids it may be used for animal food. Cleaning solution is normally collected in a tank for re-use. If the plant has a textile filter for the outlet air, it is advisable to have interlock switches on the CIP-pump and the damper for the by-pass system for the filter to ensure that the pump cannot be started unless the filter is by-passed. Access door, inspection doors, and pressure relief panels are normally opened after cleaning, and frames and gaskets are cleaned manually. Special attention must be paid to design all doors, panels and joints in such a way that they are completely leak-proof.

A condition for the success of the CIP-system is that the dryer is cleanable in-place, i.e. that the plant items are of good hygienic design with no product traps or dead space, and that ducts are at the correct slope for the removal of water. Fittings for temperature probes can, if not properly designed, provide uncleanable crevices. It is difficult to provide an existing dryer with a CIP-system, because the dryer may not be leak-proof, and the plant items may not be of hygienic design.

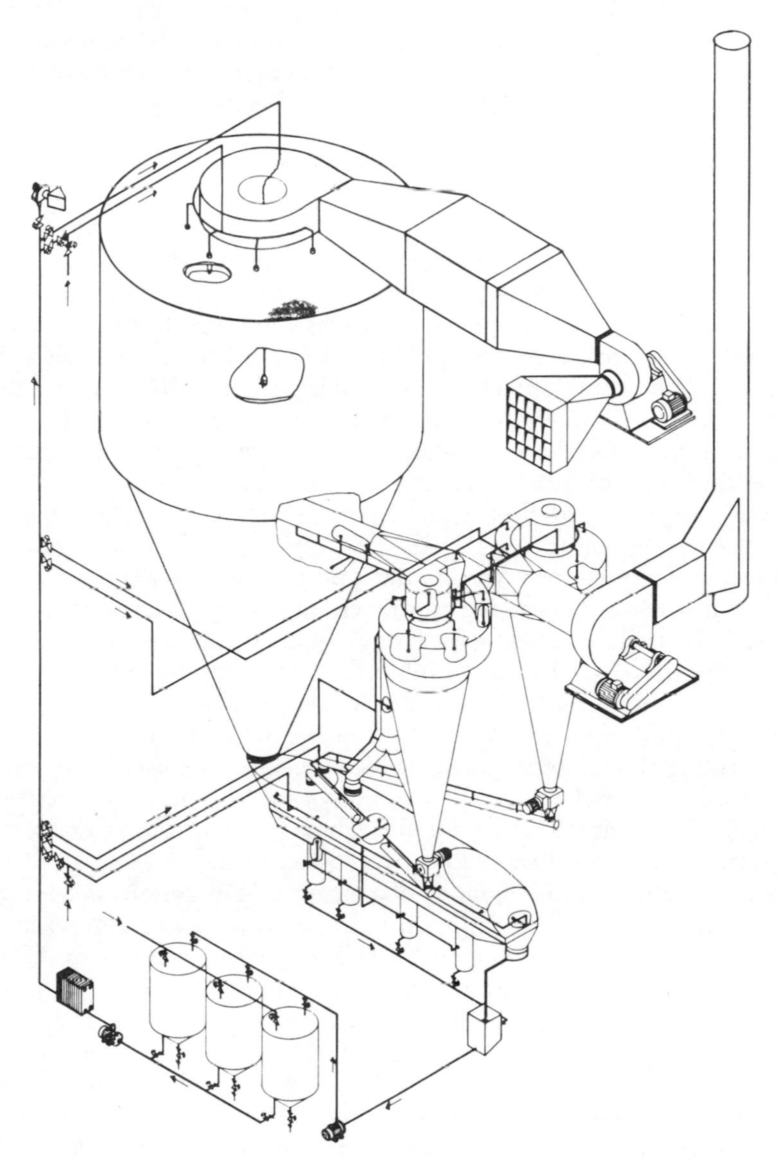

Fig. 19. Spray dryer with in-place cleaning system (courtesy APV ANHYDRO A/S, Denmark).

Protection against Fire and Explosion

The danger exhibited by milk powder with regard to fire and explosion has been recognised in recent years. As a consequence of this, authorities and insurance companies are now demanding that the design of spray dryers includes certain features to give protection against fire and explosion.

As milk powder is an organic material, it is combustible. The powder is present in the dryer in two different physical conditions — either suspended in the air as a cloud, or in the form of deposits on the interior surface of the dryer. Both present a hazard, because the conditions in the dryer may be such that the deposits can catch fire and the cloud can explode. A cloud of milk powder requires a temperature in excess of 400–450 °C to burn, and as temperatures of this order do not occur in a milk dryer, the cloud of powder does not represent a fire hazard. There have been cases where welding done inadvertently on a dryer in operation has caused a fire, but here the high temperature required to ignite the cloud of powder has come from the outside.

The position is different with regard to powder that is present as deposits (a layer of powder). Here self-ignition (spontaneous ignition) represents a real danger, because the ignition temperature of a powder layer is much lower than for a powder cloud. Self-ignition of dried milk deposits in the dryer is the most likely cause of a fire and explosion — only in rare cases will a cause other than self-ignition be the reason. Sparks from rotating parts or from discharge of static electricity will normally develop too little energy to bring the product to the combustion temperature. It is, however, advisable to establish an efficient earth connection to the dryer to eliminate any build-up of static electricity charges. The material used for filter bags in a textile filter should contain stainless steel wires to earth the bags.

The conditions for self-ignition of deposits of milk powder are present in a spray dryer, and therefore, it is imperative to avoid deposits — particularly where the temperature is high, i.e. at the chamber ceiling where the hot air is introduced. The atomiser and the air distributor will also have a high surface temperature, and therefore, deposits on these parts are critical. Improperly stopping the drying operation, for instance due to power failure, may cause a back-draught which carries milk particles into the air distributor and the inlet air duct, where they form deposits that can catch fire on re-starting the plant. The risk of self-ignition does also exist in other parts of the dryer where the temperature is lower, i.e. in the drying chamber and in the air outlet duct where the temperature is equal to the air exit temperature (80–90 °C). This is particularly critical when fat-filled powder is produced, because the temperature of the deposits will rise quickly due to oxidation of the fat.

Many materials, milk powder included, can undergo oxidation, which is an exothermic chemical process, i.e. the process generates heat. In an accumulation of material, heat will be generated throughout the volume, but can be lost only through the surface. The rate of oxidation and hence the rate of heat production is very low at ambient temperature, and self-ignition does not normally present a hazard. It is, however, a well known fact that a pile of coal can catch fire due to self-ignition caused by oxidation of iron-pyrites contained in the coal.

The rate of oxidation increases very rapidly with increase in temperature, and it is this phenomenon which gives rise to problems in a milk dryer. Oxidation readily takes place in a deposit of fat-filled powder which has a temperature of 80–90 °C equal to the exit-air temperature. The oxidation generates heat which further increases the rate of the reaction and so on, and when the heat production exceeds the heat loss through the surface, the material temperature will rise to ignition temperature. The ignition temperature for skim-milk powder deposits is 160–180 °C. It is lower for fat-filled powder and decreases with increasing fat content. The critical thickness of the powder layer at various temperatures has been determined for skim-milk powder and fat-filled powder (Beever, 1984). The quoted paper contains useful observations in relation to spontaneous ignition of milk powders, and preventative measures to be taken to avoid fires in spray dryers. A comprehensive account of dust explosions and fires is given by Palmer (1973).

Probably the biggest risk in the event of fire occurring in the dryer is that it can initiate a dust explosion of the powder suspended in the air, if the conditions are right for an explosion. As experience shows, spontaneous combustion of powder in the dryer is by far the most frequent cause of an explosion, all precautions taken to prevent fire are also precautions against explosions. The measures to be taken to prevent fire are related to the design of the dryer, but in the following are mentioned some general precautions, which apply to most dryers:

(1) Avoid powder deposits on hot surfaces. The heat will darken or char the deposited material, and particles will break off and appear in the powder. Therefore, continuous checking (at least every half hour) should be done of the cleanliness of the powder by testing to determine

scorched particles in the powder. As soon as scorched particles are found, the plant should be stopped and the reason for scorched material should be investigated.

(2) When fat-filled milk is dried, special attention should be paid to powder deposits in the drying chamber and in the air outlet duct as the product is sticky. Due to the heat generated by oxidation, deposits in low temperature zones present a hazard, as well as deposits on surfaces with a high temperature. The rate of oxidation at a given temperature varies with the grade of fat. A high content of unsaturated fatty acids will result in a high rate of oxidation, whereas a hardened fat oxidises less readily. By homogenisation of the concentrate prior to drying, the amount of free fat (surface fat) is reduced, i.e. a greater portion of the fat will be protected by the proteins. The largest number of fires in spray dryers can be traced back to the drying of fat-filled milk.

(3) It has been noted that many fires have occurred fairly soon after start-up. The reason for this can be that either some deposits have not been removed during cleaning, or the chamber walls might still be wet, encouraging deposits. Wet patches in the dryer can be due to insufficient drying after wet cleaning, or to an incorrect start-up procedure. The dryer is started with water and when the system is hot, the feed is switched to concentrate. Wet zones will occur in the dryer if this switching is done too early. Moist powder deposits are particularly dangerous, because experience has shown that the self-ignition process occurs faster in wet powder than in dry.

(4) A thermostat should be placed in the inlet-air duct that sounds an alarm and cuts off the air heating system if the air inlet temperature exceeds a certain predetermined value (for instance, 20 °C above normal working temperature).

(5) A thermostat should be placed in the air outlet duct that sounds an alarm, cuts off the air heating system, and activates the by-pass system for the textile filter, if the air outlet temperature exceeds the normal working temperature (usually the thermostat should react at 120 °C).

Reasons for the elevated outlet temperature can, for instance, be lack of feed to the atomiser or malfunction of the atomiser.

(6) A fire extinguishing system should be provided consisting of sprinklers in the chamber ceiling (shown in Fig. 19). The spray balls in the fluid-bed used for in-place cleaning should also be utilised in the fire extinguishing system. Water is automatically admitted to the spray balls and to the sprinklers at the chamber outlet temperature (normally

at 135 °C) or at the exit-air temperature from the fluid-bed. At the same time, the feed to the atomiser is automatically switched from product to water and the fans are stopped. It is, of course, possible to use the entire CIP-system as a fire extinguishing system, but normally the water supply is inadequate for that.

(7) The dryer can be equipped with fire detectors which are activated by any burning or glowing material. The purpose of the detectors is to activate the fire extinguishing system, and to prevent fire or glowing material from spreading into the fluid-bed, and into the cyclones and textile filter. The fire detectors are placed in the duct between the chamber and the fluid-bed, and in the duct between the chamber and the cyclones.

A rotary powder valve should be inserted into the duct between the chamber and fluid-bed, and the valve should stop when the fire detectors are activated. A glowing lump of powder is particularly dangerous in the fluid-bed, where there is a continuous supply of oxygen, and therefore, the fire detectors must stop all the fans. Sprinklers should be placed in the duct between the chamber and the cyclones, and the sprinklers should be activated by the fire detectors.

Fires in spray dryers present a particular danger, because often the conditions for a dust explosion in the dryer are present in which case the fire may initiate the explosion. A dust explosion is caused by the extremely rapid rise in pressure resulting from sudden oxidation of a flammable substance.

A mass of solid flammable material, which is heated to ignition temperature, will burn away slowly layer by layer as only the surface is exposed to oxidation. The energy produced is liberated gradually and harmlessly because it is dissipated as quickly as it is released. The result is quite different if the same mass of material is ground to a fine powder, and intimately mixed with air in the form of a dust cloud. Under these conditions, the surface area exposed to oxidation is very great and, if ignition now occurs, the whole of the material will burn at once. The energy is released suddenly with the evolution of large quantities of heat and a great pressure develops almost instantaneously.

There is a range of concentrations of dust in the air within which the mixture is explosive, whereas mixtures at concentrations below or above this range cannot explode. The lowest concentration of dust capable of explosion is referred to as the 'lower explosive limit', and the concentration above which an explosion will not take place as the 'upper explosive limit'. A dust explosion will occur if a flammable dust is dispersed in air (or oxygen) when:

(a) the concentration of dust in the air is within the explosive range;
(b) a suitable source of ignition is present.

The most violent explosion is produced when the amount of oxygen present is close to that which will result in complete combustion. It occurs at a powder concentration of about 1000 g m^{-3}.

In the drying chamber of a milk spray dryer, a flammable dust is dispersed in air in uncontrolled concentrations — a low concentration at the top and a high concentration at the powder outlet. It is not likely that the concentration in the entire chamber is within the explosive range, but it must be assumed that the concentration in a part of the chamber is within the explosive range. This part is referred to as the 'active volume'.

Preventive measures against explosions are aimed at avoiding the presence of a suitable source of ignition, of which fire is the most common. However, as fire can occur, and as the plant is not strong enough to withstand the pressure generated by an explosion, it is necessary to provide the spray dryer with explosion reliefs to protect persons working in the dryer room, and to prevent or reduce damage to the plant.

Explosion reliefs are provided in the form of venting panels in the chamber ceiling, or in the wall of the cylindrical part of the chamber. The pressure at which the panels open is adjustable and must be below the maximum pressure which the chamber can resist. The relief panels must be capable of operating almost instantaneously, as the pressure rise takes place extremely rapidly.

The size of the relief panels — referred to as the 'vent area' — is determined by the following parameters:

(1) The explosion intensity of the dust expressed by the group of substances to which the product belongs.
(2) The active volume of the chamber.
(3) The pressure shock resistance of the chamber, which must be greater or equal to the maximum explosion pressure reached in the chamber when it is fully vented. This pressure will decrease with increasing vent area.

Drying chambers are normally built with a pressure shock resistance up to 0·2–0·5 bar. The necessary vent area will, of course, decrease with increasing pressure shock resistance.

There are no universal rules for assessment of the active volume. Sometimes the drying chamber cone is taken as the active volume, and sometimes a greater part of the chamber volume is regarded as active volume.

In some countries, a simple, empirical rule is adopted for determination of the vent area expressed by:

$$V = \frac{1}{25} \times K$$

where V is the vent area in m^2 and K is the volume of the cone in m^3.

The relief panels must be connected by ducting to the open air. The ducts must be strong enough to withstand the maximum pressure to which they will be subjected, and must be straight and as short as possible (maximum 3 m). If venting takes place to the dryer room, a secondary explosion may occur in the building where an accumulation of settled dust may be dispersed into clouds by the primary explosion in the plant.

The required measures to be taken to protect a spray dryer against fire and explosion vary from country to country, and vary even within a country as the design must meet the local safety regulations. The Factory Inspector and possibly other authorities should be consulted at design stage to ensure approval of the installation. It is sometimes a requirement that milk powder silos are equipped with explosion relief panels.

Tall Form Dryers

The tall form dryer (Fig. 20) has pressure nozzle atomisation by means of one or more vertical nozzles. Vertical nozzles require a tall drying chamber of small diameter; the height is approximately five times the diameter, and this chamber configuration does not permit the use of a rotating atomiser.

The absence of a swirl movement of the air in the chamber is claimed to result in a true co-current flow of air and product, the so-called piston flow. This gives a short residence time, the advantage of which is somewhat counteracted by the higher outlet temperature required to achieve the desired moisture content of the powder. The elevated outlet temperature has an adverse effect on the thermal efficiency of the dryer.

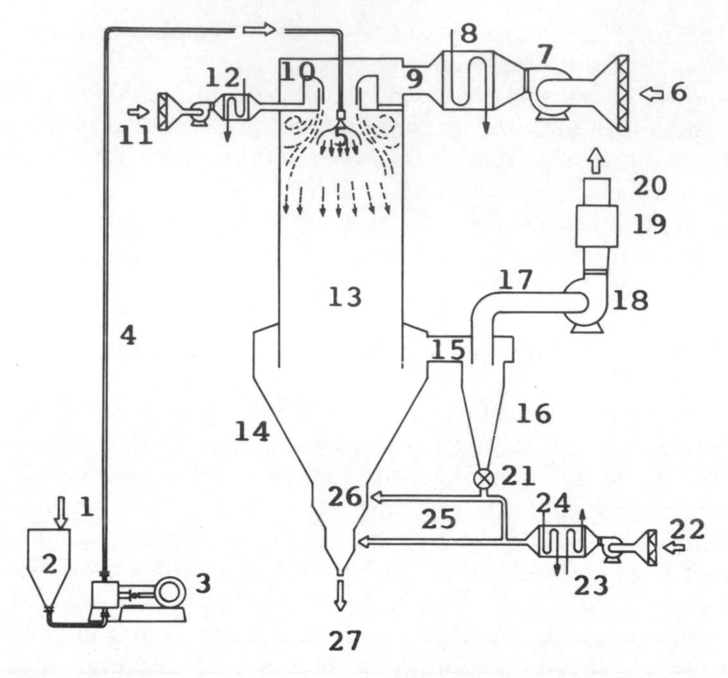

Fig. 20. Tall form dryer. 1: concentrate; 2: feed vat; 3: high pressure pump; 4: feed pipe; 5: pressure nozzle; 6: main air inlet; 7: fan; 8: air heater; 9: hot air duct; 10: air distributor; 11: secondary air inlet; 12: air heater; 13: drying chamber; 14: separating cone; 15: outgoing air duct; 16: cyclone collector; 17: duct; 18: exhaust fan; 19: silencer; 20: exhaust stack; 21: rotary valve; 22: cooling air inlet; 23: air cooler (dehumidifier); 24: air heater; 25: cooling air ducts; 26: powder cooling zone; 27: powder discharge. (Courtesy Morinaga, Japan.)

Pillsbury Dryer

The University of Minnesota and the Pillsbury Company, Minneapolis have jointly developed a dryer which is known as the Filtermat Dryer (Fig. 21). The drying chamber is rectangular (box-type) with a number of vertical pressure nozzles in the ceiling. As the height of the chamber is insufficient for drying to the final moisture content, the moist product builds up into a mat on a porous, moving belt. The drying air goes through this layer of powder, which acts as a filter for the exhaust air. The moist product is moved on the belt into following sections for further drying and cooling.

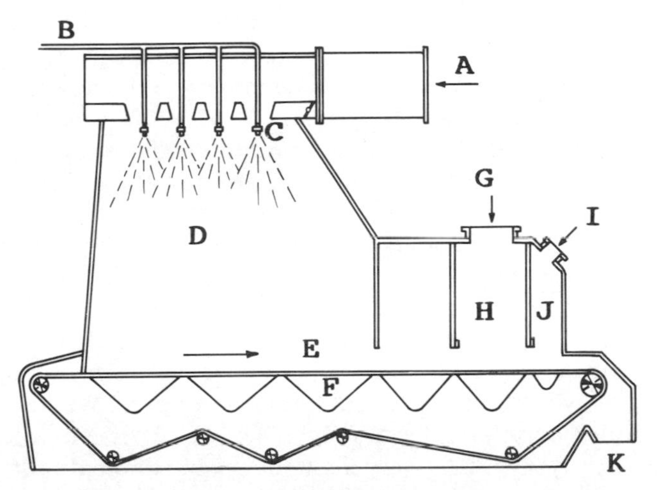

Fig. 21. Filtermat (Pillsbury) Dryer. A: hot primary air supply; B: product feed line; C: high pressure atomising nozzles; D: spray drying primary stage; E: moving belt; F: exhaust air; G: warm secondary air supply; H: bed drying secondary stage; I: cooling air supply; J: cooling stage; K: product discharge.

Compared with a fluid-bed, the air moves in the opposite direction in the Filtermat Dryer. This results in a compact cake of product which breaks-off at the end of the moving belt. Milling of the product from the belt is necessary to achieve a powder. The belt is cleaned by brushes or by water before it is again exposed to product.

The belt dryer was originally developed for the drying of food products like tomatoes, fruit and vegetables, but is also used in the dairy industry for the manufacture of special products with a high content of fat or sugar. The design is not suitable for large capacities as there is a limit to the size of the moving belt. It is not possible to use a rotating atomiser in this type of dryer.

TECHNOLOGY

Concentration

In an effort to improve the overall efficiency of the process, the tendency has been to concentrate the milk as much as possible prior to drying. In modern plants, skim and whole milk are concentrated to 48–50 per cent solids and whey to 55–60 per cent.

It has been suggested that evaporators and dryers should be developed that can handle milk of higher total solids. However, the use of higher concentrations is not an engineering problem related to the design of evaporators and dryers, it is a technological problem. Experience shows that if milk is concentrated beyond the above mentioned concentrations, the quality of the powder suffers, especially the solubility.

The stability of the casein is decisive for the solubility of the powder; poor solubility is usually due to coagulation of the casein. Casein stability is extremely sensitive to variations in pH and salt balance, and both are essentially affected by concentration of the milk. The titratable acidity increases very nearly in direct proportion to the concentration, and also the hydrogen ion concentration is increased. Whereas the pH of fresh milk is approximately 6·7, the pH of 48–50 per cent concentrate is approximately 6·1. Only a very slight increase in the acidity of the milk is enough to decrease the heat stability of the casein, as a change in pH of the milk is magnified by concentration. Although far from the isoelectric point, a pH may be reached where fine casein precipitation starts taking place, which results in poor solubility of the powder. The interaction of such factors as protein content of the fresh milk, concentration, temperature, acidity and salt balance determines the degree of casein stability.

There are indications that the concentrations used until now are not far from the optimal conditions. When these concentrations are used, seasonal variations in the composition of the milk are often sufficient to cause destabilisation of the proteins. The stability is restored by decreasing the concentration — a reduction of only 1–2 per cent in total solids is usually sufficient.

It is a well-known fact that the lactation period affects the composition of the milk. It is also changed when the cows are moved from stable to pasture; the protein content will rise which markedly increases the viscosity of the concentrate, especially when the milk has been exposed to a severe heat treatment.

It is desirable to have sufficient latitude between the evaporator and the dryer to maintain the capacity when the level of concentration has to be varied over the year. When whole milk has to be dried, it would be good policy to design the evaporator for 50 per cent solids and the spray dryer for 48 per cent. During the time where the composition of the milk permits the maximum concentration to be used, it is easy to reduce the capacity of the dryer by reducing the air inlet temperature.

The chemical and physical properties of a concentrate with high total solids affect essentially the properties of the resultant powder. Therefore, before an attempt is made to use higher concentrations, it is necessary to study more closely the factors that affect the properties of a concentrate with increasing total solids.

Concentrate Heating

The concentrate is usually discharged from the evaporator at a temperature of about 45 °C, unless backwards-flow or a separate finisher is used, in which case the concentrate temperature may be higher. The viscosity of concentrated milk is significantly affected by temperature (Fig. 22). It is desirable to feed the concentrate to the atomiser at the lowest possible viscosity, and to achieve this, the concentrate has to be heated to 70–75 °C prior to atomisation, whereby the following advantages are obtained:

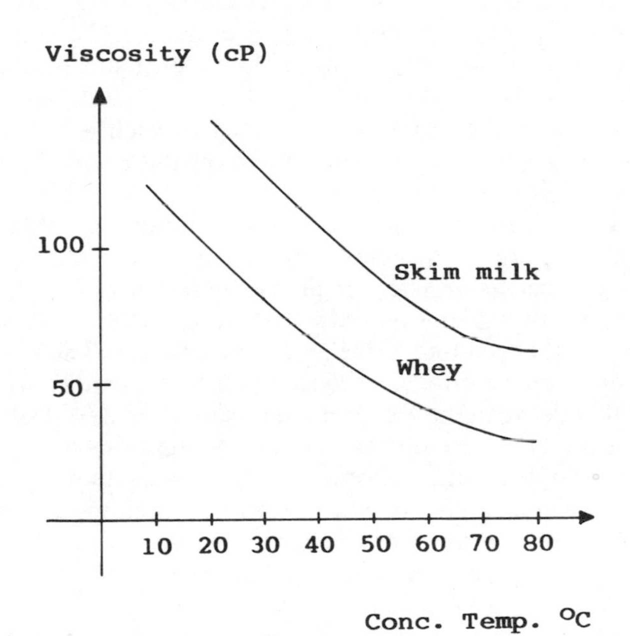

Fig. 22. The effect of temperature on the viscosity of concentrated skim-milk and whey.

(1) A saving of energy, as it is more efficient to heat the concentrate by means of steam or hot water than to use heat from the air supplied to the dryer. The evaporation capacity of the dryer is increased by approximately 5 per cent when the concentrate is supplied to the atomiser at 70 °C instead of 50 °C.
(2) The average size of the droplets into which the concentrate is atomised is reduced due to the lower viscosity. This facilitates the drying, making it possible to reduce the air outlet temperature and still maintain the specified moisture content of the powder. The lower outlet temperature results in increased capacity (i.e. lower heat consumption per ton of powder) and more gentle drying which results in improved powder quality — especially improved solubility. The last point is of particular importance when concentrate with a high total solids is dried in a single-stage spray dryer, as the outlet temperature can get near the critical temperature.
(3) As the temperature conditions in the last stages of the evaporator and in the feed tank for the dryer can be favourable for the growth of micro-organisms, heat treatment of the concentrate immediately before drying will help to safeguard the bacteriological quality of the powder.

Heating of the concentrate must be done in such a way that no holding takes place, i.e. the concentrate heater must be placed immediately before the atomiser. This is necessary in order to avoid the so-called 'age thickening' which occurs very rapidly at temperatures above 60 °C. The time-dependent thickening of hot concentrate is irreversible and results in a very high viscosity.

If the concentrate heater is placed at floor-level in the dryer room, the holding time in the feed line to the atomiser can be sufficient to cause thickening of the concentrate. The concentrate heater should be inserted in the feed line between the feed pump and atomiser, and it should be placed at the top of the drying chamber next to the atomiser. A tubular heat exchanger of the coil type, where the heating medium is vacuum-steam or hot water, can give satisfactory service when the temperature difference between product and heating medium is small, and the product has a high velocity. A scraped-surface heat exchanger may also be used, but if the concentrate heater is to be suitable for nozzle atomisation, a tubular heat exchanger can more easily be designed for the high pressure.

Manufacture of Powder to Meet a Certain Heat Classification

Heat classification refers to skim-milk powder. It is not a grading requirement, but is of practical importance in describing the suitability of powder for various, specific uses. The powder is classified according to the heat treatment to which the raw milk has been subjected before converting it to a powder. The heat classification is expressed indirectly by the level of undenatured whey protein in the powder. In order to appreciate fully what the heat classification involves, it is useful to review the composition of the milk proteins and their specific properties.

When skim-milk that has not been heated is acidified to pH 4·6, approximately 80 per cent of the proteins will precipitate. These proteins are designated the caseins, and the mentioned pH value is called the isoelectric point. It is well known that precipitation of the caseins can also take place by the action of rennet, or by saturation with sodium chloride.

The proteins remaining in suspension after precipitation of the caseins are called the whey proteins, and they represent approximately 20 per cent of the total protein content. The two main groups of whey proteins are the lactalbumins and the lactoglobulins. Unlike the caseins, they do not precipitate with acid unless they have been exposed to heat treatment to such an extent that so-called 'denaturation' takes place. The term 'denaturation' indicates that the proteins are no longer in their natural condition, and is, in this context, used to describe the change in the molecular configuration of the proteins induced by heat. When cooled they do not again assume the original configuration, i.e. the change is irreversible. There are various stages of denaturation according to the heat treatment; with excessive heat treatment, a more extended change in the configuration of the protein molecule takes place. Denatured whey proteins do not precipitate from the milk, nor do they precipitate when milk powder is reconstituted, unless the heat treatment has been excessive. Denaturation of the whey proteins changes their properties significantly. They lose biological activity and functional properties, and they precipitate when the milk or whey is acidified to pH 4·6 or saturated with sodium chloride.

The level of undenatured whey proteins in milk powder is indicated by the so-called 'Whey Protein Nitrogen (WPN) index', which is the amount of undenatured whey protein nitrogen in non-fat milk solids measured in mg g^{-1} of powder. The method for the determination of the

WPN index has been given by the American Dry Milk Institute (1971). The principle is that a sample of milk powder is reconstituted in distilled water. The caseins and the denatured whey proteins are precipitated with sodium chloride and separated-off by filtration. The level of undenatured whey proteins in the filtrate is assessed by the turbidity method using a spectrophotometer or colorimeter. The heat treatment classification for non-fat dry milk is as given in Table IV. The heat classification does not indicate the level of denatured whey proteins, because the exact content of whey proteins in the milk is usually unknown. Any estimate of the extent of denaturation depends on an assumed average value of whey protein nitrogen in the raw milk, which normally is between 8 and 9 mg g^{-1} non-fat milk solids. In milk with 8 mg g^{-1} solids, not more than 25 per cent of the whey proteins can be denatured if a low heat powder is to be produced, and at least approximately 81 per cent must be denatured if the powder is to be classified as 'high heat'. The content of whey protein nitrogen in the raw milk varies with the origin of the milk (geographical and breed), and is subject to seasonal variations as well.

It can range from 6 to 13 mg g^{-1} solids, and at certain times of the year, the level is so low that it is impossible to produce powder with a WPN index above 6. It does not seem reasonable that there may be cases where powder, according to ADMI's definition, is not classified as 'low heat' in spite of the fact that all the whey proteins may be undenatured — and this could be the case when the WPN index of the raw, unheated milk is below 6. It would be more correct to relate the heat classification to the ratio between undenatured (or denatured) whey proteins and the total amount of whey proteins in the milk.

The amount of denatured whey protein nitrogen in the powder can be assessed in the following way. The raw unheated milk is acidified to pH 4·6 whereby the caseins precipitate. Milk reconstituted from the powder is also exposed to isoelectric precipitation whereby the denatured whey protein co-precipitates with the caseins. By comparing the two results,

TABLE IV

Class	*WPN Index* (*mg* g^{-1} *powder*)
Low heat	not less than 6
Medium heat	above 1·5, but below 6
High heat	not more than 1·5

the amount of denatured whey protein is found. The ADMI test gives the amount of undenatured whey protein in the powder, and so by adding the result from the two tests, the total amount of whey protein (denatured plus undenatured) in the powder is found. The total nitrogen can, of course, also be determined by the Kjeldahl method.

In the above we have dealt only with denaturation of the whey proteins, but it should also be mentioned that the caseins can be denatured by heat. Such terms as 'destabilisation' or 'coagulation' are suggested to describe the change in the caseins produced by heat, in order to indicate that the change is of a somewhat different nature from the heat-induced changes in whey proteins. Very drastic heat treatment is required to coagulate the caseins, and the preheat conditions sufficient to denature a substantial amount of the whey proteins will normally not affect the caseins. When milk is roller-dried, the heat treatment is so severe that the caseins will coagulate, and this accounts for the poor solubility of roller powder. Whey proteins, that have been heat-denatured to such an extent that they do not redisperse into a stable suspension, also contribute to poor solubility.

In liquid milk, only the lactose and some of the inorganic salts exhibit true molecular solubility. The proteins are present in colloidal suspension, and the fat is dispersed in microscopic droplets. The so-called 'solubility' of milk powder is primarily determined by the physical and chemical condition of the proteins, i.e. the ability of the proteins to undergo redispersion into a stable suspension. Therefore, when the term 'solubility' is employed in describing the property of milk powder, it should be understood that this term implies 'suspension stability' and not true molecular solubility. The milk constituents, which are in true solution are not of importance in this context, because they do not become insoluble by conversion of the milk to powder and do not, therefore, affect the solubility index of the powder.

Denatured whey proteins will redisperse into a stable suspension when the powder is reconstituted, unless the heat treatment has been extreme. However, coagulated casein will not form a stable suspension when the powder is reconstituted, but will appear as a sediment. The solubility index is mainly a measure of the coagulated casein. By analysis of the sediment obtained when solubility tests are made, it has been confirmed that the insoluble matter mainly consists of coagulated casein with small amounts of fat and mineral complexes, whereas denatured whey proteins are only found in rare cases. Therefore, in the whole process of dried milk manufacture, it is of the greatest importance to avoid destabilisation of the casein.

Many years ago, Wright (1933) pointed out that the insoluble substance in milk powder is mainly coagulated casein, and it was found that the dangerous range of concentration is when the milk is at 84–90 per cent total solids. At this concentration, the milk exhibits maximum heat-sensitivity for insolubilisation; Wright found that at 88 per cent total solids, it took only 0·4 s to coagulate 50 per cent of the casein at 100 °C. It is, therefore, essential to avoid slow drying when the concentration reaches the dangerous level, i.e. at the end of the drying process. At this stage, the product temperature should be as low as possible, i.e. low temperature drying should be introduced at the end of the process.

It has often been suggested that the poor solubility of a milk powder is due to a high level of protein denaturation; this is not an incorrect statement, but it is inaccurate. Only if reference is made to denaturation (coagulation) of the casein is it correct that protein denaturation has an adverse effect on the solubility of the powder. Denaturation of the whey proteins does not normally affect the solubility of the powder. If that were the case, it would not be possible to make high heat powder of good solubility. This is, of course, not so, and it is equally untrue that low heat powder is synonymous with powder of good solubility. There is no relationship between WPN index and solubility index and whey protein denaturation is not synonymous with insolubilisation.

In prescribing the heat treatment required to produce powder of a certain WPN index, it must be recognised that the heat treatment during the whole process is cumulative. The heat to which the milk is subjected in the various manufacturing stages contributes to the resultant WPN index of the powder. The effect of pasteurisation, preheating before evaporation, evaporation, concentrate heating and drying on whey protein denaturation must be considered. It is easy to measure the effect of each heating stage by determining the WPN index after each stage.

The conventional pasteurisation conditions required to ensure the bacteriological quality of the product have only little effect, as a more severe heat treatment is required to cause any appreciable amount of whey protein denaturation. Only at times when the content of whey protein in the raw milk is very low, will it be necessary to use gentle pasteurisation conditions when low heat powder is to be manufactured; this may not be acceptable unless the bacteriological standard of the milk is particularly high.

The time/temperature conditions in evaporators vary according to the design of the plant. If whey protein denaturation is to be avoided during evaporation, the boiling temperature must not exceed 70 °C, and recirculation, where the product temperature is high, should be avoided.

Heating of the concentrate to 75 °C does not affect the level of denatured whey protein because, compared with unconcentrated milk, a more severe heat treatment is required to denature the whey proteins when the total solids exceed 40 per cent (O'Sullivan, 1971). It was found that for a given heat treatment, whey protein denaturation in skim-milk concentrate takes place in an inverse ratio to the degree of concentration.

In a well-designed spray dryer, the amount of whey protein denaturation is negligible. Some denaturation may take place in a counter-current dryer where partly dried particles may become in contact with the hot air. In co-current dryers operating at air inlet temperatures up to 215–220 °C, the level of whey protein denaturation in the powder is usually equal to that of the concentrate, indicating that no whey protein denaturation has taken place during drying.

The stage in the process where whey protein denaturation can take place is in the preheating prior to evaporation. The WPN index of the powder is, in practice, governed by the preheating conditions of the milk before evaporation. It is, therefore, essential at this stage to have facilities for varying time and temperature conditions over a wide range.

As the quantity of whey proteins and their response to heat treatment vary, it is not possible to prescribe the exact heat treatment required to produce powder of a certain WPN index. It is useful to measure how the complete process without preheating alters the WPN index from raw milk to powder, for with that knowledge, experience will soon be obtained as to the preheat conditions that result in powder of a certain heat classification. It is normally necessary to alter the preheat conditions over the year.

High heat powder was previously manufactured by heating the milk to 95 °C and holding it for 20–30 min, but the large throughput that is common today makes this long holding impossible, as it would require a very large holding capacity. The long holding has been substituted by ultra-high temperatures. Preheating equipment is required to heat the milk to temperatures varying from 75 to 130 °C, and a holding capacity is required to effect holdings from 15 s to 3–4 min. The design of preheating equipment capable of operating at 130 °C for 20 h before cleaning is not a simple matter. Heating by direct steam injection would, of course, meet this requirement, but it is far too expensive in running costs as regeneration is not possible. In addition, it is necessary to reduce the milk temperature to the normal feed temperature, which is a few degrees above boiling temperature in the first effect. If the milk was fed in to the evaporator at 130 °C, the rate of flashing would be far too high. A further disadvantage attached to direct steam injection in this context is that, as the steam condenses, this extra water must be removed again in the evaporator. Another problem is that steam injection can cause casein coagulation which results in powder of poor solubility.

A plate heat exchanger has a high regeneration effect and is, therefore, attractive with regard to running costs. However, a duplex installation is required which makes the initial cost high, and the change-over procedure is difficult, because it is not acceptable to stop the evaporator. If the evaporator during the change-over is starved of milk for only a short period of time, heavy burning-on of product in the tubes will occur. It is also difficult to design the change-over in such a way that cleaning solution cannot enter the product side — unless the change-over is done manually which is not suitable in an automated plant.

The most suitable preheater is a tubular heat exchanger with regeneration using steam and water as the heating medium. A low temperature difference between heating medium and product and a high velocity ensures 20 h operation without cleaning. It is recommended that the milk should be held at 90 °C for 1–2 min before it enters the high temperature section.

Only a few evaporators in Europe have sufficient preheating facilities to manufacture a heat classified powder and, therefore, most of the powder is of undefined heat classification. The demand for powder manufactured for specific uses, i.e. powder with defined properties above the normal grading requirements, will, no doubt, increase in future and, therefore, new evaporators should have the required preheating facilities. In New Zealand and Australia, practically all powder is manufactured to a defined heat classification because the purposes for which the powder is used demand specific heat induced properties.

The organoleptic properties of the reconstituted milk are affected by the heat treatment of the milk before drying. A cooked flavour is found in milk heated sufficiently to denature the whey proteins, as free sulphydryl compounds are liberated. Low heat powder is of importance to the recombining dairies to avoid a cooked flavour in the reconstituted milk. Low heat powder should also be used in cheese manufacture where fresh milk can be 'stretched' by addition of 30–40 per cent reconstituted milk. High heat powder is required for baking purposes. When powder is used for the manufacture of condensed milk and evaporated milk, a very specific heat treatment is required to obtain the right viscosity and, in the case of evaporated milk, the powder must be heat-stable to sustain the sterilisation. For these uses, the WPN is often not a sufficient criterion for the suitability of the powder. The right combination of temperature and holding time is found by testing the powder and varying the heat treatment until a satisfactory result is achieved.

Whole milk is normally given rather severe heat treatment before evaporation in order to develop the antioxidants that delay oxidation of the fat. By the heat treatment, sulphydryl compounds are released which act as antioxidants, and also a more complete destruction of oxidising enzymes takes place.

Milk Powder Quality

Standards for dry milk products based on specific analytical procedures were many years ago defined by the American Dry Milk Institute. This grading system, as well as the methods of analysis, have gained world-wide acceptance, and is equally useful to consumer and manufacturer. It contains the basic requirements the products must meet, but is not designed to describe the suitability of the product for any specific use, for which supplementary tests may be necessary; Table V shows ADMI's specification for powder to be designated 'extra grade'. This specification is widely accepted as the basic requirements for edible milk powder. Powder offered for sale to the Intervention Board of the European Economic Community must meet this specification or a more strict one, because the Intervention Board sometimes limits some of the items to increase the rejection rate. It should be mentioned that a significant quantity of powder is rejected by Intervention, not due to faults in the powder, but due to faults in the packing.

TABLE V
Quality requirements issued by ADMI for extra grade powder

	Skim-milk powder	*Whole milk powder*
Butterfat	max. 1·25%	min. 26%
Moisture	4%	2·5%
Titratable acidity	0·15%	0·15%
Bacterial estimate, per g	max. 30 000	max. 30 000
Solubility index	max. 0·5 ml	max. 0·5 ml
Scorched particles	A	A

The great value of ADMI's grading system is that detailed analytical procedures are laid down for measurement of all the specified properties. This gives an objective, consistent and uniform international measurement of quality which essentially facilitates the purchase and sale of the products. The moisture content is determined by the toluene distillation method. The infra-red heating method is frequently used in dry milk factories for quick, moisture determinations. It performs a useful function in processing control, but can yield erroneous results due to improper selection of temperature/time conditions. It is strongly recommended to frequently check this method against the toluene method, and always use the latter where moisture tests are required for specification purposes.

The bacteriological estimate (or standard plate count) is the estimated total number of organisms per gram of powder, and is determined as laid down by ADMI. Specifications for powder for particular purposes often call for a very low plate count. The bacteriological quality should, in all cases, be specified in more detail to verify the absence of pathogens. Special analysis are required for coli, salmonellae and staphylococci. A special microbiological specification would be a low spore count. It is, in general, regarded as unnecessary to establish standards for spore counts when the bacteriological quality of the powder is under control, but for powder that, in its use, offers favourable conditions for spore germination and cell multiplication, it can be significant to limit both aerobic spore formers and anaerobic spore formers. Since the heat treatment during processing generally does not destroy spores, milk must come from farms having a history of low spore counts.

By the solubility test, the amount of insoluble matter is determined in the reconstituted milk when distilled water is used. The solubility index is the volume of sediment (in ml) from 50 ml of reconstituted milk. In some countries, the solubility is reported on a volume/volume percentage basis. Thus, for a milk powder with a solubility index of 0·5 ml, the insoluble residue is 1·0 ml per 100 ml, and this would be reported as '99 per cent solubility'. Similarly for a powder with a solubility index of 0·25, the insoluble residue is 0·50 ml per 100 ml, which would be reported as '99·5 per cent solubility'. When expressing the limits, the designation 'maximum' is used in relation to ADMI's index, whereas the same limit is expressed as 'minimum' when reference is made to the volume in per cent.

The test for scorched particles (also called the sediment or purity test) is the only one where there may be room for a subjective judgement, as the sediment pad containing the impurities is graded visually against standard pads ranging from A to D.

So far we have dealt with the powder properties on which ADMI's grading system is based, but individual powder users have their own items included in the specifications.

The most common, non-grading item is the bulk density of the powder. Unfortunately, there are different methods for determination of the bulk density, and as the method used has a marked effect on the results obtained, the method should be prescribed together with the desired value. According to British Standards, the density is measured after the sample has been dropped ten times from a height of 150 mm onto a soft pad (a folded duster). The density is expressed in g ml^{-1} (g cm^{-3}). Sometimes the reciprocal value is used which is referred to as the 'specific volume' expressed in ml g^{-1} (cm^3 g^{-1}). The powder density is obviously of considerable interest from an economic point of view, as it influences the cost of storage, packing and transport.

The amount of free fat in the powder is of particular importance when instant powder is produced. The free fat is defined as the portion of the fat extracted from whole milk powder by washing with a solvent, such as carbon tetrachloride, for a defined period, normally 15 min. The free fat is expressed as a percentage of the powder, or as a percentage of the total fat. Normally a free fat content of 1 per cent (maximum) of the powder is required for instant powder which equals very nearly 4 per cent of the fat, as the fat content of whole milk powder comprises nearly one quarter of the powder. It is possible to reduce the level of free fat in the powder by homogenising the concentrate immediately before atomisation.

Further powder requirements are the heat treatment classification, which has been dealt with, and 'instant' properties, which will be covered in the following passage.

Factors Affecting Powder Properties

It is obviously important for the powder manufacturer to know which factors affect the powder properties. It may be raw milk quality, and it may be process conditions. In the latter case, it must be known where in the process the powder properties are influenced, and how variations in operation conditions affect powder quality.

Of the quality requirements for extra grade powder, according to ADMI listed in Table V, the fat and moisture content are entirely a function of processing. For whole milk powder, it is, of course, a condition that the ratio of fat to non-fat milk solids in the raw milk is such that the dried product will contain a minimum 26 per cent fat.

The characteristics which are influenced by the raw milk quality are the acidity, the microbiological count, and — indirectly through the acidity — the solubility. The bacteriological quality of the raw milk affects the quality of the powder, both directly and indirectly. In the direct sense, organisms which occur in the raw milk and survive the process, will appear in the powder. The presence of thermophilic and thermoduric organisms in the raw milk are, in this context, of greatest importance, because whereas other micro-organisms can be controlled to an acceptable level, the thermoresistant organisms will survive the process and contribute to the plate count of the powder.

Also indirectly the number and type of bacteria in the raw milk are of importance for the quality of the powder. The natural level of lactic acid in milk direct from the cow is very low, with a titratable acidity of approximately 0·11 per cent. However, during the time interval between milking and processing, lactic acid may develop as a result of bacterial activity. One of the most numerous micro-organisms in normal raw milk is *Streptococcus lactis,* which grows well between 21 °C and 35 °C producing lactic acid using the lactose as a nutrient. The developed acidity can lead to serious defects in the finished product. Not only may the titratable acidity exceed the permissible level, but the same may apply to the solubility. A high lactic acid content will result in an increased hydrogen ion concentration, so causing problems with protein stability during processing. It must be remembered that the acidity problem is magnified by concentration of the milk, because the titratable acidity increases very nearly in direct proportion to the concentration. The level may be reached where the heat stability of the proteins is seriously affected, which results in poor solubility of the powder.

In practice, the greatest problem attached to a high bacterial content of the milk is normally not a high plate count of the powder, but a high developed acidity, which can be reached very quickly. It is not uncommon for the total titratable acidity at the processing point to reach 0·16 per cent lactic acid and, on hot summer days, values of 0·17 per cent or even 0·18 per cent are not unknown in milk which has travelled far and has been too warm. In this condition the milk becomes highly unstable, causing problems in the processing and yielding an unsatisfactory finished product. Although the strength of the acid expressed by the pH value — and not the volume expressed by the titratable acidity — controls the heat stability of the milk, it is common practice to express the acidity as titratable acidity which is the number of ml of sodium hydroxide solution required to neutralise 100 ml of milk using phenolphthalein as indicator. Different units are used for titratable acidity. The difference lies in the strength of the NaOH solution used and Table VI indicates the relationship between the different units and the corresponding content of lactic acid.

Soxhlet Henkel degrees are the number of ml of N/4 NaOH required to neutralise 100 ml of milk. Thörner degrees are the number of ml of N/10 NaOH required to neutralise 100 ml of milk. Dornic degrees are the number of ml of N/9 NaOH required to neutralise 100 ml of milk.

1 ml N/9 NaOH = 0·01 g lactic acid, i.e. 1 Dornic degree corresponds to 0·01 g lactic acid per 100 ml of milk or 0·01 per cent. Therefore, the number of Dornic degrees divided by 100 gives the content of lactic acid in per cent.

The acidity of the milk at the processing point should preferably be about seven Soxhlet Henkel in order to produce a first class powder, and should be rejected as a raw material for a first class, edible product if the acidity exceeds eight Soxhlet Henkel. The acidity of the powder (i.e. of the milk reconstituted from the powder) cannot be derived directly from the acidity of the milk, as there is an initial drop in titratable acidity when the milk is heated, due to the release of carbon dioxide. However, raw milk is far from the only cause of quality defects in the powder. Many of the problems arise in and around the process, and efforts to improve powder quality as a whole necessitate close attention to both milk quality and the process.

TABLE VI
Designation for titratable acidity of milk

Soxhlet Henkel	*Thörner*	*Dornic*	*% Lactic acid*
1	2·5	2·25	0·0225
2	5·0	4·5	0·045
3	7·5	6·75	0·0675
4	10·0	9·0	0·0900
5	12·5	11·25	0·1125
6	15·0	13·5	0·1350
7	17·5	15·75	0·1575
8	20·0	18·0	0·1800
9	22·5	20·25	0·2025
10	25·0	22·5	0·2250
11	27·5	24·75	0·2475
12	30·0	27·0	0·2700

The conventional approach is to blame only the dryer for process-related defects in the powder. The fact is, however, that only one property of the powder — the moisture content — is entirely governed by the dryer. Some of the other properties, which are a function of the process, may be determined by the dryer, by the evaporator, or by other stages in the process. Some properties are governed by conditions outside the dryer. This is the case when whey is dried, as the desired non-hygroscopic nature of the powder is chiefly governed by crystallisation of the lactose between the evaporator and the dryer. Free fat and bulk density are, to a great extent, determined outside the dryer, and the heat classification entirely outside the dryer.

Some of the process-related, powder properties are dealt with in the following:

Too high a plate count of the powder is more often caused by poor sanitation somewhere in the process, rather than by poor bacteriological quality of the raw milk.

Other powder properties which are a function of the process are solubility and scorched particles.

Solubility is probably the most difficult of all the problems in connection with milk drying, as all the factors affecting solubility are not known. The air outlet temperature is critical. Assuming the acidity of the milk is within the acceptable range, the problem may be related to too high a total solids in the concentrate, which reduces the heat stability of the proteins and causes too high a viscosity. The high viscosity can also be a consequence of an unusually high protein content of the milk, or a high heat treatment of the milk. The high viscosity results in large droplets when the concentrate is atomised, which requires an increase in outlet temperature to reach the desired residual moisture. The high outlet temperature may overheat the milk solids and cause insolubilisation. The answer is to increase the rotational speed of the atomiser (or the pressure of the nozzle atomiser), or to reduce the degree of concentration, or to heat the concentrate.

It can be very useful to apply the solubility test to the concentrate to find out if it contains insoluble matter. If the solubility index of the concentrate is equal to the solubility index of the powder, the fault is not in the dryer, and the opposite is the case if there is no insoluble matter in the concentrate. It can also be useful to analyse the sediment obtained when the solubility test is done.

Scorched particles usually originate from milk solids that have been held up somewhere within the process, and thereby have been overheated or burnt. The most frequent cause of scorched particles is deposits in the drying chamber, atomiser or air distributor. The heat will darken or char this material, and particles which break off, get mixed with the product, and appear in the powder as dark specks. Scorched particles may be present in the concentrate if burning on of the product occurs in the evaporator. By applying the test for scorched particles to the concentrate, it may be possible to locate where in the process the contamination has taken place. Sometimes a filter for the concentrate is inserted between the evaporator and the dryer. Another cause of scorched particles may be inadequate filtration of the drying air or poor maintenance of the filter, whereby impurities are introduced to the product by the air. The air may contain milk powder dust, which becomes scorched when passing through the air heater.

The bulk density of the powder is chiefly governed by the total solids of the feed to the atomiser, in that the density increases with increasing total solids. The density is very sensitive to variations in the concentration and it is, therefore, desirable to have an evaporator with automatic control of the total solids in the concentrate. The temperature of the drying air affects the density slightly, in the way that the density is reduced with increasing air temperature. It could be expected that particle size would affect the density, but practice shows that it alters insignificantly with variations in particle size. Another matter is agglomeration, which is part of the instantising process. Agglomeration has a marked effect on the density. A heavily agglomerated powder is very light, and the density of agglomerated powder is often used to express the degree of agglomeration.

The density of individual particles is determined by the density of the solids, i.e. the product composition and by the amount of entrapped air within the particle. 'Nozzle powder' contains less entrapped air than powder from a rotating atomiser, but the benefit of this on the density is, to some extent, offset by the fact that a rotating atomiser can accept concentrate of a higher total solids than a nozzle atomiser. The amount of entrapped air in the powder from a rotating atomiser can be reduced by using a so-called 'steam swept atomiser'. By admitting steam to the product in the atomiser disc, steam is incorporated into the particles instead of air. The steam condenses practically instantaneously, and the particles shrink and become more solid. The tapped density of skim-milk powder is normally 0·58–0·60 g cm^{-3}; with steam atomisation, it is possible to increase the density by approximately 20 per cent to 0·70–0·72 g cm^{-3}.

Effect of variations in certain process conditions on powder properties

Experiments have been made at the Dairy Research Institute at Melle in Belgium with the purpose of finding out how variations in one processing parameter — keeping all other conditions constant — affect the properties of the powder. The results have been published by De Vilder *et al.* (1976, 1979) and are summarised below.

In the first experiment (Table VII), the influence of raising the inlet temperature of the drying air was studied. The inlet temperature was raised from 170 °C to 225 °C, whereas the outlet temperature was kept constant at 90 °C by gradually increasing the feed rate. The results show that the solubility is affected very little by raising the air inlet temperature when the outlet temperature is kept constant. The moisture content increased because the air outlet temperature was kept constant. When the air inlet temperature is increased, the outlet temperature will also have to be increased if the moisture content is to remain constant. As a rule, the outlet temperature must be increased by 1 °C for each 5 °C increase in inlet temperature. The free fat decreases with raising inlet temperature — probably because a hard particle surface is formed as a result of the high drying temperature. The hard surface prevents the fat dissolving agent from penetrating into the particles, so that less fat is extracted when the free fat is determined. The density decreases with increasing inlet temperature in spite of the increased moisture content — probably because of an increase in entrapped air in the powder particles.

In the second experiment (Table VIII), the influence of raising the air outlet temperature was studied. The inlet temperature was kept constant

TABLE VII
Influence of raising inlet temperature at constant outlet temperature 90 °C. Homogenised whole milk concentrate, 48% solids

Inlet temp. of the air (°C)	*Moisture content (%)*	*ADMI-solubility index*	*Free fat content % of total fat*	*Bulk density* ($g\ cm^{-3}$)
170	2·58	0·05	6·62	0·68
180	2·69	0·05	5·55	0·66
187	2·92	0·05	4·72	0·66
195	2·99	0·10	4·14	0·65
205	3·05	0·15	4·10	0·63
215	3·19	0·30	4·00	0·63
225	3·30	0·30	4·05	0·62

TABLE VIII

Influence of raising outlet temperature at constant inlet temperature 195 °C. Homogenised whole milk concentrate, 48% solids

Outlet temp. of the air (°C)	*Moisture content (%)*	*ADMI-solubility index*	*Free fat content % of total fat*	*Bulk density (g cm^{-3})*
75	4·75	0·05	2·15	0·64
80	4·41	0·05	2·64	0·63
85	3·93	0·05	3·04	0·61
90	3·23	0·50	2·78	0·60
95	2·59	0·50	4·49	0·58
100	2·18	2·90	4·38	0·57
105	1·76	3·30	5·43	0·55

at 195 °C, whereas the outlet temperature was raised from 75 °C to 105 °C by gradually reducing the feed rate. The moisture content is, of course, reduced with increasing outlet temperature. The solubility is significantly affected by high outlet temperatures. At outlet temperatures of 100 °C and above, the solubility is very poor. The free fat increases with increasing outlet temperature, which may be ascribed to crack formation in the surface of the particles due to overheating. It is a normal feature that the density falls with falling moisture content. The low outlet temperatures (75, 80 and 85 °C) correspond to the outlet temperature range when two- or three-stage drying is used. The results clearly illustrate the beneficial effect of multistage drying on the solubility and free fat.

The third experiment (Table IX) was done to illustrate the influence of the degree of atomisation at constant air inlet and outlet temperatures. The rotating speed of the atomiser was varied from 19 600 to 31 300 rpm, corresponding to a peripherential speed of the disc of 106–169 m s^{-1}. The inlet temperature was constant at 195 °C, and the outlet temperature was kept constant at 90 °C by gradually increasing the feed rate.

By increasing the degree of atomisation, the concentrated milk is given a large surface area, which has the effect that the water is more readily evaporated. Therefore, a lower moisture content is achieved at the same outlet temperature. The solubility shows a marked improvement with increased speed of the atomiser, even where the powder is over-dried. But the low moisture content is not produced by high outlet temperature, as was the case in Test two, and therefore, it has no adverse effect on the solubility. The improvement in solubility is due to the short drying time of the milk when it is atomised into smaller particles. One would expect the free fat to rise due to the increased surface of the powder. However, by increasing the speed of the atomiser a certain homogenising effect is obtained in the disc, which may explain the modest drop in free fat. The density is little affected, but declines slightly due to the lower moisture content.

The last experiment (Table X) shows the influence of heating the concentrate. The air inlet temperature was constant at 195 °C, and the outlet temperature was kept constant at 90 °C by gradually increasing the feed rate. The atomiser speed was constant at 25 000 rpm.

The effect of heating the concentrate is lower viscosity, and, therefore, atomisation into smaller droplets. This change was, in the previous experiment, achieved by increasing the atomiser speed, but the result is,

TABLE IX
Influence of rotating speed of the atomiser at constant inlet temperature 195 °C and constant outlet temperature 90 °C. Homogenised whole milk concentrate, 48% solids

Revolutions of the atomiser (min^{-1})	*Peripherential speed of atomiser disc* ($m\ s^{-1}$)	*Moisture content* (%)	*ADMI-solubility index*	*Free fat content % of total fat*	*Bulk density* ($g\ cm^{-3}$)
19 600	106	3·27	0·90	7·52	0·62
22 250	120	2·82	0·74	5·06	0·59
25 000	135	2·65	0·05	5·57	0·59
28 300	153	2·55	$<$0·05	6·48	0·58
31 300	169	1·99	$<$0·05	4·22	0·58

TABLE X
Influence of concentrate heating at constant inlet temperature 195 °C, constant outlet temperature 90 °C, and constant atomiser speed 25 000 rpm. Non-homogenised whole milk concentrate, 49·5% solids

Concentrate temperature (°C)	*Moisture content (%)*	*ADMI-solubility index*	*Free fat content % of total fat*	*Bulk density* ($g\ cm^{-3}$)
50	3·18	1·20	13·8	0·58
60	2·87	0·60	14·5	0·57
70	3·04	0·30	15·9	0·55
80	3·02	0·20	21·1	0·54

of course, the same, and therefore, there is good agreement between the results in Tables IX and X with regard to the change in moisture content, solubility, and bulk density. The free fat started at a higher level, because the concentrate was not homogenised. The increase in free fat with increase in concentrate temperature is due to the increase in powder surface, the effect of which has not been off-set by increased homogenisation in the atomiser disc, as in the previous test, because the disc speed was kept constant.

Summarising the results of the last two experiments the following recognition can be recorded. When milk is atomised into smaller droplets, the outlet temperature can be reduced and the solubility is improved. Smaller particles are obtained by increasing the speed of the atomiser or, in case of nozzle atomisation, by increasing the pressure. Smaller particles can also be obtained by reducing the viscosity of the feed — either by reducing the total solids or by heating the concentrate; concentrate heating has the greatest effect at high concentrations. The tendency of the free-fat content in whole milk powder to increase due to the larger powder surface can be counteracted by increased atomiser speed (or increased nozzle pressure), or by homogenisation of the concentrate. Homogenisation does, however, increase the viscosity, and therefore, these two contradictory parameters have to be optimised when whole milk is dried.

The experiments clearly illustrate that a slight change in process conditions significantly affects the characteristics of the powder.

Instant Milk Powder

Instant milk powder is powder that has been manufactured in such a way that it has better reconstitution properties than normal powder. It is not unusual for instant powder to be described as a product with improved solubility; this is, of course, not correct. The instantising process will never improve the solubility — in fact, the process can have an adverse effect on the solubility if correct conditions are not used.

The purpose of instantising is to improve the rate and completeness of the reconstitution of the powder. This is of special importance for domestic use, because the industry normally uses mechanical equipment for reconstitution, and hence the 'instant properties' are less important. The concept of an instant powder is that when it is placed upon the surface of unheated water, the powder will quickly sink into the water and disperse without stirring. The powder must have a good wettability (short wetting time), good sinkability, and good dispersibility. The solubility should be as for normal powder.

Dispersibility, as defined and determined in IDF Standard 87 (1979), is probably the best single criterion for assessing the 'instant properties', because a condition for dispersion of the powder in a defined time is that the particles have been wetted and have sunk into the water. The temperature of the water influences significantly the rate of reconstitution, with a maximum reached when the water is 70–80 °C. The water temperature used for the determination of dispersibility and wettability according to IDF's standard is 25 °C, which is assumed to match the temperature used in households, although the temperature of tap water is generally lower.

Powder suitable for reconstitution in hot beverages, such as tea or coffee, does not have to be manufactured with distinct 'instant properties', because the rate of reconstitution will be fast in any case, due to the high temperature. It is more important that the powder is heat-stable, as the combination of high temperature and the acidity of the coffee or tea will severely test the stability of the milk protein. If the powder is not heat-stable, the phenomenon known as 'feathering' occurs, which gives the beverage the same appearance as that produced by sour milk.

The instantising process is different for skim-milk and whole milk, and the reconstitution rate of skim-milk powder is governed by the texture of the powder. The instantising process for skim-milk powder consists of agglomeration of the particles into porous aggregates of sizes up to 2–3 mm. By the agglomeration, the amount of interstitial air (i.e. the air between the particles) is increased, and the reconstitution commences when the interstitial air is replaced by water. As a result of the agglomeration, the powder particles are wetted and dispersed before the actual dissolution begins.

Agglomeration is, in practice, accomplished by wetting the powder — usually in a fluid-bed — with steam or water, sufficiently to cause the surface to be 'tacky', whereby the particles will agglomerate. During the wetting, the moisture content may rise to 6–10 per cent, and the agglomeration is followed by re-drying by means of hot air to the normal moisture content. The re-wet method is a separate process divorced from the drying process. Low to medium heat powder should be used for two reasons. The first is that, as the instantising process includes further heat treatment of the milk solids, it is essential to use low heat powder to avoid a cooked flavour of the reconstituted milk. Secondly, it has been found that the stability of the agglomerates is poor when they are manufactured from a high heat powder; less shattering of the agglomerates takes place, during subsequent handling, when a low heat powder is used. The bulk density of the agglomerated powder is as low as 0·35–0·40 g cm^{-3}.

The so-called 'straight-through' method for the manufacture of instant skim-milk powder consists of discharging powder with a high moisture content from the drying chamber into a fluid-bed for agglomeration and drying to final moisture content. The fine powder fraction separated from the drying air, and the air from the fluid-bed, is fed back to the atomisation zone to agglomerate with the moist product from the atomiser. However, the benefit is limited in this context, as discharging the powder with a high moisture content from the chamber with the intention that it should agglomerate in the fluid-bed, means that the surface of the particles will be dry and the interior wet. Exactly opposite conditions are required for agglomeration. If attempts are made to take the powder out with a moisture content sufficiently high for the particles to agglomerate, i.e. with a 'tacky' surface, great difficulties are encountered in getting the powder out of the chamber.

The 'straight-through' process does not render a product sufficiently agglomerated to be designated 'instant'. The fact that the bulk density of the powder is very similar to the bulk density of normal powder indicates a low level of agglomeration. The powder is classified as 'semi-instant' or 'dust-free' powder due to the absence of fines. For industrial use, the 'non-dusty' property is often more important than any 'instant properties'.

The manufacture of instant, whole milk powder is more complicated, and was not commercially successful until many years after the introduction of instant, skim-milk powder. The basic method is the same as for skim-milk — to convert the powder to porous agglomerates to permit the single particles to be wetted and dispersed before dissolution starts. This is the essence of all processes for instantising milk powder. However, due to the hydrophobic nature of the fat, reconstitution of the agglomerates will not take place in water at temperatures below 45 °C (about 10 °C above the melting point of the fat), unless the powder particles have been coated with a surface-active or wetting agent; the use of lecithin as a surface-active agent is accepted world-wide. A special grade is used, consisting mainly of the natural phospholipid fractions extracted from soyalecithin (available under the trade name 'metarin').

The instantising process for whole milk powder starts with the manufacture of a suitable base powder, i.e. a powder, which after lecithination has 'instant properties' in water of 25 °C. The base powder must meet the following specifications, above the normal grading requirements:

(1) The free fat content should be as low as possible, and not exceed 4 per cent of the total fat (1 per cent of the powder). This can normally be achieved by homogenisation of the concentrate immediately prior to atomisation;
(2) The particle density should be as high as possible to improve the sinkability. Concentrate of high total solids should be used, and the amount of entrapped air in the particles should be reduced to a minimum. A decrease in the drying air temperature to 170–180 °C will contribute to that;
(3) The powder should consist of porous agglomerates with an absence of fines. The greatest portion should have particle sizes between 100 and 250 μm, with a maximum of 15–20 per cent below 90 μm. The bulk density should be 0·45–0·50 g cm^{-3}.

In order to meet the last requirement, the concentrate should have a high viscosity, i.e. an unheated concentrate with high total solids. Atomisation should take place using the lowest atomiser speed that is consistent with drying efficiency. These conditions will produce large particles which require extended drying time, so that the chance of partly dried (and hence 'sticky') particles colliding and adhering together is increased. To off-set the adverse effect of the extended drying time, two- or three-stage drying should be employed, whereby the temperature in the drying chamber will be lower than when single-stage drying is used, and the free fat content will also be lower.

The powder is discharged into a fluid-bed for final drying, and for removal of the fines. The amount of fine particles blown-off is governed by the quantity of air passing through the perforated plate. The fines separated from the drying air, and from the fluid-bed air, are fed back to the atomisation zone to agglomerate with the moist product from the atomiser. When the powder has been dried to its final moisture content in the fluid-bed, it is ready for lecithination, either in the fluid-bed while the powder is still hot, or later in a separate fluid-bed after the powder has been cooled and stored intermediately. The method to use depends, to a great extent, on how the powder is to be packed.

A 1 : 1 mixture of lecithin and butter oil heated to 70 °C is sprayed onto the powder in the fluid-bed by means of nozzles, and in a quantity to give a lecithin content of 0·2 per cent in the powder. The temperature of the powder during lecithination should be approximately 50 °C. The best result is obtained when the powder is packed at this temperature. The lecithin will continue to flow round the particles, and any non-lecithinated surface exposed by particle breakdown during packing will become coated by the continuing spread of the lecithin. The hot powder is highly exposed to oxidation of the fat and must, therefore, be packed with inert gas reducing the oxygen level in the packing to a maximum of 2 per cent. If the lecithination process is linked to the dryer (the so-called 'straight-through' process), the lecithin is sprayed onto the powder immediately after it has reached its final moisture content.

In the case of hot packing, it is obviously necessary to synchronise the packing operation with the dryer, but with can packing, it will be very costly to install a packing line to match the capacity of the dryer when a large dryer is employed. The powder can be packed hot into tote-bins for later consumer packing, but then the tote-bins must be evacuated and filled with inert gas. This will hardly be justified, because the benefit of the initial hot packing in tote-bins will be lost by the further handling, which will inevitably result in some breakdown of the agglomerates. If the powder is lecithinated as it comes out of the dryer for later packing in a cold condition, it must be held in the fluid-bed after lecithination for at least 5 min at 50 °C before it is cooled. Intermediate storage must be in tote-bins or drums, but as the powder is cooled, gas packing is not necessary at this stage provided the powder is stored for only a few days. Pneumatic transport and silo storage will cause particle breakdown and must, therefore, be avoided.

In practice instant, whole milk powder is often manufactured in two stages. First the base powder is produced in the spray dryer without lecithination. The base powder is cooled and stored intermediately in tote-bins or drums. Lecithination takes place in a separate fluid-bed completely divorced from the drying operation. The capacity chosen is the most suitable, taking the packing line into consideration. The capacity of the packing line dictates the size of the fluid-bed. The powder is heated in the fluid-bed to 50 °C before lecithination and packed hot.

IDF Standard 87 contains methods for the determination of the dispersibility and wettability of instant skim-milk powder and whole milk powder, but gives recommended values only for dispersibility; no values are given for wettability. When the IDF method is used for determination of the wettability, it is realistic to require a wetting time for skim-milk powder not exceeding 15 s, and for whole milk powder not exceeding 20–30 s in order to designate the powder as 'instant'.

Packing and handling of instant powder is critical. The packing must protect against any pick-up of moisture, and against physical breakdown of the agglomerates. In the case of whole milk powder, the packing must also protect against oxidation of the fat. As agglomeration is used as a means of achieving good instant properties, it is of vital importance to preserve the high level of agglomeration until the powder is reconstituted. Even a small amount of agglomerate breakdown results in the formation of fine particles which decreases both wettability and dispersibility. In addition to that, agglomerate breakdown in whole milk powder will expose powder surfaces not coated with the wetting agent, which will further reduce the wettability.

Baby Food Powder

The manufacture of baby food powder based on cows' milk has shown a steady increase since the beginning of this century. Baby food powder is nowadays a highly developed product, and manufacture of infant foods is an important part of the dry milk industry.

The industry has been accused of advertising that baby food powder is better than human breast milk. These accusations are unfounded. The manufacturers of baby food powder emphasise that breast milk is the best food for infants, but some women are unable to feed their children, and to these mothers, to-day's highly developed powdered baby foods are invaluable. When it is reported from developing countries that infants have died from artificial milk, it is not caused by the powder, but is due to the use of contaminated water for reconstitution, or due to poor bottle hygiene.

The incapacity of some mothers to feed their children is not something that only belongs to our time — it has always existed. Records of feeding infants by others go as far back as the 17th century, when sometimes another woman — the so-called wet nurse — was called upon to breast feed the child. Wet nurses were very carefully selected, as there was a widespread belief that the milk of the nurse would transmit her character to the child. During the 19th century, artificial feeding with cows' milk became more accepted, but there was still widespread objection to the use of boiled milk. Whole milk diluted with water was used, and sometimes sugar was added.

At the beginning of the 20th century, roller-dried milk was produced for infant feeding, with the result that the death rate of artificially-fed babies decreased dramatically — probably not due to the formulation of the roller-dried powder, but more due to its improved bacteriological quality. Roller-dried, full cream and half-cream milk with minor additives were used successfully in many countries until about the middle of this century. One of the pioneers was the British company, Cow & Gate Ltd, whose products became known in many parts of the world. In records of the company is described the excitement caused by an urgent order in the 1930s from a famous Maharajah in India for a consignment of baby food powder. The company's export manager visualised the tremendous potential in India, but was somewhat disappointed when it later was learned that the powder was required for the Maharajah's racing stables.

Since the middle of the 20th century, emphasis has been placed on the advice from paediatricians that baby food powder should approximate to the composition of breast milk, as nearly as is practicable. Medical authorities advise against feeding infants below four months of age with unmodified cows' milk, and this has led to the development of the so-called 'humanised' baby food powder, suitable for babies from the time that they are born. From four months of age, the infants can be fed on cows' milk that has been modified only slightly.

The total solids of breast milk and cows' milk is about the same, but there is a considerable difference in the composition. It is, therefore, necessary to modify cows' milk to obtain a product similar in analysis to breast milk. However, breast milk is not well defined, and it varies widely in composition, particularly during the first 14 days after parturition. In practice, humanised baby foods are produced to match the average analysis of breast milk produced some 14 days after parturition — the so-called matured milk. This analysis should not be interpreted too rigidly, because, due to the natural variations, many mothers produce milk 'out of specification'. Table XI shows the composition of average, matured breast milk and cows' milk. It will be seen that when infants are fed with normal full cream, cows' milk instead of breast milk, they will receive a higher protein intake with a different composition, different fatty acid composition, less carbohydrate, and a higher mineral intake.

The high content of protein and minerals in cows' milk puts a high load on the infants' kidneys. It is essential that humanised baby foods do not have too high a protein content and have a low mineral level, as the capacity of the kidney of a newly born child is very limited; this becomes particularly critical when loss of water is increased by fever or diarrhoea.

The simplest modification of cows' milk is the addition of carbohydrate,

TABLE XI
Composition of breast milk and cows' milk

		Average matured breast milk per 100 ml	*Cows' milk per 100 ml*
Total protein	g	1·5	3·3
Protein composition:			
Whey proteins	%	70	20
Casein	%	30	80
Fat	g	3·3	3·5
Fat composition:			
Unsaturated fats	%	52	43
Saturated fats	%	48	57
Carbohydrate	g	7·0	4·7
Minerals	g	0·21	0·72
Energy	kcal	72	66
Protein calories	%	6	21

which will dilute the proteins and minerals. The protein composition is changed by means of whey solids, whereby also the level of lactose is increased. Whey which has been ultrafiltrated should not be used, because the lactose in the whey will be removed from the protein fraction by ultrafiltration. The whey must originate from cheese manufacture where additives such as nitrate or sodium chloride, have not been used. The whey proteins must be undenatured, i.e. the heat treatment of the original milk and the whey must be gentle, and the whey must be at least 90 per cent demineralised by electrodialysis or ion-exchange. With regard to the fat it is possible to change the fatty acid composition of cows' milk by the addition of vegetable fat, whereby the level of unsaturated fats is increased. An addition of lactose is normally used to adjust the carbohydrate level.

As far as the minerals are concerned, it should be mentioned that calcium, magnesium, and phosphorus are desirable in certain quantities, whereas potassium, sodium, and chlorine should be avoided. The low mineral level can be achieved by using demineralised milk as well as demineralised whey. Breast milk is low in iron, and iron deficiency can be avoided by the addition of iron to baby foods. With regard to vitamins, the only addition to those being naturally present in cows' milk are vitamin A, vitamin D, and vitamin C.

Possible contaminants in cows' milk, such as trace metals, traces of pesticides, herbicides, and antibiotics must be kept under strict control, and it is, therefore, not unusual to specially select the milk to be used for the manufacture of infant foods to ensure the absence of contaminants.

In summary, the position with regard to infant foods is that modern techniques have made it possible to modify cows' milk to produce a baby food, which has properties similar to breast milk, and therefore, is safe to use from when the child is born. The only shortcoming of the manufactured baby food is that it does not possess the immunological properties of breast milk. It is well known that breast milk has anti-infective properties, and no method exists today to give manufactured baby foods these properties. Therefore, although the manufactured baby foods give the infant all necessary nutrients, breast feeding should always be preferred when it is possible.

Great care must be taken in the manufacture of infant foods to safeguard the bacteriological quality of the finished product. It is not unusual to require a plate count as low as 50 organisms per g of powder with pathogens absent. Dry mixing of ingredients into the powder after spray drying should be avoided, as it can lead to bacteriological contamination. Further, by dry mixing it is difficult to obtain a uniform distribution, and the added ingredient, e.g. lactose, may segregate out in the cans when they are subjected to vibrations.

Carbohydrate should be added to the milk before evaporation, or to the concentrate in the form of a 50 per cent solution, which must be pasteurised before it is mixed with the product. The fat-soluble vitamins (vitamin A and vitamin D) should be added to the milk before evaporation, and the same applies to the vegetable fat and other possible ingredients, e.g. dextrin-maltose. The mix should be homogenised, pasteurised at 110 °C with 1 min holding, and evaporated to 47–48 per cent total solids. The concentrate should have a temperature of 60–70 °C, and should be spray dried using an air inlet temperature of 180 °C.

The heat-sensitive vitamins (vitamin C) should be added to the concentrate. It is necessary to overdose with vitamin C as a portion is destroyed during drying. However, it is more costly to dry-mix the vitamin into the dried product, and it will be difficult to obtain a uniform distribution. Further, by adding vitamin C before drying, the manufacturer can guarantee the end-user that the child receives the correct amount of vitamins, because the portion with the lowest heat resistance, which in the case of dry mixing may be lost when the powder is reconstituted in hot water, has already been lost in the dryer and substituted by overdosing the concentrate.

Improvement of reconstitution properties of baby food powder by means of the instantising process is normally not necessary, because the powder is reconstituted in hot water, and the high carbohydrate content gives the powder 'instant properties'.

Buttermilk

Buttermilk can be evaporated and spray dried, but there is a great difference between sweet buttermilk and sour buttermilk. Sweet cream buttermilk behaves like skim-milk, and can be evaporated and dried under the same conditions.

Sour cream buttermilk is difficult to deal with due to the high content of lactic acid. It can only be evaporated to 26–28 per cent solids, as the concentrate becomes very viscous. The flow in the evaporator should be backwards so that the end-concentration is reached in the effect where the temperature is highest, which will reduce the viscosity. Spray drying is difficult due to the stickiness and hygroscopicity of the product. A low temperature should be used and, as deposits in the chamber cannot be avoided, frequent cleaning is necessary. It is useful to equip the drying chamber with a rotating air sweep which continuously will blow-off wall deposits. It consists of a tube which follows the vertical and conical shape of the chamber. It rotates at slow speed at a short distance from the interior surface of the chamber. Air of high pressure is supplied through nozzles attached to the tube and facing the chamber wall.

Whey

One of the problems associated with cheese making is the disposal of whey. For each ton of cheese produced, an outlet has to be found for about eight tons of whey. Whey is also the by-product from the manufacture of casein and quarg, but cheese whey is by far the most important in quantity. Over many years, the traditional methods for the disposal of whey have been to return it to the farmers for animal feeding or to dump it, either by spreading it on fields or by discharging it to natural water courses; these methods are no longer possible. The concentration of cheese manufacture in large factories has aggravated the whey problem, making it imperative to dispose of the whey at the point of manufacture.

The polluting effect of whey is illustrated by the fact that the biological oxygen demand (BOD) for the degradation of 1 litre of whey is equal to the BOD of the effluent water generated during 24 h by one person. As the quantity of whey from a large cheese factory is in the order of 500 000 litres per day (corresponding to manufacture of about 60 tons cheese per day), it will be seen that an effluent plant for biological treatment of the whey corresponds to a plant for the biological treatment of the effluent water from a population of half a million people. This clearly illustrates that biological degradation of the whey solids is out of the question, and then there is only one alternative left, and that is to recover the solids.

Utilisation of whey solids has attracted great attention in recent years. Robinson and Tamine (1978) describe a number of proposals for employing whey solids in the production of human food, and emphasise that investments should be spent on the recovery of the solids rather than on disposal plants for the whey. The recognition of the value of the undenatured proteins in whey is of fairly recent date, as the recovery of the undenatured whey proteins was first made possible by the introduction of ultrafiltration. However, ultrafiltration alone is not an answer to whey disposal, because the manufacturer is left with the permeate which has a BOD-value nearly as high as the whey. Evaporation and spray drying is the generally accepted method for the disposal of large quantities of whey, and a whey drying plant is often an integral part of a cheese factory.

In the following, the various constituents of whey solids will be examined, and the conditions will be explained for the conversion of whey into an edible powder of high quality.

Whey is not a product of defined chemical composition. It varies considerably with the product from which it is derived, and with the manufacturing process. Different kinds of cheese yield different types of whey, and often various substances are added during cheese manufacture, and as these are water soluble, they will be present in the whey. The whey contains normally 6 per cent solids, which is half of the total solids of the original milk. The most valuable constituent of the whey solids are the proteins, which amount to 20 per cent of the total protein content of the milk. The whey contains almost all the vitamins and minerals of the original milk. Table XII shows a typical composition of cheese whey solids. As much as 70 per cent of the whey solids is lactose. The drying process and the properties of the whey powder are chiefly governed by the amount of lactose that has been converted into crystals prior to drying. The best results are obtained when nearly 100 per cent of the lactose is present in the powder as crystals, rather than in the glassy or amorphous condition.

In skim-milk powder, 51 per cent is lactose, but the protein content is

TABLE XII
Typical composition of cheese whey solids (%)

Lactose	70
Proteins	14
Minerals	9
Fat	4
Lactic acid	3

36 per cent, which is nearly 2·6 times the protein content of whey solids. The high protein content of skim-milk solids acts as carrier for the lactose, and therefore, the condition of the lactose is of no importance either for the drying process, or for the properties of skim-milk powder, in spite of the fact that more than half of the powder is lactose.

The lactose in whey is present in two forms, as α-lactose and β-lactose. The relative proportions are that 40 per cent is α-lactose and 60 per cent β-lactose. α-Lactose will crystallise from a super-saturated solution, and the rate of crystallisation will increase with decreasing temperature of the solution. The α-lactose crystallises as an α-lactose monohydrate, which is a non-hygroscopic form. As much of the lactose as possible must be present in the powder in the crystalline hydrate form to obtain a non-hygroscopic or non-caking whey powder. Amorphous lactose in the powder will render it highly hygroscopic, thereby obstructing its use.

By concentrating the whey and cooling the concentrate, some of the α-lactose will crystallise. However, the two forms of lactose are in equilibrium in solution, and therefore, when some of the α-lactose crystallises, a quantity of β-lactose will be converted to the α form, so that a constant proportion between the two forms of lactose in solution is preserved. The conversion of β-lactose to α-lactose is called mutarotation, and the rate of mutarotation is influenced by the temperature of the solution. The optimal temperature for mutarotation is about 30°C; at low temperatures (from 0 to 5°C) the process is extremely slow.

As a result of mutarotation, the solution of α-lactose again becomes super-saturated, and the crystallisation process continues until the amount of lactose in solution has been reduced so much that the solution is no longer super-saturated. If, however, the solution is cooled, super-saturation occurs again as the solubility of lactose decreases with decreasing temperature of the solution. However, if the temperature is too far below the optimal temperature for mutarotation, the crystallisation approaches zero due to lack of transformation of the beta form into the alpha form. Thus the conditions for promoting both fast and rich crystallisation are contradictory, and it requires considerable experience to choose the right conditions to obtain maximum crystallisation.

The development of nuclei can be promoted by adding small crystals of lactose to the super-saturated solution. This is known as 'seeding', the purpose of which is to provide the solution with a sufficient number of small crystals on the surface of which the crystallisation can continue. The seeding material is either fine-ground lactose or well-crystallised whey powder.

With regard to the proteins, lactoglobulin and lactalbumin account for 90 per cent of the proteins in whey. Nutritionally they are superior to most other proteins in human and animal nutrition, due to their favourable amino-acid composition. They should be preserved in the undenatured form, i.e. the heat treatment of the whey must be gentle.

The fat is removed by centrifugation, and the resultant whey cream can be churned into butter.

The lactic acid content of the whey, at the point where it is processed, depends on the storage conditions after the whey is discharged from the cheese vats. The whey should be pasteurised as soon as possible after separation from the curd, and cooled before storage in order to inhibit acid development. The image that whey has as a waste product often governs the treatment it receives, and still today it is not uncommon for whey to be neither pasteurised nor cooled. This results in a very high content of lactic acid, which makes the whey difficult to dry, and renders the powder highly sticky and hygroscopic; lactic acid is a liquid at room temperature. When whey is collected from smaller cheese factories to be dried at a central factory, it is often in very poor condition, after having travelled far at high temperatures. Disturbances in the drying operation often coincide with the change-over from processing of whey of own manufacture to processing of whey brought in from other factories. The pH of sweet, continental whey may be 6·1, of cheddar cheese whey 5·9, and of cottage cheese whey 4·5.

The conversion of whey to powder consists of evaporation, treatment of the concentrate to effect maximum lactose crystallisation and finally spray drying. The correct handling of the whey is to pass it from the cheese vats over a screen to remove suspended curd particles, and then to pasteurise it. From the pasteuriser, the whey goes to a separator for removal of the fat. If the whey has to be stored before evaporation, it should be cooled to 4°C or heated to 65°C.

The whey should be evaporated to 55–60 per cent solids. If it has a high content of lactic acid, it is important to evaporate to the highest possible concentration. The treatment of the concentrate to obtain maximum lactose crystallisation is not only the key to whey drying, but is also the key to obtain high powder quality.

The lactose is crystallised by cooling the concentrate, and to obtain maximum lactose crystallisation, it is essential to cool in two stages. First the concentrate must be cooled quickly to 28–30 °C, taking advantage of fast mutarotation when the solution is still highly supersaturated. The quick cooling results in many small crystals; the crystal size should be below 50 μm. The best way of effecting the quick temperature drop is to use flash cooling by pumping the concentrate into a vessel which is under vacuum. The product is cooled by the sudden expansion. The absence of a heat exchanger, the surface of which may foul, is an advantage. Seeding may take place in the flash cooler ensuring good distribution of the seeded material.

The concentrate is pumped at 28–30 °C into crystallisation tanks as shown in Fig. 23. The tanks are made of stainless steel and provided with a jacket through which water at 10–12 °C is pumped. The water may be recirculated if chilled water is added to keep the temperature down. Sufficient tank capacity must be available to hold the concentrate in the crystallisation tanks until the temperature has dropped to approximately 15 °C. A temperature drop of approximately 3 °C per hour can normally be achieved, i.e. the temperature has come down to 15 °C in the course of 5 h. The concentrate should be held at this temperature for 3–4 h before drying.

The agitation system in the tanks is decisive for obtaining maximum crystallisation, and for holding down the viscosity of the concentrate. The agitator must be capable of breaking-up the concentrate and effecting a thorough mixing. The speed must be slow (10 rpm), and the direction of rotating should be reversed every 30 s. The degree of lactose crystallisation should be checked, and when it has reached about 85 per cent, the concentrate can be released for drying. Whey concentrate, where 85 per cent of the lactose is present as crystals, is highly abrasive.

The feed pump should be low speed, as high speed pumps quickly wear out. The atomiser disc should have interchangeable inserts, as shown in Fig. 24, to avoid wear of the disc. The inserts are made of aluminium oxide, and as they get eroded only at the side where the material leaves the disc, they can be used in different positions and finally replaced by new ones. The concentrate may be preheated to

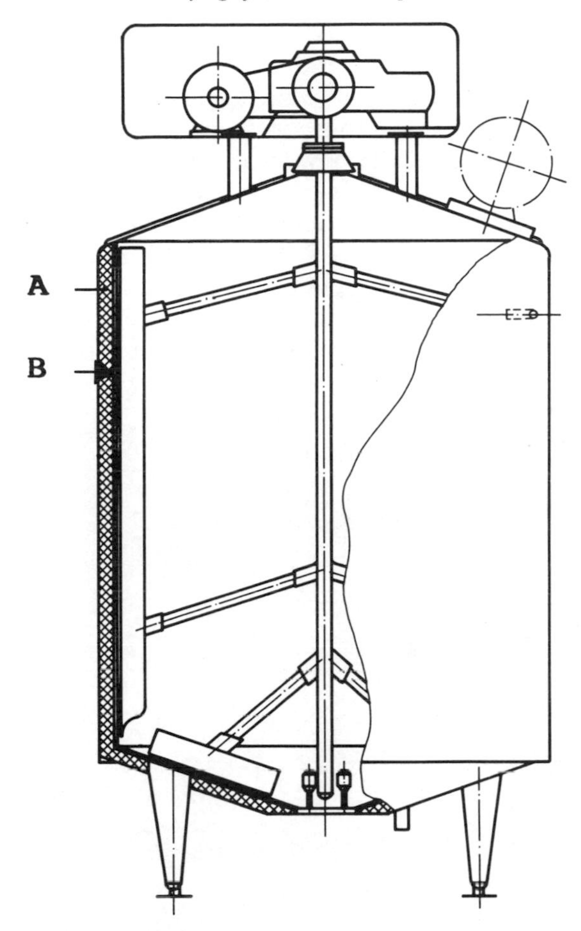

Fig. 23. Crystallisation tank. A: insulation; B: water jacket.

40–45 °C, which will not affect the crystal structure, but will reduce the viscosity, and also reduce the heat to be supplied by the drying air.

Under certain conditions during drying, the semi-dried material may stick to the hot, interior metal surface of the dryer, and the powder particles may also stick to each other and form lumps. This phenomenon is known as thermoplasticity or sticking, and the problem was first explained by Pallansch (1972), who designed a so-called stickometer to determine the sticking temperature of whey powder under certain conditions. Pallansch found that the sticking temperature depends on the composition of the whey solids, and that it is a function of the content of amorphous lactose, lactic acid and moisture. Figure 25 shows the increase in sticking temperature with increased degree of lactose crystallisation, with all other conditions being constant. It shows that the sticking temperature of the powder is raised from 60 °C to 78 °C when the level of lactose crystallisation is increased from 45 to 80 per cent. Figure 26 shows the drop in sticking temperature as the content of lactic acid in the powder is increased, with all other conditions constant. The gradual increase in lactic acid up to 16 per cent in the powder shifted the sticking point linearly down from 98 °C to 57 °C. A graph similar to Fig. 26 illustrates how the sticking point decreases in direct

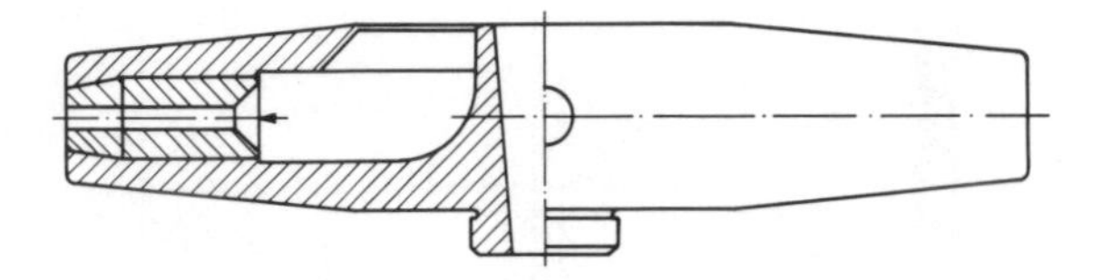

Fig. 24. Atomiser disc with inserts.

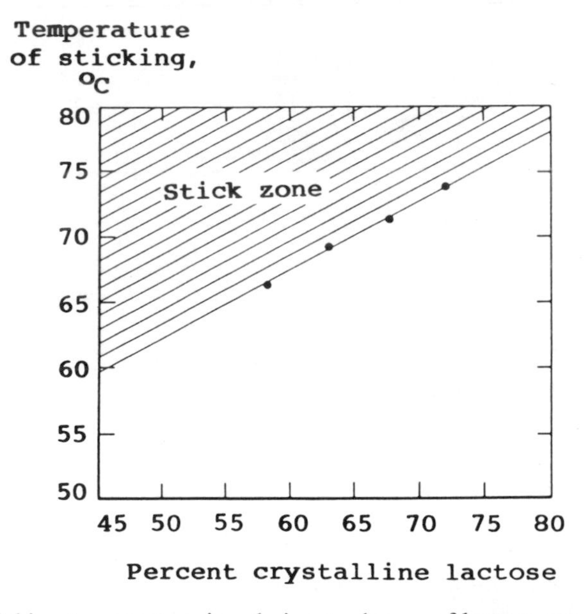

Fig. 25. Sticking temperature in relation to degree of lactose crystallisation.

proportion to the increase in the moisture content of the powder. At 2·5 per cent moisture, the sticking point was 92 °C, but dropped to 70 °C at 4·5 per cent moisture in the powder.

To achieve trouble-free drying, the powder temperature should be below the sticking point. The lactic acid content can often be reduced by changing the process after the cheese vats, i.e. by treating the whey in such a manner that the developed acidity is reduced to a minimum. If that does not bring the sticking point sufficiently up, the degree of crystallisation should be increased, if necessary, by increasing the concentration. A reduction in moisture content increases both sticking point and product temperature, and an increase in moisture content decreases both sticking point and product temperature. It is, therefore, difficult to obtain better conditions in the dryer by variation of the moisture content.

It is, however, clear from the above that the combination of a high content of amorphous lactose and a high content of lactic acid is particularly difficult and should, therefore, be avoided.

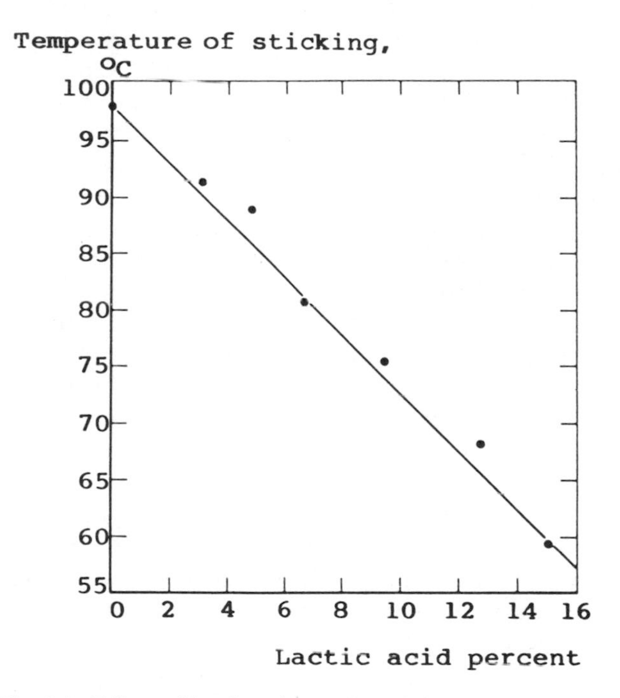

Fig. 26. Effect of lactic acid on the sticking temperature

The best whey powder is produced by the double crystallisation method, i.e. crystallisation both before and after spray drying. The more lactic acid the whey contains, the more necessary it is to have double crystallisation to get a non-hygroscopic powder. The powder is dried to 12 per cent moisture with an air inlet temperature of 150–160 °C and an outlet temperature of 50–60 °C. The powder drops from the base of the drying chamber onto a conveyor belt to give it a residence time of approximately 3 min. This intermediate holding of the product with high moisture content allows lactose crystallisation to continue in the damp powder mass. Further, any lactose anhydride is converted to the non-hygroscopic α-lactose hydrate by taking up one molecule of water. In spite of the low outlet temperature, the product temperature may be slightly above the sticking point, as the sticking temperature is low due to the high moisture in the powder. The effect of this is that the chamber wall gets coated with semi-dry powder, which drops-off as the moisture content is reduced. Further, the powder particles stick together to form agglomerates, which results in a coarse powder structure.

The after-crystallisation process is completed in a few minutes, and the conveyor belt feeds the product into a secondary dryer for removal of the excess moisture. The secondary dryer can be a rotary drum dryer, i.e. a kiln through which hot air is blown, or it can be a fluid-bed dryer. The conditions in the secondary dryer must be such that the chemically bound water in the lactose is not removed. In this way, a fully non-hygroscopic powder is obtained, as 95 per cent of the lactose is in the crystalline form. Further, a coarse, free-flowing, non-dusty powder with excellent wetting properties is obtained. These powder properties are required for whey powder to be used in the food industry.

By the so-called 'straight-through' drying method, the after-crystallisation on the conveyor belt is omitted. A higher drying temperature is used, and secondary drying is effected in a fluid bed placed under the drying chamber. The moisture content in the powder discharged into the fluid bed is too low for any after-crystallisation to take place, and also too low for any agglomeration to take place. The texture of the final powder is as normal milk powder, and the free-flowing properties, the non-dusty properties, and the reconstitution properties are inferior to the powder obtained by the low temperature/double crystallisation method. The advantage of 'straight-through' drying is the use of a higher drying temperature and a cleaner drying chamber.

CLOSING REMARKS

The developments in milk drying since the Second World War have been very significant, and dry milk has achieved an importance similar to the traditional dairy products.

Milk drying has developed into a highly sophisticated industry, technically as well as technologically. The manufacture of special powders, i.e. powders with specific properties may be of increasing importance in the future, and will no doubt be more rewarding than the manufacture of standard skim-milk and whole milk powders.

The energy required to convert milk to powder must, in the future, be reduced as much as possible due to the high energy costs. An obvious method would be to concentrate more before drying, but means must then be found to counteract the adverse effect from the physical and chemical changes that occur in concentrates of high total solids.

The safety of the equipment must be further improved, and the same applies to protection of the environments from effluent, noise and dust.

REFERENCES

American Dry Milk Institute, Inc., Chicago (1971). *Standards for Grades of Dry Milk including Methods of Analysis,* Bulletin 916.

Beever, Paula (1984). *Journal of the Society of Dairy Technology,* **37**, 2, 68–71

Challis, B. C. *et al.* (1982). *Amine Nitration and Nitrosation by gaseous Nitrogen Dioxide,* IARC Scientific Publications 41, pp. 11–20.

De Vilder, J., Martens, R. and Naudts, M. (1976). *Milchwissenschaft,* **31**, 7, 396–401.

De Vilder, J., Martens, R. and Naudts, M. (1979). *Milchwissenschaft,* **34**, 2, 78–84.

Hall, C. W. and Hedrick, T. I. (1966). *Drying Milk and Milk Products,* The AVI Publishing Co., Westport, Conn.

Harding, F. and Gregson, R. (1978). Nitrates and Nitrites in Skim Milk and Whey Powders in the United Kingdom, *Proceedings of the XXth International Dairy Congress,* pp 356–7.

International Dairy Federation, IDF Standard 87 (1979). Square Vergote 41, 1040 Bruxelles.

Leaver, G. H. (1984). *Gas Fired Low NO_x Combustion Systems and Air Heaters for Direct Food Processing,* Urquhart Engineering Co. Ltd, Greenford, Middlesex, UK.

Muir, D. D., Abbot, J. and Doyle, B. (1981). *Journal of the Society of Dairy Technology,* **34**, 1, 15–17.

O'Sullivan, A. C (1971). *Journal of the Society of Dairy Technology,* **24**, 1, 45–8.

Pallansch, M. J. (1972). *Proceedings of the Whey Products Conference,* Dairy Products Laboratory, Agricultural Research Service, US Department of Agriculture, Washington, D.C. 20250. Eastern Region Research Laboratory Publication No. 3779.

Palmer, K. N. (1973). *Dust Explosions and Fires,* Powder Technology Series, Chapman & Hall, London.

Robinson, R. K. (1982). *Dairy Industries International,* Dec. 1982, 19–23.

Robinson, R. K. and Tamine, A. Y. (1978). *Dairy Industries International,* March 1978.

Rothery, G. B. (1968). *The Australian Journal of Dairy Technology,* **23**, 76–81.

Wright, N. C. (1933). *Journal of Dairy Research,* **4**, 122–41.

Chapter 5

Modifications to the Composition of Milk

F. A. Glover*

formerly National Institute for Research in Dairying, Shinfield, Reading, UK

The modifications described in this chapter are those which can be achieved by the new membrane processes, namely ultrafiltration, reverse osmosis and electrodialysis. It is only in the last 20 years that membrane technology has advanced sufficiently to permit operation on an industrial scale.

Ultrafiltration (UF) is a fractionation process separating molecular species on the basis of size, using a membrane which retains the large molecules and allows water and the smaller molecules to pass. Thus from milk, some of the water, salts and lactose can be filtered out, leaving a concentrate richer in protein and fat. Reverse osmosis (RO) is similar in operation to UF, but employs a tighter membrane able to pass water only; RO is, therefore, a concentration process. Both UF and RO are pressure driven. The essential component in electrodialysis is also a membrane, but the driving force is a potential difference and ions pass through the membrane according to their charge. Electrodialysis is, thus, a deionisation process. For the dairy industry, the potentialities of UF are greater than those of RO, and electrodialysis has few uses.

A very clearly written, general article on the concentration of proteins by UF has been published recently by Lewis (1982), and Glover *et al.* (1978) reviewed RO and UF specifically for dairying; the same subject has been updated and considerably extended by Glover (1984).

*Present address. 39b St Peter's Avenue, Caversham, Reading, Berkshire, UK.

PROPERTIES OF MILK RELEVANT TO MEMBRANE PROCESSING

For an understanding of the application of membrane processing to milk, it is necessary to note some of its physical properties. The course of ultrafiltration is governed by molecular size, dimensions being more relevant than the more commonly known molecular weights. Table I gives some important weights and diameters, and shows the large step in size between lactose and α-lactalbumin. The pore sizes of UF membranes fall in this gap, which demonstrates how UF can be so conveniently applied to milk for fractionation purposes. Osmotic pressure is important in reverse osmosis but does not concern ultrafiltration, since the small molecules which give rise to the osmotic pressure pass through the UF membranes. Milk has an osmotic pressure of about 0·7 MPa due mainly to lactose and salts, and since RO concentrates all the components, osmotic pressure increases directly with the concentration factor. Viscosity is significant in UF. Milk has a viscosity of about 1 mPa s at 50 °C, due largely to the protein. As concentration proceeds the viscosity increases, at first slowly, and then very rapidly in the region of 14–18 per cent protein, so that the viscosity can reach 200 mPa s even at 50 °C at which a UF plant is commonly operated. This causes high pressure drops within a plant, reduces the rate of flow, reduces the rate of diffusion of molecules in the concentrate and causes heating due to the extra input of pumping energy. Viscosity thus imposes a limit on the degree of concentration, namely about six-fold, when the protein concentration is in the region of 18 per cent.

TABLE I
Molecular sizes of milk components

Component	*Molecular weight*	*Diameter* (nm)*
Water	18	0·3
Chloride ion	35	0·4
Calcium ion	40	0·4
Lactose	342	0·8
α-lactalbumin	14 500	3
β-lactoglobulin	36 000	4
Blood serum albumin	69 000	5
Casein micelles	10^7–10^9	25–130
Fat globules	–	2 000–10 000

*Taking the molecules as approximately spherical.
With acknowledgement to Kessler (1981).

THE PRINCIPLES OF ULTRAFILTRATION

UF is a pure sieving process, through a membrane having pores in the range 5–35 nm. The driving force is an applied pressure, generally about 500 kPa (5 bar), and the principle of operation is known as cross-flow filtration, since the feed flows over the membrane parallel to its surface at a velocity in the region of 5 m s^{-1}. The performance of the membrane is described by two quantities; the retention *R* of molecular components in the feed, and the rate of filtration known as the permeate flux *J*. For any particular molecule

$$R = \frac{(C_f - C_p)}{C_f} 100$$

where C_f = concentration of the molecule in the feed
C_p = concentration of the molecule in the permeate.
For very large molecules R = 100 per cent, for very small molecules $R = 0$, while intermediate sized molecules have values of R between 0 and 100 per cent, since membranes have a pore size distribution and hence diffuse cut-off levels. Permeate flux is expressed in units of litres per square metre of membrane per hour (litres m^{-2} h^{-1}), and it is mainly by this quantity that any plant is judged. In practice, fluxes from whole milk start at about 60 litres m^{-2} h^{-1} and then fall as concentration increases.

Transport through the membrane is by viscous flow, but in reality, the membrane is not the only barrier. As permeate is extracted from the feed, solids are left behind at the membrane surface, causing a local increase in concentration. The concentration of solids, therefore, increases in the direction looking towards the membrane. It is the phenomenon known as concentration polarisation (Fig. 1), which is all

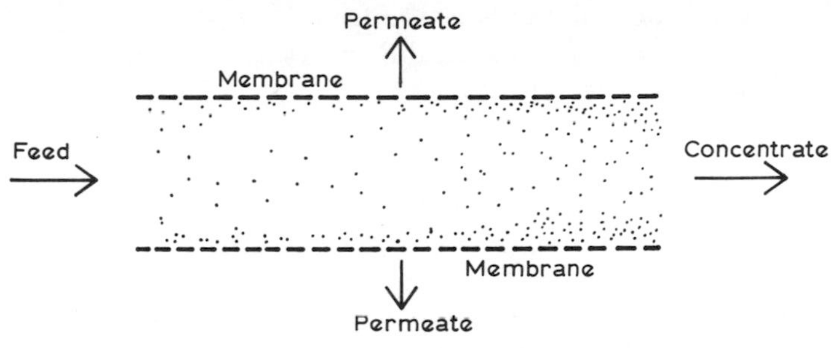

Fig. 1. Concentration polarisation.

important in both UF and RO, since it takes over control of the processes, setting up a barrier to filtration and reducing flux. At a constant concentration of the feed, an equilibrium is set up between the collection of solids near the membrane and the diffusion of those molecules back into the main stream assisted by the cross flow. All systems of plant operation aim to minimise concentration polarisation, either by turbulence or high rates of shear in the flow channels between membranes. As filtration proceeds, the concentration of solids in the feed increases, and concentration polarisation becomes more serious, until finally solid deposits form on the membrane and in its pores. The membrane has become fouled, and by this time, the flux has fallen very low.

The following expression illustrates the factors which determine the rate of filtration.

Let J = permeate flux;
A = area of membrane;
Δp = pressure difference across the membrane;
v = kinematic viscosity of the permeate;
R_m = resistance of the membrane;
R_d = resistance of the layer of solids at the membrane, the gel layer, due to concentration polarisation and fouling;

then

$$J = \frac{A\,\Delta p}{32\,v}\,\frac{1}{R_m + R_d}$$

Since equilibrium is reached through diffusion of the solids back into the feed, the permeate flux can also be described in terms of:

D = diffusion coefficient of solids in the feed;
C_f = concentration of solids in the feed;
C_g = concentration of solids in the gel layer;
d = thickness of the gel layer;

then

$$J = \frac{D}{d}\ln\left(\frac{C_g}{C_f}\right)$$

This equation is useful in understanding the limit to which concentration can be taken during an ultrafiltration process. When the concentration of solids in the feed reaches the concentration in the gel layer, $C_f = C_g$, $\ln(C_g/C_f) = 0$ and the flux becomes zero.

Yield of Components

The yield of any component is the amount retained in the concentrate as a fraction of that in the initial feed.

If y = yield;
f = concentration factor
R = retention coefficient, $\dfrac{C_f - C_p}{C_f}$

in batch operation $y = f^{(R-1)}$
and if C_0 = initial concentration;
C_1 = final concentration;
$C_1 = C_0 f y$;
i.e. $C_1 = C_0 f^R$

This calculation applies only if the R remains constant throughout ultrafiltration, which is rarely the case in practice due to concentration polarisation and membrane fouling. An average retention must then be taken, putting

$$R = \frac{\log\left(\dfrac{C_1}{C_0}\right)}{\log f}$$

Milk is a multi-component system, and this simple analysis does not apply directly to all components. The proteins and fats are completely retained by the membrane and form a considerable proportion of the concentrate, other components are partly retained and others pass freely through the membrane. The reduction in volume during ultrafiltration is from the water phase only. Hence there is a greater loss of water phase components than the concentration factor indicates, resulting in lower concentrations of these components when expressed in the usual analytical terms as fractions of the whole. As ultrafiltration proceeds, the concentration of a component with zero retention will appear to fall. However, if calculated in terms of the water phase only, the concentration will remain constant, consistent with the behaviour of a component having zero retention.

Diafiltration

Diafiltration is a modification of ultrafiltration in which water is added to the feed as ultrafiltration proceeds in order to wash-out feed components able to pass through the membrane. With the continued removal of more permeate, more of the small molecules are removed from the concentrate. Diafiltration may thus be regarded as a purification process. Together with ultrafiltration, it is a way of producing milk enriched in protein and fat, and very low in lactose and salts content.

The most efficient diafiltration is carried out by adding water at the same rate as permeate is removed. The volume of water to be added to achieve a required level of purity is calculated as follows, taking lactose as the component to be removed.

Let U = volume of water to be added;
R = retention factor for lactose, in practice in the region of 0·1;
m_w = mass of water in the feed before diafiltration;
l_1 = mass of lactose in the feed before diafiltration;
l_2 = mass of lactose in the feed required after diafiltration;

then

$$U = \frac{m_w}{1-R} \ln\left(\frac{l_1}{l_2}\right)$$

In terms of processing time, the most efficient method for milk is to ultrafilter first to a two-fold concentration, then to start the diafiltration, finishing off with more ultrafiltration if further concentration is required.

MEMBRANES FOR ULTRAFILTRATION

The first successful membranes on a large scale were made of cellulose acetate. Treatment of cellulose with acetic acid, such that an average of 2·5 out of every 3 of the OH groups on the cellulose is replaced with acetyl, CH_3COO-groups, produces a material with good membrane forming properties, namely an ability to form films, able to take up water and having a high wet strength. A thin layer of this cellulose acetate is prepared, then one surface is subjected to a controlled time–temperature treatment which produces a thin, tight skin on a much thicker and more porous layer. The tight skin, 0·1–1 μm thick having pores 2–20 nm diameter, is the effective ultrafilter, the thicker layer 100 μm thick serves as a support.

However, cellulose acetate membranes had severe limitations, particularly for dairy processing. Because cellulose acetate is subject to hydrolysis, its use was confined to a pH range of 3–7, and it would not tolerate temperatures above 35 °C. This placed severe restrictions on cleaning operations so vital in dairying. Polyamide membranes with improved tolerance to temperature and pH were used for some time, but more successful, and now widely used in practice, are the polysulphone membranes. The basic chemistry of this is the diphenylsulphone repeating unit $(C_6H_5)_2\ SO_2$

It forms a membrane with high mechanical strength, capable of withstanding temperatures up to 80 °C and a pH range of 2–12. The structure of a polysulphone membrane is shown in Fig. 2, which is an electron micrograph of the cross section.

The latest addition to the long list of membrane materials is zirconium oxide. It is an inorganic membrane supported on a graphite backing in the form of narrow tubes. Its advantages lie in its great mechanical strength, withstanding pressures up to 2 MPa without creep, tolerating the whole pH range 0–14 and temperatures up to 400 °C.

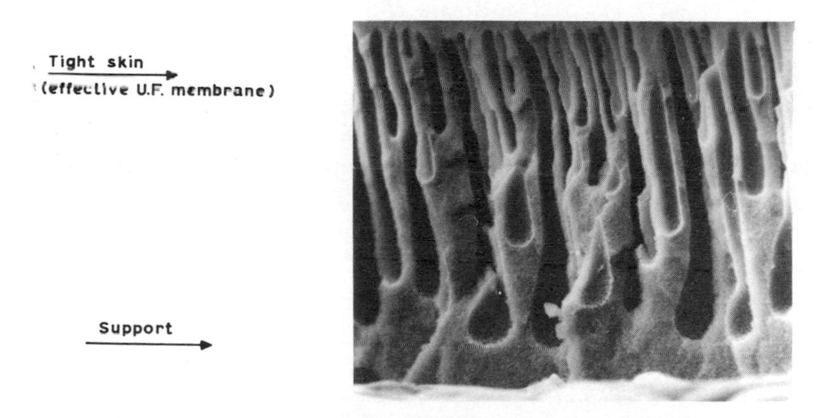

Fig. 2. Ultrafiltration membrane structure. Electron micrograph-cross section of polysulphone membrane magnification ×400. (Micrograph by courtesy of Dr R. F. Madsen, Danish Sugar Corporation.)

Membrane Geometry

Membranes are tubular, flat or spirally wound. The essential difference between these arrangements is the space between adjacent membranes, termed the flow channel, which ranges from 25 mm in the tubes down to about 0·5 mm for the flat and spirally-wound types. Membranes are assembled in modules, which are units in the engineering sense used in the building of plants. A module will contain up to 30 m^2 area of membrane. Wide channel systems rely on high degrees of turbulence to minimise concentration polarisation: the narrow channels achieve this by having shorter lengths, high rates of shear and mixing in the wider sections of the assemblies and the entrances and exits of the channels. The advantages of the tubular systems are that they are easily cleaned and are tolerant of particulate matter, whilst the flat assemblies have the advantage of a low hold-up volume per unit of membrane area, plant occupies a smaller space per unit of membrane area and membrane faults can be detected and replacements made within smaller membrane areas. Spirally-wound systems may be classed with the flat assemblies. In energy consumption, the flat and spirally-wound systems have the advantage over the tubular arrangements. Another version of the tubular system has membrane tubes less than 1 mm in diameter, known as hollow fibres rather than tubes. It is less common in large scale plants than the other types.

Membrane Modules

Tubular

The example of the tubular system is the design of Paterson Candy (PCI) (Fig. 3). Membranes are cast onto the internal surface of a synthetic fibre tube, 12·5 mm internal diameter in lengths up to 3·6 m. The fibre tubes are inserted into supporting, perforated stainless steel tubes, and assembled into bundles of 18 tubes fitted into end-caps which direct the flow into two groups, each of 9 tubes. Within each group, the tubes are connected in series, the two groups are in parallel. One module (3·6 m long) has a membrane area of 2·6 m^2. The feed flows along the inside of the tubes and permeate passes radially through the fibre backing, through the perforations in the stainless steel tubes, and is collected in stainless steel surrounding shrouds. In the event of a membrane failure, it is not possible to detect an individual faulty tube due to their close assembly inside the shroud. All 18 tubes, 2·5 m^2 area of membrane must be replaced.

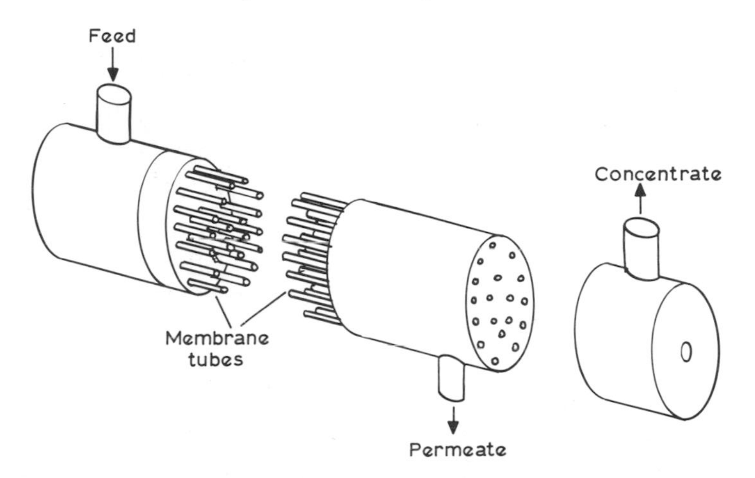

Fig. 3. Ultrafiltration module, tubular membranes. Paterson Candy International (PCI). Membranes 12·5 mm diameter, 3·6 m long; each module has 18 tubes, total area 2·5 m^2. Module outside diameter 10 cm. (Illustration from PCI technical literature.)

Flat sheet

The modules made by the Danish Sugar Corporation, De Danske Sukkerfabrikker (DDS) are examples of the flat sheet arrangement (Fig. 4). The membranes are in oval sheets, 42 cm × 31 cm, area 0·1 m^2, clamped in pairs between plates which have a pattern of curved ribs, so that when two plates with membranes in between are pressed together the ribs touch, forming the flow channels in between. The depth of the channels is 0·7 mm, and the flow along the channels is in parallel. Plates are 0·5 cm thick, made of polysulphone and have a cavity inside to receive the permeate through fine slits cut in the surface. Permeate is led out through fine, steel tubes (1 mm internal diameter) from the edge of the plates. Many membranes and plates are clamped together in a horizontal frame to form a module containing 27 m^2 area of membrane.

Recently DDS have improved their modules considerably (Kristensen *et al.*, 1981). By changes to the inlet and outlet ports and pipework, a 50 per cent reduction in energy consumption has been achieved. Even more important has been a modification of the flow channels between adjacent membranes. In order to get a more uniform distribution of flow over the whole membrane area, the shortest centre channels have been removed, thus reducing the ratio of the longest to shortest channels, and channel heights have been increased to 1·3 mm near the centre and to

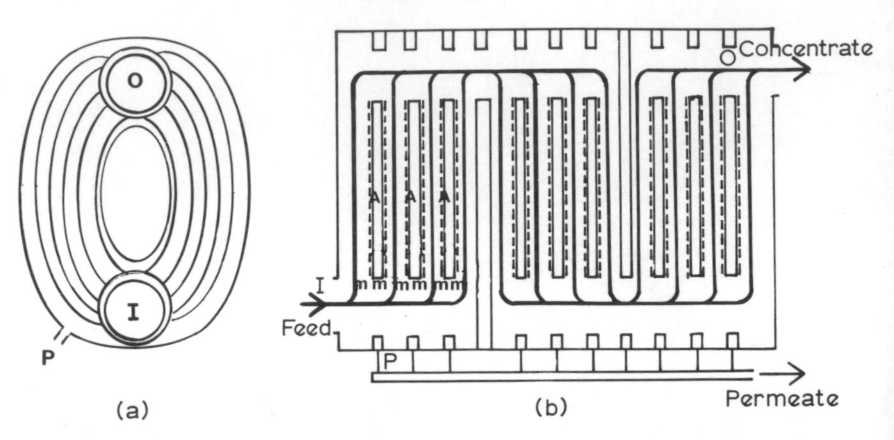

Fig. 4. Ultrafiltration module, flat sheet membranes. Danish Sugar Corporation (DDS). (a): membrane support plate for module 37; I, inlet port; O, outlet port; P, permeate outlet, dimensions 38 cm × 31 cm, channel height 1·3 mm at centre, 1·8 mm at periphery. (b): membrane assembly; A, support plates; m, membranes. (Illustration from DDS technical literature.)

1·8 mm at the periphery. Rates of flow in the outer channels are now higher, concentration polarisation is less and permeate flux is improved. Whereas the limiting total solids in the concentrates was 42 per cent, the new modules enable 51 per cent to be reached, and even 58 per cent if the initial fat content of the milk is first increased from the standard 3·8 per cent up to 6·1 per cent. Modules of this new design have been produced in three sizes, 4·5 m^2, 9 m^2 and 13·5 m^2 membrane area.

Spirally wound

A layer of plastic mesh (1 mm thick), a flat membrane and a layer of absorbent material to collect and transport the permeate are rolled into a spiral to form a cylindrical module (Fig. 5). At the ends, the spaces between the membrane and absorbent material are sealed, leaving the ends of the mesh open. A perforated, stainless steel tube is inserted along the centre of the roll. The flow of feed is parallel to the axis of the module between membranes, the mesh acting as a spacer and turbulence promoter. Permeate flows radially through the membrane and into the absorbent material, then tracks along it towards the centre of the roll, finds its way through the perforations in the central tube and out at the ends. The feed, now concentrated, leaves the module at the further end.

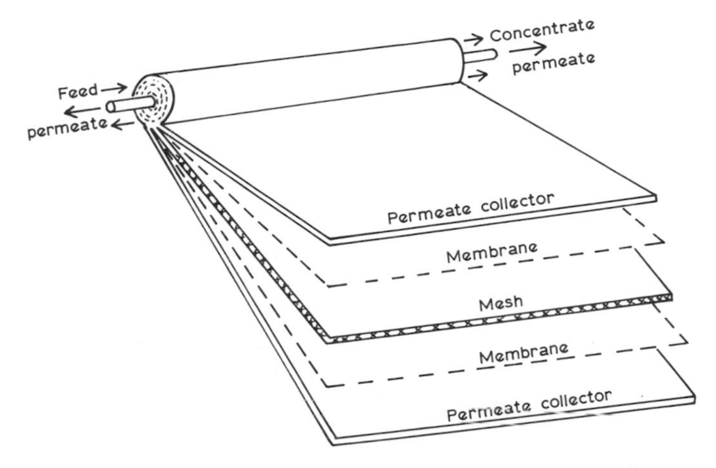

Fig. 5. Ultrafiltration module, spirally wound membrane, Abcor. Length 84 cm, diameter 11 cm, membrane spacing 0·7 mm, area 5 m^2. Three such sections are housed in series in a stainless steel tube to make one module, 3 m long, membrane area 15 m^2. Schematic diagram to show the formation of the spiral. (Illustration from Abcor technical literature.)

Hollow fibre

A hollow fibre is a narrow tube, internal diameter 0·5 or 1 mm/length 1 m, made of non-cellulosic polymer. The structure is dense on the inner surface, 0·1–1·5 μm thick, and much looser towards the outer surface in a layer 50–250 μm thick. This tube is itself the membrane; the feed passes along the inside of the tube and permeate moves radially outwards. Fibres are assembled in bundles of 1000 or more, and sealed in a clear plastic cartridge to form a module containing 2·5 m^2 area of membrane.

ULTRAFILTRATION PLANT

The factors which determine the design are:

(i) Membrane performance — permeate flux in relation to the concentration required.
(ii) Flow velocity — compromise between high flow velocity to control polarisation and energy afforded for pumping.

(iii) Pressure drop along the plant — choice of flow velocity, arrangement of modules and pumps to ensure an adequate working pressure is available over all membrane.
(iv) The volume of feed to be processed.
(v) The retention time of the liquid in the plant in relation to bacterial growth.
(vi) Capital and maintenance costs of plant and membranes.

The basic design of an ultrafiltration plant is shown in Fig. 6. The essential components of a plant are:

(i) Feed tank.
(ii) Pumps — feed and circulation, usually centrifugal.
(iii) Flow control valve.
(iv) Flow meter.
(v) Thermometer.
(vi) Pressure gauge — inlet.
(vii) UF module assembly.
(viii) Pressure gauge — outlet.
(ix) Pressure retaining valve.

A heat exchanger is usually included, either to heat the milk to the desired operating temperature or to prevent overheating due to the pumping. Fully automated plants are fitted with controls for pressure, flow and temperature and safety devices to stop the pump should the feed cease or the system become blocked.

To achieve a useful degree of concentration, the feed must pass many times, probably 100 or more times over the membrane. Passing once only would require far too big an area of membrane. Plant may then be operated in two ways:

(i) Continuously — recirculating round an inner loop.
(ii) As a batch process — recirculating back to the feed tank.

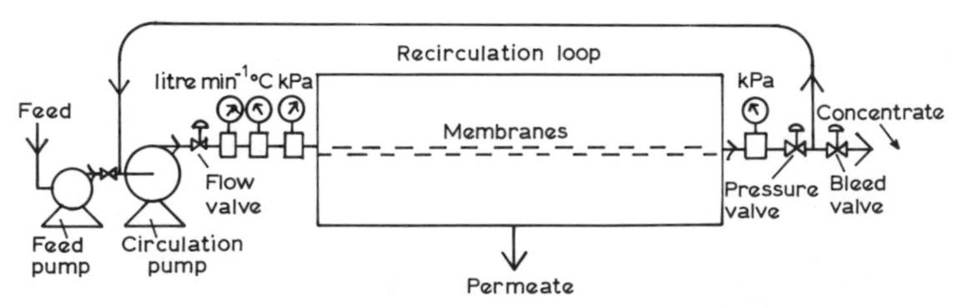

Fig. 6. Ultrafiltration plant design, continuous internal recycle or feed and bleed.

Continuous Operation

The design in Fig. 6 is for this mode, known as internal recycle, since the feed is pumped many times round an inner loop which consists essentially of the circulation pump and the membrane assembly. The plant is operated so that, at the outlet end of the membrane, the required concentration is reached and a little of the concentrate is bled-off; the rest is diverted back to the beginning of the loop at the entrance to the circulation pump, where more fresh feed is added from the feed pump to make up the volume of concentrate removed. An alternative name for this method is, therefore, the feed and bleed mode. It is the most appropriate mode for large commercial plants where the usual pattern is to run continuously for 20 h and clean for 4 h.

The advantages of the feed and bleed operation are short retention times in the plant, measured in minutes rather than the hours of the batch process, and economy of pumping energy since the whole pressure is not lost on each recycle as the internal loop is closed. Its disadvantage is that the concentration in the loop at the membrane is permanently high, close to the final required level, hence the permeate flux is low.

Commercial size plants may contain several hundreds of square metres of membrane area, and one pump cannot provide sufficient pressure for all the necessary modules. Plants are, therefore, made up of repeating units, each with its own pump and array of modules which are then all provided with a working pressure. An example of such a plant fitted with tubular membranes is shown schematically in Fig. 7. In large

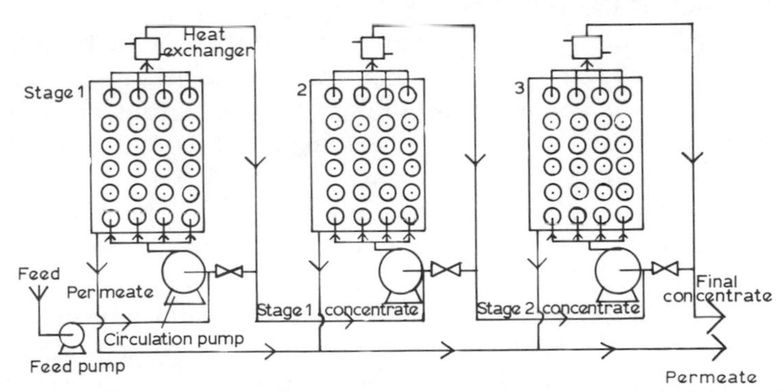

Fig. 7. Ultrafiltration plant — commercial scale, Paterson Candy International. Each stage contains 24 modules, 4 banks in parallel each containing 6 modules in series. 1 module 2·5 m^2 area of membrane, 1 stage 62 m^2, whole plant 186 m^2. (Illustration from PCI technical literature.)

plants, the size of successive units decreases to maintain the flow of the declining volume of concentrate.

Batch Operation

Instead of recycling the feed back to the entrance of the recirculation pump, it is returned to the feed tank, so that it circulates round the whole plant. This mode of operation is suitable for processing small volumes which can be contained wholly within the plant, as in laboratory and pilot-plant trials. No concentrate is withdrawn until the end of the process, it is all left in the plant until the required concentration is reached. The method has the advantage of a simple plant with only one pump and the removal of a large volume of permeate while the concentration of the feed over the membrane is low and hence flux is high. Against the method are the long retention times of the feed in the plant and, of course, it is possible to process only the volume which the plant will hold. A modification in operation will extend this capacity by topping-up the tank with fresh feed as ultrafiltration is going on, but the limit here is reached when the whole plant is filled with concentrate.

Engineering Equipment

Beyond the specialised component — the membrane assembly, the engineering equipment comprises pumps, pipes, valves and controls, all standard items familiar to the dairy industry. The main component is the pump which must be capable of running continuously. Delivery rates depend on the membrane geometry and size of plant, tubular arrangements requiring the higher flow rates because of their wider flow channels. For example, the tubular membrane (diameter 1·2 cm) requires a flow of about 20 litres min^{-1} to give a velocity over the membrane of 2·7 m s^{-1}, so that through one module containing 18 tubes arranged in parallel, the pump must deliver 360 litres min^{-1} and through one stage of the commercial plant in Fig. 7, the flow must be 1440 litres min^{-1}. The delivery pressure has to be in the region of 0·5 MPa. Centrifugal pumps perform this duty very well, they are easy to control and are not expensive. However, at the very high concentrations of milk now being aimed at for cheesemaking, centrifugal pumps are less able to contend with the very high viscosities of the concentrates, hence it is proposed that positive pumps should be used in the final stages of UF plants.

Valves can be of the diaphragm type, one at the outlet of the centrifugal pump to control the flow, another at the outlet of the membrane assembly to maintain the pressure or bleed-off the concentrate. Pressure gauges must be of the diaphragm or capsule type, so that milk cannot enter the Bourdon tube. Pressure safety devices are not required in ultrafiltration since pressures are low, and centrifugal pumps are not damaged if inadvertently operated against a closed valve. For the measurement of flow rate, rotameters are used though they are difficult to read through concentrated milk; turbine flow meters are much better, but more expensive. Temperature control is necessary for both product and plant. UF is commonly carried out at 50 °C, a compromise between 37 °C where maximum bacterial growth occurs, and 60 °C where the whey proteins start to denature, and a high temperature favours the rate of the process mainly on the grounds of lower viscosities of the feed and permeate. Operation at low temperature causes serious concentration polarisation.

An impression of UF in practice is given in Fig. 8, which is a photograph of a large commercial plant using flat membranes for the manufacture of whey protein concentrate. The International Dairy Federation (1979) has conducted a survey entitled 'Equipment available

Fig. 8. Ultrafiltration plant — commercial scale, flat sheet membranes, DDS. Ultrafiltration of 1 million litres of whey per day for the manufacture of whey powder. (Photograph by courtesy of the Danish Sugar Corporation, DDS-RO Division, 4900 Nakskov, Denmark.)

for Membrane Processing'. This is the most comprehensive collection of data on membranes for UF, RO and electrodialysis, modules, operating parameters, cleaning and disinfection.

ULTRAFILTRATION PLANT PERFORMANCE ON MILK

Performance is usually represented by a permeate flux curve, i.e. a plot of flux against concentration under stated conditions. Few details on the performance of large-scale industrial plants in operation with firms are available, but an appreciation of possible processing rates may be obtained from the many research papers published. They are accounts of experiences mainly on the pilot-plant scale. Figure 9 is a flux curve obtained with the new zirconium oxide membranes by Goudedranche *et al.* (1981). These are the best fluxes yet seen in the literature, favoured by the new membranes, a small plant and a batch operation. From polysulphone membranes, fluxes commonly start at about 40–50 litres m^{-2} h^{-1}; fluxes in commercial plants operated in the continuous mode will be lower than this. The improved performance of the zirconium membranes is a result of their higher mechanical strength over that of the former polymer types. The method of operation is to start with a low pressure in the initial stages of concentration, then as flux decreases with concentration, the applied pressure is increased to keep up the flux.

The Effect of Pressure, Flow Velocity and Temperature

As the operating pressure is increased from its lowest value, the flux increases until a region is reached where further pressure increases are of no advantage. As more pressure attempts to increase the flux, the gel layer on the membrane increases, nullifying the effect of the additional pressure. The point at which the flux becomes independent of pressure is related to the velocity of the feed over the membrane, the higher velocities making it profitable to use higher pressures and obtain still higher fluxes.

Increasing flow velocity is more beneficial than increasing pressure, as the action is to assist in dispersing the polarised layer. Continued gain in flux is limited by the energy which can be afforded for pumping, and in the case of whole milk, there is a danger of damage to the fat globules by excessive pumping.

Flux increases more than linearly with temperature. Increase in tem-

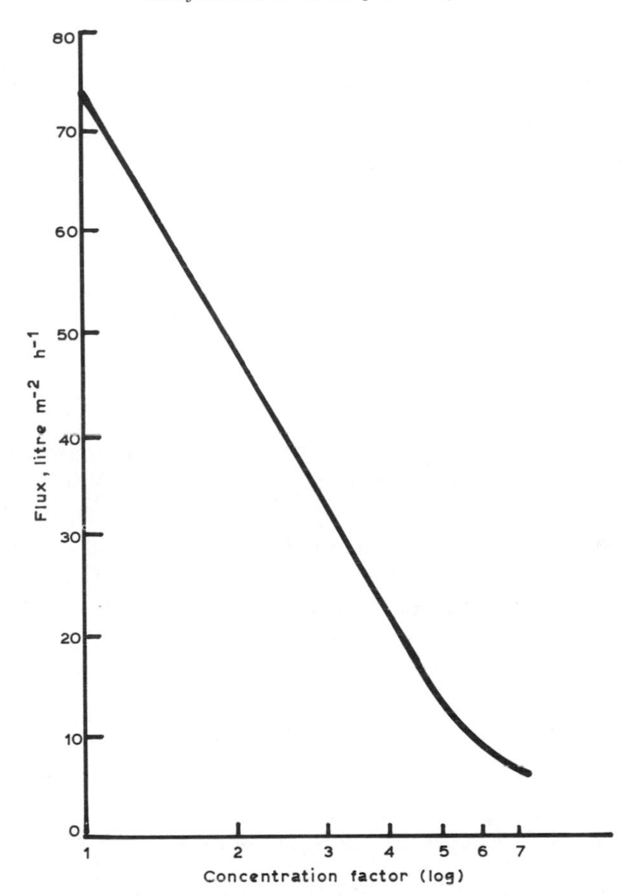

Fig. 9. Ultrafiltration of whole milk — flux against concentration factor. Membrane ZrO_2, pilot plant, 1·6 m^2 membrane area, batch operation, at 52 °C. (With acknowledgement to Goudedranche *et al.* (1981) and reproduced by courtesy of Technique Laitière.)

perature has the dual effect of lowering the viscosity which assists flow rate, and of increasing diffusion which assists dispersion of the polarised layer.

Composition of the Product

Membranes retain all the fat and practically all the protein from milk. Retention coefficients of the non-protein N-compounds are generally

20–40 per cent, and higher for the high concentration factors. The losses through the membrane are mainly urea and amino acids. Retentions of lactose may be up to 10 per cent; since some minerals are partly bound to protein, namely calcium, magnesium, phosphate and citrate, they will be retained with the protein, but others are free and will pass through the membrane. Likewise the fat soluble and protein bound vitamins are completely retained.

An example of the compositions of concentrates and permeates in the ultrafiltration of whole milk is given in Table II.

Ultrafiltration of Skim-Milk

Skim-milk behaves in exactly the same way as whole milk in ultrafiltration. The permeate flux from skim is slightly higher than from whole milk by about 20 per cent, indicating that the fat in whole milk, though it forms a large proportion of the total solids and is present in the largest particle sizes, is not a great hindrance to filtration. More important is the fact that skim- and whole milk have the same protein content, and the protein, being in smaller particle sizes than the fat, is more liable to denser packing at the membrane surface. Of all the components in milk, protein exerts the greatest control over the rate of ultrafiltration.

TABLE II
Ultrafiltration of whole milk — example of compositions of the products — concentrate and permeate (Concentrations in per cent w/v)

Volume concentration factor	*Total solids*	*Fat*	*Protein*	*Non-protein nitrogen compounds*	*Lactose*	*Ash*
Concentrate						
×1	12·9	3·9	3·1	0·18	4·7	0·77
×3	28·6	12·6	9·8	0·18	4·1	1·3
×5	43·3	21·8	16·1	0·18	3·2	1·9
Permeate						
×1	5·7	0	0	0·18	4·8	0·53
×3	6·1	0	0·06	0·19	5·1	0·53
×5	6·7	0	0·49	0·19	5·2	0·54

System: tubular membranes, small pilot-plant, batch operation, 50 °C.
Note. See p. 239 on the concentration of components with low retentions.

Growth of Bacteria during Ultrafiltration

Microbial growth is not a serious problem if the process is carried out at a high temperature (50 °C) and on the feed and bleed principle, such that retention times are below the lag phase of the growth. Some plants claim a retention time as low as 10 min, which is well below lag phases of around 2 h at 50 °C (Kiviniemi, 1974). Design features, such as low hold-up volume, low pressure drops and optimum flow velocity, all help to keep growth in check, and provided the starting milk is of good hygienic quality, bacterial numbers do not increase beyond that due to the concentration of the milk.

Deposits on Membranes

Deposits from milk have not been studied, but an examination of membranes after processing whey will illustrate the nature of deposits from milk. The whey proteins, being smaller than the casein micelles, are the main fouling agents. The electron microscope shows deposits up to 1 μm thick, and particulate matter penetrating the membrane to a depth of 2 μm. On the surface are seen micro-organisms, granules, thin strands, coarse fibres and sheets of various thicknesses, and they are clearly visible within 10 min of the start of filtration, increasing in thickness thereafter. The worst offender is β-lactoglobulin, since it is able to form sheets, as does blood serum albumin; γ-globulin and α-lactalbumin exist as granules. It is postulated that salts could form bridges between the proteins and the membrane, so building up a fouling layer (Lee *et al.,* 1975).

Cleaning Membranes

Polysulphone and other non-cellulosic membranes are cleaned with highly alkaline detergents and chlorine to break down protein deposits, and acids to remove minerals. Chlorine released from sodium hypochlorite at pH 10·5 rapidly attacks protein. As the chlorine is consumed, it must be made up to the recommended level of 250 ppm until no further decline occurs. Proprietary cleaning solutions containing organic and inorganic sequestrants are available specifically for ultrafiltration plants. Cleaning is assisted by a high flow rate, low pressure and high temperature (50 °C), and it is important to use pure water as recommended by the manufacturer of the cleaning agent. A membrane is clean when the permeate flux from pure water is restored to the level at which it stood before the UF process.

USES OF UF MILK CONCENTRATES

The first use of ultrafiltration in the dairy industry was to fractionate whey, concentrating the protein for food processing, and reducing the pollution load of the carbohydrate portion by making use of the permeate; up to 10 million litres of whey are now filtered daily. Of more direct use to industry is the ultrafiltration of milk for cheesemaking, which promises to become the widest application. Lesser applications are milk standardisation and making fermented milk products. Ultrafiltration of milk on the farm immediately after milking is being investigated.

Cheesemaking

The principle of making cheese with ultrafiltered milk is to include the whey protein in the cheese, and the attraction is an increase in yield of product. Thus, better use is made of the whey proteins than has been done hitherto, when their loss meant that about 20 per cent of the protein in milk was largely wasted. If milk is ultrafiltered first to a concentration such that during cheesemaking no further liquid drains, the extraction of permeate during the ultrafiltration may be regarded as similar to the drainage of whey in the traditional process. The difference is that permeate does not contain protein; this has been retained in the cheese and has increased the yield by as much as 15–30 per cent of the traditional yield according to the type of cheese. Since the milk concentrate, termed the 'pre-cheese', must be similar in composition to the final cheese, the process is more easily applied to those types having the lower solids content, for example cottage cheese (21 per cent T.S.), Ricotta (28 per cent T.S.) and Feta (40 per cent T.S.). Concentration of milk up to five-fold is sufficient for these varieties, and commercial processes have been set up successfully. France is the leader in this field because of the popularity of soft cheeses amenable to manufacture by this process, which is known as the MMV process — Maubois, Mocquot and Vassal — in whose names the original patent was taken out in 1969. It was reported in 1980 that more than 60 000 tons of cheese per year was being made by this method (Maubois, 1980).

Camembert cheese can be made from ultrafiltered milk concentrates, with the advantage of a 20 per cent increase in yield and improved uniformity in the weight of cheese taken from the moulds. The Alfa-Laval Company has combined ultrafiltration with their 'Camatic' process to make Camembert cheese in a continuous process (Hansen, 1981). The largest production using ultrafiltration is of Feta cheese, mainly in Denmark for export to the eastern Mediterranean countries (Hansen, 1977). For this type, an extra yield of cheese of 30–35 per cent is obtained, only 20 per cent of the normal quantity of rennet is required, losses of fat in the whey are avoided, continuous production is possible and the quality of the cheese is good.

Many other soft cheeses have been made on a trial basis from ultra-filtered milk concentrates, namely cottage cheese, cream cheese, Danish Blue, Edam, Herve, Mozzarella, Quarg, Queso Blanco, Queso Fresco, Ricotta and Teleme (see Glover, 1984).

However, to make Cheddar cheese, which has a solids content of 62 per cent, would require ultrafiltration to a concentration factor of 8, and though this has been achieved experimentally, it is not a practical reality because of the very low permeate fluxes at this level. The greatest advance towards making Cheddar cheese using UF has been made by the Australian CSIRO Division of Food Research in Sydney. (Sutherland and Jameson, 1981). Steps in their process include ultrafiltration, diafiltration and special care in handling the curd to avoid losses of fat and casein. Further development is required to achieve the required quality of the final cheese. Related to this process is the production of a cheese base and a processed cheese, in which milk concentrates are taken to a final 60 per cent T.S. by evaporation. By complete drying, it is possible to make a cheese powder for subsequent reconstitution and making cheese in an importing, under-developed country.

Fermented Milk Products

The use of milk concentrates generally improves the yield and body of fermented products. There is a report on the production of yoghurt and ymer by Morgensen (1980).

Trials on Other Uses of UF Milk Concentrates

Adjustment of protein levels in milk up to 6·5 per cent are reported to have no effect on the organoleptic properties (Roenkilde Poulsen, 1978).

The use of UF concentrates in making sweetened condensed milk has been found to avoid sandiness in the texture (Serpa Alverez *et al.,* 1979). If milk is first concentrated by UF, it is possible to prepare low-fat, high-protein whipping cream, which is attractive to the producer since the yield of product per unit volume of milk is increased (Scurlock and Glover, 1983). Such a product could be regarded as nutritionally preferable when the consumption of milk fat is questioned. Trials on the ultrafiltration of the milk of other species, namely the buffalo, goat and sheep, have been carried out, again for the purpose of making cheese.

USE OF THE PERMEATE

The permeate from the ultrafiltration of milk is a solution, mainly of lactose, in the same concentration as in the water phase of the original milk which is about 4·8 per cent. Such a solution has a high biological oxygen demand and, therefore, cannot be discarded as waste; far better that it should be regarded as a valuable source of carbohydrate. Three uses are possible, after different treatments, for human food, for animal food and as an industrial fuel (Coton, 1980). For human food, the lactose can be hydrolysed enzymically or by ion exchange into a glucose/galactose sweet syrup; this is finding uses in the brewing and confectionery industry. By fermentation the lactose may be converted to alcohol, or lactic acid, or antibiotics also for human use. Other fermentations will produce a biomass or ammonium lactate for animal food, or methane for fuel. Evaporation and crystallisation produces a food for human or animal use. Of these the most economically viable are the production of the glucose/galactose syrup and the alcohol, and indeed both have reached commercial practice.

THE PRINCIPLES OF REVERSE OSMOSIS

Consider the arrangement in Fig. 10 where a membrane is placed between a solution and a solvent. The membrane is permeable only to the solvent, no other molecules in solution or suspension can pass. In an attempt to equalise the concentrations on both sides of the membrane, the solvent, usually water, will pass through the membrane from left to right. The driving force is the gradient of the chemical potential across the membrane. This is a thermodynamic quantity, depending on

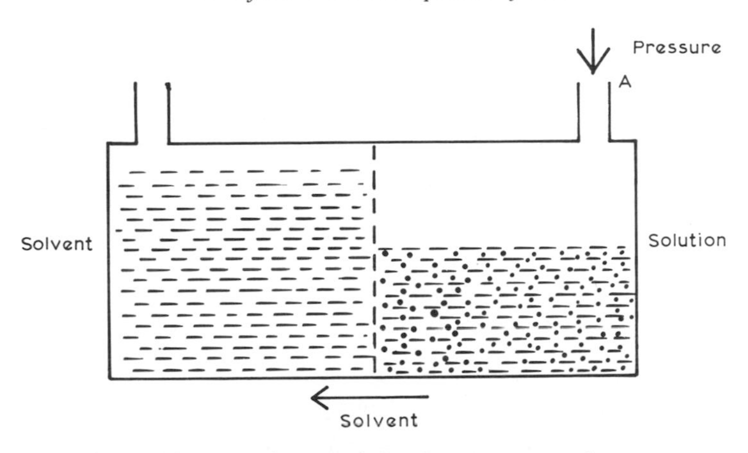

Fig. 10. The principle of reverse osmosis.

concentration, pressure and temperature, whose gradient determines the movement of matter. The concentration of water is higher on the left than on the right, water, therefore, moves down the gradient from left to right causing the liquid level at A to rise, thus setting up a pressure. Solute molecules cannot move since they are restrained by the membrane. Equilibrium is reached when the chemical potentials, made up of concentration and pressure on both sides of the membrane, become equal. Temperature is not involved since it is the same on each side. The pressure set up by the rise of water at A is the osmotic pressure of the solution in the right hand compartment. It is directly proportional to the concentration, and inversely proportional to molecular weight.

If a pressure in excess of the osmotic pressure is applied at A, the above process may be reversed, and water may be driven from right to left through the membrane achieving a concentration of the solution. This is reverse osmosis, which can be used for concentration or for purification depending on whether the concentrate is the required product as in milk processing, or the permeate as in water purification.

The favoured theory of the passage of water through the membrane considers that the solvent dissolves in the membrane, then moves through by diffusion. The solute is far less soluble in the membrane and is, therefore, held back. This theory does not require the membrane to have pores; some small molecules do get through, in practice, due to defects in the membrane.

The rate of transport of any component through the membrane is a function of the gradient of its chemical potential. At a constant temperature, chemical potential is a function of concentration and pressure.

$$\text{Flux} \propto f\,(\text{concentration gradient}) + f(\text{pressure gradient})$$

Since the solvent, water, passes easily through the membrane, the difference in the concentrations of water across the membrane is small, and hence, in the above expression, the concentration term may be neglected, and the pressure term becomes predominant.

If J_w = water flux;
ΔP = pressure applied for reverse osmosis;
$\Delta \Pi$ = osmotic pressure of the solution;

the driving pressure will be the applied pressure minus the opposing osmotic pressure.

Then

$$J_w \propto f(\Delta P - \Delta \Pi), \text{ or } J_w = A(\Delta P - \Delta \Pi)$$

where A is the solvent flow constant which contains the membrane area, thickness and void volume fraction, the diffusion constant and the solubility of the solvent in the membrane, the molar volume of the solvent in the solution, and the densities of the solvent and solution.

For the solute which is largely retained by the membrane, the difference in concentration across the membrane is relatively large. Hence for the solute, the concentration term is predominant.

If J_s = solute flux;
ΔC_s = difference in solute concentration across the membrane;
then $J_s \propto \mathrm{f}\,(\Delta C_s)$;
or $J_s = B\,\Delta C_s$;

where B is the solute flow constant, similar in nature to the solvent flow constant A.

From this analysis it should be noted that:

(a) Water is driven through the membrane by pressure, solute is driven through by a concentration gradient, i.e. the transport of solvent and solute are not connected.
(b) An increase of pressure increases the water flux, but has no effect on the passage of solute. Hence the concentration of solute in the permeate decreases with an increase in pressure.
(c) An increase in the feed concentration reduces the permeate flux

since the osmotic pressure of the feed is increased, and it increases the passage of solute because of a higher concentration gradient.

(d) An increase in temperature increases the rate of diffusion of both water and solute, both at a rate of about 3·5 per cent $°C^{-1}$.

(e) Ageing of the membrane causes the constants A and B to reduce due to creep and tightening under the applied pressure, hence both permeate flux and passage of solute fall with age. The fall is logarithmic with time, most noticeable in the early stages of membrane life.

This theoretical treatment was developed to fit dilute monocomponent systems, and is offered here to provide an understanding of the fundamental aspects of reverse osmosis. Milk has many components covering a great range of molecular sizes, and some in very high concentrations. The theory could not be expected to fit the practical behaviour of milk.

As with ultrafiltration, the main hindrance to the process is concentration polarisation, which must again be minimised either by high degrees of turbulence or high rates of shear.

MEMBRANES FOR REVERSE OSMOSIS

Cellulose acetate formed the first successful RO membrane as it did for UF. Although nylon membranes appeared for water purification, it was almost 20 years before a successor to cellulose acetate became available for milk processing. Paterson Candy International Ltd has its non-cellulosic membrane ZF99 which is superior to cellulose acetate in permeability, retention, pH and temperature tolerance. It can be operated safely over a pH range of 3–11, and up to a temperature of 80°C. The Danish Sugar Company (DDS) has its new RO membrane, designated HR, and described as a thin-film, composite membrane consisting of a UF membrane coated with a very thin, polymer layer. The polymer is the effective RO membrane, laid on a polysulphone layer which in turn is supported on a polypropylene backing. It has properties similar to the ZF99. These new membranes are susceptible to damage by chlorine and must be cleaned only with approved agents.

Membrane Geometry and Modules

Membrane geometry for RO is the same as that for UF, namely tubular, flat and spirally wound, and the same comments on the different types apply. There is no hollow fibre design suitable for milk as there is for UF, and the Du Pont RO system of bundles of many thousands of very fine, nylon fibres only 90 μm diameter could not take the high solids content of milk.

Modules for RO are also very like UF modules. For the tubular membrane type of Paterson Candy (Fig. 3) all membrane tubes are connected in series. The flat membrane arrangement of DDS (Fig. 4) has circular plates for RO (30 cm or 40 cm diameter) built into vertical stacks containing 19 m^2 or 28 m^2 membrane area.

REVERSE OSMOSIS PLANT

RO plants look exactly like UF plants, comprising pumps, membrane modules, controls and gauges. The same modes of operation may be used, namely continuous internal recycle and batch operation, and also since concentration factors for RO are lower, single-pass operation has sometimes been possible. Single-pass plants are designed as a series of stages in tapered formation, successive stages having few modules to ensure that high rates of flow are maintained and not reduced by the extraction of permeate in the earlier stages.

Recently, centrifugal pumps with high pressure delivery have become available providing higher flow velocities, and hence more efficient use of membranes. Plant configuration is shown in Fig. 11 for the multistage recycle system in which modules are arranged in a number of stages, each with a short path length and booster pump at the entrance of the stage. Throughout the plant higher pressures and higher velocities than in single stage recycle are then maintained. In the plant of Fig. 11 working on whey the pressure drop across each stage is only 0·9 MPa from 4·5 MPa to 3·6 MPa and the flow velocity nowhere falls below 2·2 m s^{-1} which is well above the critical velocity of 1·6 m s^{-1} for such tubular membranes at which experience has shown that membrane fouling in whey starts.

Engineering Equipment

There are differences in the engineering equipment for RO and UF due to the much higher pressure demanded by RO. For the tubular system working on milk, it is required to deliver about 20 litre min^{-1} into a single module at a pressure of about 5 MPa. Positive displacement pumps (constructed in stainless steel) of the three piston type, as in the

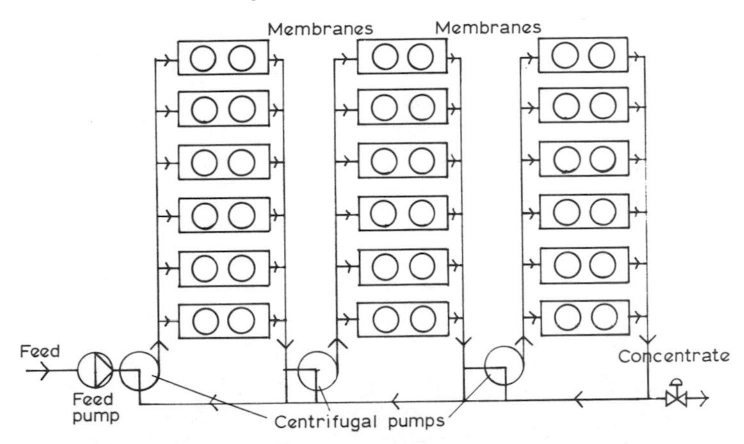

Fig. 11. Reverse osmosis plant — multistage recycle system (PCI), 3 stages each with 6 parallel rows of 2 modules, total area 92 m^2. (With acknowledgement to Pepper and Orchard (1982) and reproduced by courtesy of the Society of Dairy Technology.)

normal dairy homogeniser, serve very well. They are fitted as feed pumps to raise the pressure for entry to the module assembly, then booster pumps are used between stages as described above. Any pressure pulsation can be smoothed out by including a nitrogen-filled accumulator in the pipeline.

The Archimedian-screw type of pump is capable of raising the pressure for RO, but cannot maintain high rates of flow due to slipping of fluid backwards past the screw; it can, therefore, be used only on small capacity RO plants. The pitot pump has a number of attractions for RO and has been used. It has few moving parts, its action is rotary, it can operate against a closed valve and the flow is pulse-free. Its disadvantage is that its efficiency is low.

Since RO feed pumps are the positive type, they will not necessarily require flow controls or flow meters; they will, however, need protection against overload. The normal, spring-loaded, relief valve is not acceptable for dairy processing on hygienic grounds due to the dead space in the valve, and stainless steel, bursting discs are much more suitable. In all other respects, the auxiliary equipment for RO is the same as that for UF.

A commercial RO plant using tubular membranes is illustrated in Fig. 12.

Fig. 12. Reverse osmosis plant — commercial scale, tubular membranes (PCI). Area 327 m^2, concentrating cheese whey, 10 000 litre h^{-1} to 24 per cent T.S. (×4). (Photograph by courtesy of Paterson Candy International, Laverstoke Mill, Whitchurch, Hants. RG28 7NR England.)

REVERSE OSMOSIS PLANT PERFORMANCE ON MILK

In comparison with ultrafiltration, additional factors come into play to govern the rate of reverse osmosis. The most important of these is osmotic pressure, and its increase both in the mainstream of the feed and, more particularly, in the polarised layer at the membrane surface as concentration proceeds, both considerably reduce the effectiveness of the applied pressure. Secondly, because RO membranes retain all solids, calcium phosphate, which is in saturated solution in normal milk, can be precipitated on the membrane if hot milk is concentrated. This precipitation of calcium phosphate may be reduced by pretreating the milk before RO, such as reducing the pH to increase the solubility of the calcium phosphate, preheating to transfer some of the calcium to the casein micelles, or removing some of the calcium by ion exchange. The most practical of these methods is the preheating. Nevertheless, as in UF, the main fouling agent is the protein, mainly because of the lower diffusion coefficients of the large molecules and micelles. The diffusion coefficient of casein is one third of that of the whey proteins and one twentieth of that of lactose, even at normal concentrations in milk.

The reverse osmosis of milk is not yet practised on a large scale. There is some activity in the yoghurt and ice cream business, and some trials on hard cheesemaking, but nothing for the preparation of concentrates for the liquid market. The only information on the behaviour of milk during RO comes from research reports, which are few indeed compared with the vast number on UF in the scientific literature and advertising material. An impression of the rate of RO for whole milk is given in Fig. 13; the results were obtained in experiments using a small pilot-plant. These data show that a concentration factor of ×2 is easily attainable and ×3 is possible though scarcely practicable, corresponding to total solids contents of approximately 25 per cent and 37 per cent

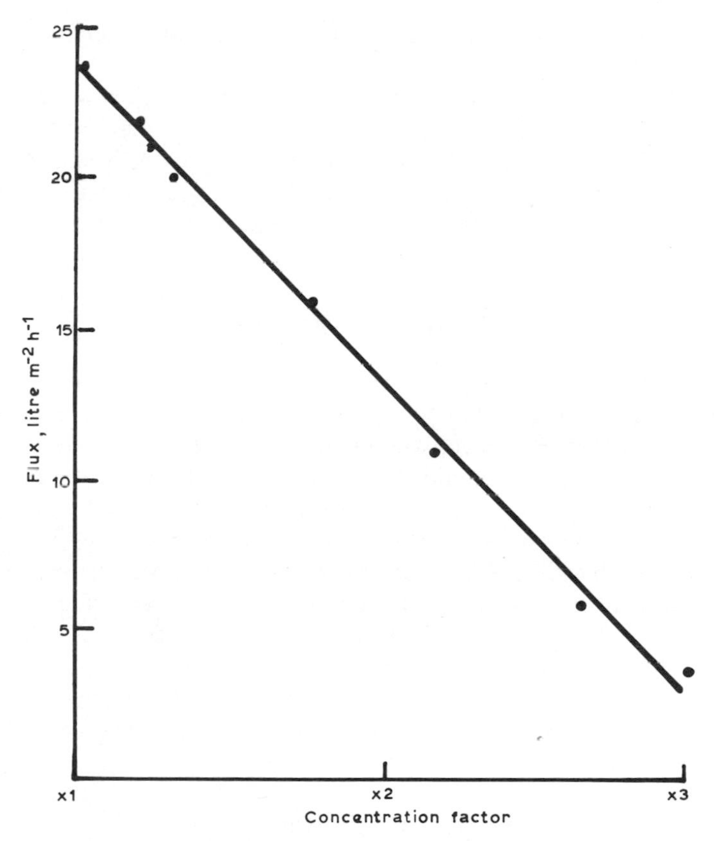

Fig. 13. Reverse osmosis of milk — flux against concentration: membrane — cellulose acetate type, pressure 4·8 MPa, flow velocity 2·4 m s^{-1}, 30 °C, small pilot plant, batch operation. (With acknowledgement to Abbot *et al.* (1979) and reproduced by courtesy of the *Journal of Dairy Research.*)

respectively. For skim-milk concentrated under the same conditions, the fluxes are only about 3 litres m^{-2} h^{-1} higher than those from whole milk. More realistically, skim-milk has been concentrated two-fold in a larger plant as a continuous operation by the Paterson Candy, multi-stage recycle plant fitted with 16 m^2 area of cellulosic membrane (Pepper and Orchard, 1982). The operating conditions were: pressure 3·8 MPa, flow velocity 2·5 m s^{-1}, temperature 30 °C and fluxes at the first and second stages (13·5 and 18 per cent T.S.) were maintained at 20 litres $m^{-2}h^{-1}$ and 10 litres $m^{-2}h^{-1}$ respectively over a period of 6 h. A larger plant of the same type having a membrane area of 250 m^2 is concentrating 11 250 litres h^{-1} of skim-milk by a factor of 1·5.

Effects of Pressure, Flow Velocity and Temperature

The effects of changes in the operating parameters follow the same trends as in UF, with the difference that as pressure is increased the flux passes through a maximum, so that beyond this point, it is definitely disadvantageous to apply more pressure as it only decreases the flux. The level of the flux at the maximum rises as the flow velocity is increased.

Composition of the Product

For practical purposes, all the components of milk are retained in the concentrate, and only a small proportion of the smallest ions escapes into the permeate. Retention of the whole mineral content of milk exceeds 99 per cent; individual ions with retentions below this are Na^+ 95 per cent, K^+ 98 per cent and Cl^- 94 per cent. The permeate contains 1000 ppm total solids made up mainly of lactose 250 ppm, Na^+ 110 ppm, Cl^- 200 ppm (Skudder, 1978); it can only be presumed that defects in the membrane caused the loss of lactose into the permeate. In the experience of Pepper and Orchard (1982), using a skim-milk feed initially with a BOD of 65 000 mg $litre^{-1}$ concentrated continuously to 18 per cent T.S. in the multi-stage recycle, RO plant, the permeate BOD was only 100 mg $litre^{-1}$; there was also very little bacterial growth in the concentrate.

Deposits on the Membrane

Solids collect at the membrane in two forms: a compacted layer in contact with the membrane, and a gelatinous layer between this and the mainstream of the feed. The thickness of the layer of solids may be up to a few micrometres, and consists mainly of proteins. Electron microscopy of the deposits shows that they are granular in structure, more compact nearer the membrane surface. The granules are mainly casein, some of them linked by bridges forming a network. A few fat globules are trapped in-between the granules, but it is evident that the compacted casein forms the greatest obstruction to the process. In the deposited layer, the concentration of the casein can be up to 10 times that in the original milk (Skudder, 1978).

Cleaning Reverse Osmosis Membranes

For their cellulose acetate membranes, Paterson Candy recommend cleaning by circulating a 1 per cent enzyme detergent solution for 90 min at 35 °C, and then sterilising with 50–100 mg litre^{-1} free chlorine for 30 min. Recommended cleaning materials for their new ZF99 membranes are:

0·1–0·3 per cent w/w nitric or phosphoric acid at 45 °C;
0·1–0·5 per cent w/w sodium hydroxide at 50 °C; and
0·1–0·5 per cent proprietary alkaline detergent.

Sterilisation is with hydrogen peroxide or formalin. The introduction of ZF99 membranes has halved the cost and cleaning time of membranes.

USES OF RO MILK CONCENTRATES

The main application of reverse osmosis in dairying is the concentration of whey to facilitate handling, transport and storage. On milk its use is not nearly so widespread, although there is some activity for making ice cream and yoghurt, and trials are in progress for cheddar cheese making, recovery of milk solids from rinse water, and the consideration of RO as an initial step in drying from the point of view of energy saving.

Ice cream

Skim-milk powder is used in making ice cream. Since skim-milk RO concentrate is cheaper to make than the powder, the possibility arises of substituting concentrate for powder. The DDS Company have prepared skim-milk RO concentrate to 15·4 per cent T.S. for this purpose, and report that the quality of the ice cream is as good in all respects as the product made by traditional methods.

Yoghurt

Skim-milk concentrates are used for yoghurt as they are for ice-cream — as a substitute for powders. Some yoghurt has been made by this method for several years now, and involves first concentrating milk by RO by a factor of 1·4–15 per cent T.S.

Cheese

There has been some exploratory work on making cheese from RO milk concentrates. Using 15 per cent T.S. concentrates, increases in the yield of Cheddar cheese of 1–3 per cent are reported, and this improvement is attributed to the higher retention of casein fines, fat and whey solids. Cottage cheese and pâte frâiche have also been made from RO concentrates. The quality of all the cheese has been pronounced good.

Recovery of milk solids from dairy rinse water

Experiments in Sweden have shown that in a dried-milk factory, rinses of pipelines obtained in the first few seconds contained as much as 8 per cent solids. This level could be concentrated to 18 per cent T.S. by RO, and the material dried and sold for animal feed, so reducing pollution and effecting a considerable saving.

Milk powder

Milk powder has been made experimentally in a process in which RO was used to concentrate milk prior to evaporation and spray-drying. The whole milk powders had high free fat levels due to damage suffered by the milk in the RO pumps and valves, but the skim-milk powder was just as good as that made conventionally.

Energy saving

In the initial stages of the concentration of milk, RO consumes less energy than evaporation since RO demands pumping energy only; evaporation involves a phase change. Over the range of concentration for which RO can be applied, for example in the case of concentrating whey from 6 per cent T.S. to 28 per cent, the energy required by RO is approximately ¼ of that of the latest mechanical, vapour recompression evaporator. Unfortunately for the drying of milk, RO cannot concentrate to the level of 50 per cent T.S. where spray-drying starts, hence RO is not a total replacement for evaporation.

THE PRINCIPLE OF ELECTRODIALYSIS

Electrodialysis is a process for the removal of ions from a solution by driving them through a membrane in an electric field. Anions (−ve) move towards the anode (+ve), cations (+ve) move towards the cathode (−ve). Three types of membrane are used:

(A) anion permeable membranes; they contain cationic groups and repel cations;
(C) cation permeable membranes; they contain anionic groups and repel anions;
(N) non-selective membranes, permeable to both anions and cations.

The selective membranes are made of styrene-divinylbenzene containing quaternary ammonium and sulphonic groups for the anionic and cationic groups respectively. Non-selective membranes are formed from regenerated cellulose. Pore sizes are in the range 1–2 nm, and the sheets of membrane are reinforced mechanically with synthetic fibres. It is important that the membranes should have a low electrical resistance, that they should be highly selective between anions and cations and highly permeable to one of them, that they should have a high capacity to hold the required ions, that they permit little unionised solvent to pass and finally, that they should have good resistance to chemical attack. Details of electrodialysis membranes, their properties and manufacturers are given in the IDF Document 115 (1979).

The basic operation of electrodialysis is illustrated in Fig. 14 for removing salt from impure water (Young, 1974). A and C membranes are arranged, in parallel, 1 mm apart in an assembly like a plate heat exchanger. At the ends of the stack are the anode and cathode, to set up a potential difference across each pair of membranes of 1–2 V and operate at a current density of the order of 10 mA cm^{-2}. A typical electrodialysis stack could contain 200 pairs of membranes, each measuring approximately 1 m × 0·5 m giving an effective membrane area in the stack of nearly 200 m^2. In the illustration, it can be seen that sodium and chlorine ions are removed from compartments 2 and 4 and transferred to compartments 1, 3 and 5. Desalted water is then led-off at D and the brine at B. In practice, the feed has to be circulated many times to effect a useful degree of demineralisation; for example when processing whey, a single pass through a stack may remove only 10 per

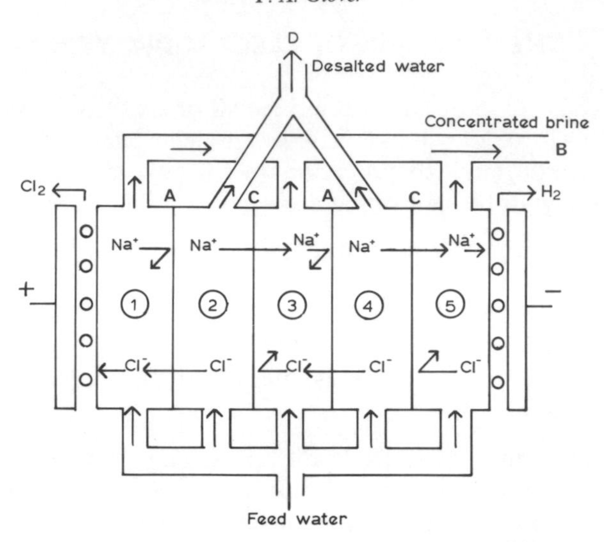

Fig. 14. The principle of electrodialysis. (With acknowledgement to Young (1974), and reproduced by courtesy of the Journal of the Society of Dairy Technology). A, anion permeable membranes; C, cation permeable membranes.

cent of the minerals, to reach 90 per cent demineralisation some 30 passes are required.

The main problem in electrodialysis, as in UF, is concentration polarisation. Where the transport number of ions in solution is less than that in the membrane, at any one membrane there will be a depletion of ions on one side of the membrane and an increase in concentration on the other. This unequal distribution of ions must be minimised by employing high degrees of turbulence between the membranes, and since high current densities increase this inequality, densities must be limited. When processing milk or whey, calcium phosphate collects on the cathode side of a cation membrane, but more serious is the deposition of large organic molecules on the face of anion membranes; denatured protein components with large negative charges are particularly obstructive in this respect.

The system used for whey is a slight variation on the basic process. Whey is fed into compartments 1, 3 and 5, and a dilute salt solution is fed into 2 and 4 to maintain electrical conductivity across the stack, and to flush-out the mineral ions extracted from the whey. During the process, the concentration of ions rises in the flushing solution which must then be diluted. To protect the electrodes from chemical attack by components of the feed, a separate electrolyte solution is fed into the compartments adjacent to the electrodes. Since electrodialysis operates most effectively with solutions of high electrical conductivity, the whey is usually concentrated first to 28 per cent T.S. The operation is then carried out as a batch process achieving 90 per cent demineralisation.

A modification of the classical electrodialysis, known as transport depletion, uses non-selective membranes instead of the usual anion membranes, the purpose being to avoid concentration polarisation. However, with this advantage must be accepted lower efficiency, since the cations are not repelled by the non-selective membranes.

Another variant of electrodialysis is termed ion substitution, in which one species of ion is removed and another substituted. Instead of the simple, alternate A-C-A membrane arrangement, the sequence A-C-C-A is used, and a solution containing the ions for substitution is fed into the C-C compartment.

Demineralised milk is required for the preparation of baby food and special adult diets. However, the greatest use of electrodialysis in dairy technology is for the preparation of demineralised, whey protein.

POSSIBLE FUTURE DEVELOPMENTS IN MEMBRANE PROCESSING

Ultrafiltration has more potential applications than reverse osmosis, and the use of UF is more widespread in France where soft cheese is popular, and in Denmark where the Danish Sugar Corporation has been so progressive in the technology. In England and other countries where hard cheese is the predominant variety, UF waits on the development of a method for making the Cheddar type from UF concentrates. Work in this direction is in progress, with the cheesemaker modifying his process, and the manufacturers of plant constantly trying to increase the total solids of the final concentrate. It has even been suggested that all cheese will eventually be made from UF concentrates.

REFERENCES

Abbot, J., Glover, F. A., Muir, D. D. and Skudder, P. J. (1979). Application of reverse osmosis to the manufacture of dried whole milk and skim milk. *Journal of Dairy Research,* **46,** 663–72.

Coton, S. G. (1980). The utilisation of permeates from the ultrafiltration of whey and skim milk. *Journal of the Society of Dairy Technology,* **33** (3), 89–94.

Glover, F. A. (1984). Ultrafiltration and Reverse Osmosis for the Dairy Industry. Technical Bulletin 5, National Institute for Research in Dairying, Reading, England.

Glover, F. A., Skudder, P. J., Stothart, P. H. and Evans, E. W. (1978). Reviews of the progress of Dairy Science: reverse osmosis and ultrafiltration in dairying. *Journal of Dairy Research,* **45,** 291–318.

Goudedranche, H., Maubois, J. L., Ducruet, P. and Mahaut, M. (1981). Utilisation de nouvelles membranes minerales d'ultrafiltration pour la fabrication de fromages du type Saint Paulin. *Technique Laitière,* (950), 7–13.

Hansen, R. (1977). Feta cheese production by ultrafiltration. *Nordeuropaeisk Mejeri Tidsskrift,* **43,** (9) 304–9.

Hansen, R. (1981). Ultrafiltration Camembert manufactured on the new Camatic. *Nordeuropaeisk Mejeri Tidsskrift,* **47,** (5), 147–52.

International Dairy Federation (1979). Equipment available for membrane processes. International Dairy Federation Bulletin — Document 115.

Kessler, H.G. (1981). *Food Engineering and Dairy Technology,* Verlag A. Kessler, Freising, FRG, p. 86.

Kiviniemi, L. (1974). Microbial growth during the ultrafiltration of sweet whey and skim milk. *Kemia-Kemi,* **1,** (12), 791–5.

Kristensen, S., Nielsen, W. K. and Madsen, R. F. (1981). New DDS ultrafiltration module for high concentration of milk products in connection with production of cheese. *Northern European Dairy Journal,* **47,** 268–74.

Lee, D. N., Miranda, M. G. and Merson, R. L. (1975). Scanning electron microscope studies of membrane deposits from whey ultrafiltration. *Journal of Food Technology,* **10,** 139–46.

Lewis, M. J. (1982). Concentration of proteins by ultrafiltration. In: *Developments in Food Proteins — 1* (Ed. B. F. J. Hudson), Applied Science Publishers, London, pp. 91–130.

Maubois, J. L. (1980). Recent developments of membrane filtration in the dairy industry. In: *Ultrafiltration membranes and applications. Polymer Science & Technology,* Vol. 13 (Ed. A. R. Cooper), Plenum Press, New York, pp. 305–18.

Morgensen, G. (1980). Production and properties of yoghurt and ymer made from ultrafiltered milk. *Desalination,* **35,** (1–2–3), 213–22.

Pepper, D. and Orchard, A. C. J. (1982). Improvements in the concentration of whey and milk by reverse osmosis. *Journal of the Society of Dairy Technology,* **35,** (2), 49–53.

Roenkilde Poulsen, P. (1978). Feasibility of ultrafiltration for standardising protein in milk. *Journal of Dairy Science,* **61,** (6), 807–14.

Scurlock, P. G. and Glover, F. A. (1983). Use of high protein creams for the production of cream with a reduced fat content. Annual Report. p. 130, National Institute for Research in Dairying, Shinfield, Reading, England.

Serpa Alverez, D., Bennasar, M. and De la Fuente, B. (1979). Application of ultrafiltration to the manufacture of sweetened condensed milk. *Le Lait,* **59,** (587), 376–86.

Skudder, P. J. (1978). Concentration of milk by reverse osmosis. Ph. D. thesis, University of Reading.

Sutherland, B. J. and Jameson, G. W. (1981). Composition of hard cheese manufactured by ultrafiltration. *Australian Journal of Dairy Technology,* **36,** 136–43.

Young, P. (1974). An introduction to electrodialysis. *Journal of the Society of Dairy Technology,* **27,** (3), 141–9.

Chapter 6

Utilisation of Milk Components: Whey

J. G. Zadow
Division of Food Research, CSIRO, Dairy Research Laboratory, Highett, Victoria, Australia

Whey may be defined, broadly, as the serum or watery part of milk remaining after separation of the curd that results from the coagulation of milk by acid or proteolytic enzymes. Its composition will vary substantially, depending on the variety of cheese produced or the method of casein manufacture employed. On average, whey contains about 65 g kg^{-1} of solids, comprising about 50 g lactose, 6 g protein, 6 g ash, 2 g non-protein nitrogen and 0·5 g fat. The protein fraction contains about 50 per cent β-lactoglobulin, 25 per cent α-lactalbumin and 25 per cent other protein fractions including immunoglobulins. However, there will be wide variations in composition depending on milk supply, and the process involved in production of the whey.

Wheys can be conveniently classed into groups:

Sweet wheys: titratable acidity 0·10–0·20 per cent, pH typically 5·8 to 6·6.
Medium-acid wheys: titratable acidity 0·20–0·40 per cent, pH typically 5·0–5·8.
Acid wheys: titratable acidity greater than 0·40 per cent, pH less than 5·0.

In general, wheys produced from rennet-coagulated cheeses develop low levels of acidity, while the production of fresh, acid cheeses, such as Ricotta or Cottage cheese, yields medium-acid or acid wheys. Whey from caseins produced by acid addition is classed as high-acid whey, whereas whey from rennet casein is sweet whey. Typical pH values and titratable acidities for a range of wheys are shown in Table I.

TABLE I
pH values and acidity of various wheys

	pH	*Titratable acidity*
Rennet types		
Edam	6·5–6·6	0·09
Gouda	6·5–6·6	0·10–0·12
Tilsit	6·5	–
Gruyère	6·5	0·07
Mozzarella	6·3	0·11
Emmental	6·2–6·5	–
Camembert	5·8–6·0	0·18–0·25
Cheddar	5·7–6·3	0·15–0·19
Danbo	5·3	–
Fresh acid types		
Ricotta (whole-milk based)	5·3	0·14
Ricotta (whey based)	5·3	–
Cottage	4·5–4·6	0·50–0·55
Cream-Neufchatel	4·5	0·60
Quarg	4·5	0·70
Caseins		
Rennet	6·5	0·10
Lactic acid	4·5	0·64
Hydrochloric or sulphuric	4·0–4·4	–

Data from Kosikowski (1977).

Compositional and Seasonal Variations

The mean composition of dried, sweet and acid whey samples manufactured in the United States is shown in Table II. Standard errors for many of the components are also shown, indicating possible variations.

As would be expected, there are considerable variations in the composition of wheys, depending on their source and on seasonality factors. Seasonality is of particular importance in countries, such as New Zealand, Australia, and Ireland, where milk production varies widely throughout the year. Matthews (1978) has examined the changes in the composition of New Zealand rennet, sulphuric and lactic casein wheys over a twelve month period. Seasonal changes in the composition of the wheys all followed similar patterns, with the main change being an increase in protein content and decrease in lactose concentration

TABLE II
Composition of dried sweet and acid wheys

Component	*Content (per kg powder)*			
	Sweet whey	*Samples*	*Acid whey*	*Samples*
Proximate				
Water (g)	31·9(0·887)	151	35·1(1·74)	52
Food energy (kJ)	14 760		14 190	
Protein (Nx6·38) (g)	129·3(0·874)	165	117·3(1·819)	57
Total lipid (fat) (g)	10·7(0·334)	148	5·4(0·31)	49
Carbohydrate (total) (g)	744·6		734·5	
Ash (g)	83·5(0·607 6)	147	107·7(2·226)	48
Minerals (mg)				
Calcium	7 960(314)	66	20 540(1 055)	
Iron	8·8(0·33)	41	12·4(1·78)	19
Magnesium	1 760(51·8)	41	1 990(96·4)	19
Phosphorus	932(39·9)	65	13 480(825·1)	20
Potassium	20 800(500·4)	60	22 880(976·4)	20
Sodium	10 790(671·3)	62	9 680(853)	18
Zinc	19·7(1·56)	43	63·1(5·49)	19
Vitamins				
Ascorbic acid (mg)	14·9(3·35)	41	9·0(3·3)	13
Thiamin (mg)	5·19(0·141)	44	6·22(0·720)	13
Riboflavin (mg)	22·08(0·605)	44	20·60(1·301)	13
Niacin (mg)	12·58(0·506)	43	11·60(1·952)	10
Pantothenic acid (mg)	56·20		56·32	
Vitamin B6 (mg)	5·84(0·159)	41	6·20(0·514)	10
Folacin (mcg)	120(0·94)	40	330(46)	10
Vitamin B12 (mcg)	23·71(0·971)	48	25(1·747)	10
Vitamin A (IU)	440		580	
Lipids				
Saturated, total (g)	6·8		3·4	
4:0	0·6(0·6)	4	0·2	
6:0	0·1(0·007)	4	0·1	
8:0	0·1(0·005)	4	Trace	
10:0	0·2(0·01)	4	0·1(0·02)	4
12:0	0·1(0·008)	4	0·1(0·02)	4
14:0	1·0(0·054)	4	0·5(0·05)	4
16:0	3·3(0·16)	4	1·5(0·13)	4
18:0	1·0(0·058)	4	0·6(0·04)	4
Mono-unsaturated, total (g)	3·0		1·5	
16:1	0·3(0·02)	4	0·3(0·07)	4
18:1	2·5(0·098)	4	1·1(0·09)	4

(*contd.*)

TABLE II—*contd.*

Component	*Content (per kg powder)*			
	Sweet whey	*Samples*	*Acid whey*	*Samples*
Polyunsaturated, total (g)	0·3		0·2	
18:2	0·2(0·03)	4	0·2(0·03)	4
18:3	0·1(0·02)	4	Trace	
Cholesterol (mg)	60			
Amino acids (g)				
Tryptophan	2·05		2·41	
Threonine	8·17		5·9	
Isoleucine	7·19		5·81	
Leucine	11·86		11·16	
Lysine	10·30		10·08	
Methionine	2·41		2·21	
Cystine	2·53		2·11	
Phenylalanine	4·07		3·86	
Tyrosine	3·63		3·00	
Valine	6·97		5·79	
Arginine	3·75		3·27	
Histidine	2·37		2·30	
Alanine	5·98		5·06	
Aspartic acid	12·69		11·49	
Glutamic acid	22·48		20·96	
Glycine	2·80		2·11	
Proline	7·86		6·99	
Serine	6·22		5·41	

Bracketed values are standard errors of means.
Data from Posati and Orr (1976).

towards the end of the season. The true protein content of rennet whey varied from 110 to 140 g kg^{-1} of solids, and the lactose content from 750 to 810 g kg^{-1} solids throughout the season. The major differences between the wheys were the lower calcium and phosphate levels and higher amounts of protein and lipids in rennet whey; lower lactose and lipid contents, higher non-protein nitrogen contents, and the presence of lactate in lactic whey. Roeper (1971) has also examined seasonal changes in whey composition in New Zealand. It was reported that from August to February, when more than 80 per cent of the whey is produced, there was little change in the composition of Cheddar whey, apart from a gradual decrease in lactose. Ash content, true protein and non-protein nitrogen began to increase from February onwards, peaking in May–June. Drought conditions also had a significant effect on whey composition, with protein and lactose levels changing to typical winter values. However, the seasonal changes in whey composition reported in both Australia and New Zealand are not likely to pose major difficulties for most whey processors. The seasonal variation of a number of components in wheys from New Zealand and Australia is shown in Table III.

PRODUCTION

Statistics

The full range of unit processes, such as concentration, drying, fermentation, demineralisation, membrane processing, and lactose hydrolysis, may be applied to all wheys. However, the commercial viability of many processes will depend on specific compositional factors, for example, the use of wheys with high mineral content for the manufacture of demineralised products may prove uneconomical.

Within the past twenty years, there has been a general realisation of the economic potential of whey. In the past, it has been treated as a waste product, to be disposed of in rivers or on land. Pressures by environmentalists, and the realisation that whey contains about half the solids of milk and much of its nutritional value, have substantially changed the attitude of the industry. Whey is now recognised as a valuable raw material, with its potential for economic returns to the dairy industry still largely undeveloped.

Strobel (1972) pointed out that there were few statistics on whey production and whey solids utilisation, and this remains the case. Still, cheese whey solids production can be assessed at approximately 4 500 000 tonnes per annum, based on a world cheese production in 1980 of 11 000 000 tonnes (Food and Agriculture Organisation, 1981). Of the total cheese whey manufactured, approximately 50 per cent is produced in Europe, 18 per cent in North America, 2·5 per cent in Australia and New Zealand, 15 per cent in other dairy-developed countries, and the remainder in under developed countries (Allum, 1980). The major European producer of whey is France (0·5 million tonnes of solids per

TABLE III
Seasonal variation in composition of New Zealand and Australian wheys

Component	*Whey variety*				
	Rennet (1)	*Sulphuric (1)*	*Lactic (1)*	*HCl (2) casein*	*Cheddar (2) cheese*
Total solids†	64–67	60–64	62–65	58–61	61–66
True protein*	110–140	85–120	84–110	96–124	99–110
NPN:TN ratio$	0·24–0·28	0·24–0·29	0·30–0·35	0·18–0·24	0·22–0·30
Ash*	74–78	120–132	117–123	116–194	76–91
Lactose*	750–810	680–760	620–690		744–810
Calcium*	6–8	21–24	24–27		6–7
Inorganic phosphate*	10–16	29–34	30–34		8–30
Potassium*	18–26	8–24	20–24		19–25
Sodium*	6–16	4–14	6–12		6–11
Chloride*	14–21	12–18	13–19		19–20

†Value expressed as g solids/kg whey.
*Values expressed as g kg^{-1} solids.
$ Non-protein nitrogen to total nitrogen ratio.
(1) Data extracted from Matthews (1978).
(2) Unpublished data, CSIRO Division of Food Research.

annum), followed by West Germany (0·3 million tonnes) and Italy (0·2 million tonnes). Casein production in the major western countries amounted to 160 000 tonnes in 1982 (Australian Dairy Corporation, 1983), resulting in production of an additional 320 000 tonnes of whey solids.

The overwhelming use of whey powder in Europe is as animal feed, with calf milk replacers accounting for 90 per cent of all usage in the EEC. Similar consumption patterns are observed in Australia and New Zealand. By contrast, in the USA, approximately 66 per cent of dried whey is incorporated into human food, with major users being the dairy and baking industries (Delaney, 1979). Detailed information on worldwide utilisation of whey solids is very sparse, although some details have been reported by Clark (1979) for the USA.

Throughout the world there has been a slow trend for whey products to become more important in the economics of the dairy industry. There is, however, little doubt that the general trends in whey utilisation have changed dramatically over the past few years. The adoption of membrane processing by many manufacturers in dairy-developed countries has resulted in the production of significant quantities of whey protein concentrates with functional properties of considerable interest to the food industry. In 1978, modified whey products (demineralised, delactosed wheys and whey protein concentrates) accounted for 23 per cent of the total whey processed in the USA for animal feed, and 11 per cent of the whey processed for human consumption (Allum, 1980). The potential of whey protein concentrates, in particular, as ingredients for food-stuffs remains high, and a rapid increase in production can be expected over the next few years to meet demand. With the increasing economic importance of whey products to the dairy industry, it is probable that, in the future, more detailed information will be available on production and utilisation.

Economics

The basic decision by a dairy company on the alternative approaches to whey utilisation should be determined by the relative economics of the various options. However, the interplay of these options is, as can be seen from Fig. 1 (after Ryder, 1980), very complex, and there is little detailed information available concerning the relative economics of these process options. Ryder (1980) has reported some aspects, examining in detail the costs of dumping, spraying on land and animal feeding,

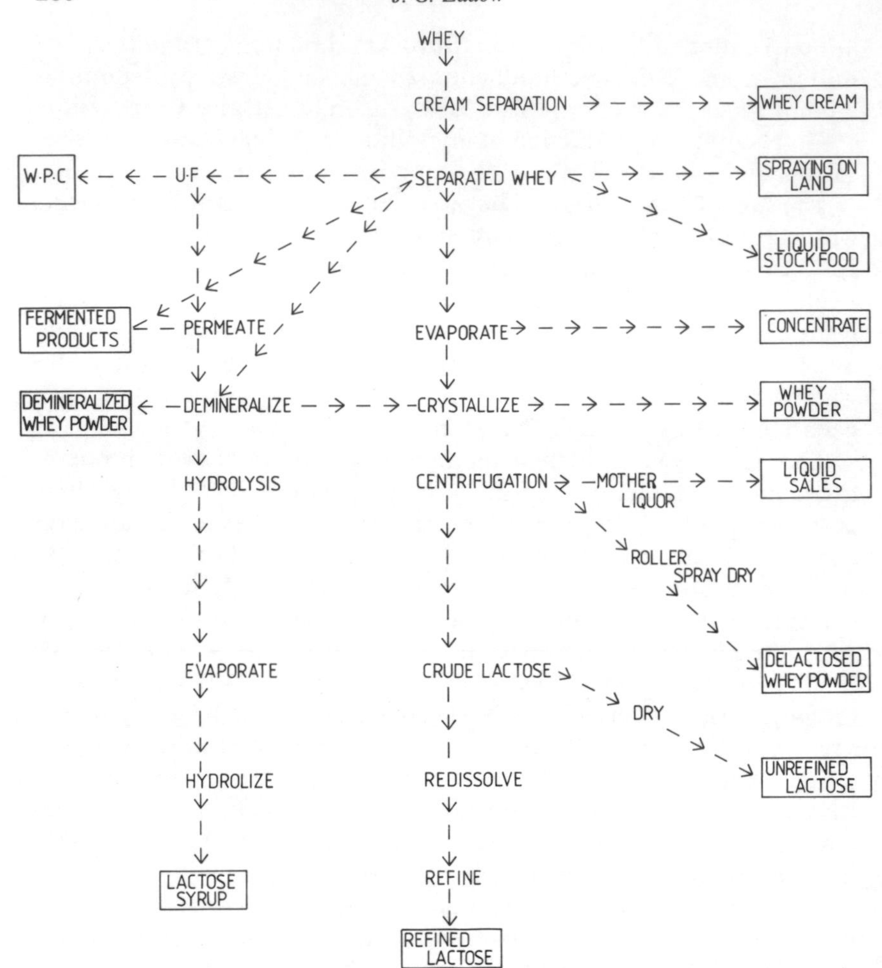

Fig. 1. Process options available to whey processors.

but an estimation of processing costs for the various options is, however, particularly difficult.

Aspects of the economic utilisation of permeate have been discussed by Coton (1980), and a useful comparison of the costs of concentration of whey by reverse osmosis (RO) and evaporation has been reported by Pepper (1981). Preconcentration of the whey by reverse osmosis, followed by conventional evaporation, proved to be the most economic combination. A case history for a plant processing 100 000 tonnes of whey per annum showed that the annual savings of this process were greater than the capital cost of the RO equipment. Muller (1979) has detailed the savings to be made by the RO concentration of whey, based on a case study involving the supply of whey to a central processing depot from three small factories. It was concluded that RO concentration offered satisfactory economies, even if only small quantities of whey are available.

WHEY CHARACTERISTICS

Analytical

Harper (1979) has reviewed recent studies on the analysis of whey and whey products, covering most aspects of the determination of whey components.

There has been particular interest in the measurement of lactose. Some recommended procedures are the Munson–Walker method involving the formation of cuprous oxide, polarimetry, the Chloramine T method (involving back titration of liberated iodine), the phenol/sulphuric acid method, and the picric acid method involving measurement of a coloured product at 520 nm (Jenness and Patton, 1959). Recently, high performance liquid chromatography techniques have been employed (Euber and Brunner, 1979), as well as an enzymatic technique based on the use of a cryoscope (Zarb and Hourigan, 1979). This method involves treatment of the product with lactase in buffer, then determination of the difference in freezing point of the sample on mixing, and after incubation for 1 h at 37 °C. This difference is directly proportional to the concentration of lactose present. Removal of the protein from the sample before analysis is not required, and the method is specific regardless of the presence of other sugars.

Whey Properties

Physical and chemical properties

There is comparatively little data available on the detailed physical properties of whey.

The surface tension of cheese wheys has been shown to vary between 40 000 and 84 000 N m^{-1}, increasing with increasing total solids, and decreasing with temperature (Zaetz *et al.,* 1982).

The viscosity characteristics of whey and concentrated whey are important, not only in terms of evaporation efficiency, but also for operations, such as lactose hydrolysis, which might be expected to function more efficiently if applied to concentrates. Studies on the viscosity of whey and whey concentrates have been reported by Chebotarev *et al.* (1983); lactose hydrolysis reduces viscosity and the degree of non-Newtonian behaviour. Lipatov and Chebotarev (1981) have also examined the viscosity and density characteristics of whey over a temperature range of 20 to 50 °C, and have suggested that whey is a heterogeneous system, with the fat globule size following a logarithmic distribution, and the casein particles being of an irregular shape up to 1 mm in diameter. Their results provide for the use of a 'separability index' as a characteristic of the fat–whey system; the results may be used as the basis of a special separator design.

Flavour

Studies on the flavour of acid whey, based on eight flavour characteristics, have been reported by McGugan *et al.* (1979). When increasing concentrations of whey were added to skim-milk, 'brothiness' was first noted when 20 per cent whey was added, and diacetyl, bitterness and sweetness at 40 per cent addition. Volatile acidity, non-volatile acidity, saltiness and astringency were only noted in 100 per cent whey. Neutralisation of the whey resulted in a change in all flavour characteristics.

Effect of heat

Hillier *et al.* (1979) have examined protein denaturation, using gel electrophoresis, in wheys heated over the range 70–150 °C. The rate of denaturation of α-lactalbumin appeared to be first order, but was actually considered to be, more likely, second order displaying pseudo-first order kinetics. The rate of denaturation of β-lactoglobulin A and B followed second order kinetics, whilst that of serum albumin was more complex. Below 95 °C, the rate of denaturation of β-lactoglobulin A was faster than that of β-lactoglobulin B.

A number of studies have reported the effect of total whey solids on the denaturation of whey proteins. In general, it has been found that, with higher total solids, the denaturation of β-lactoglobulin A and B slows, but the denaturation of α-lactalbumin increases. Increased lactose concentrations reduced denaturation of both proteins, perhaps as a result of the formation of heat-induced complexes. Increased calcium contents, up to 0·4 mg ml^{-1}, tended to slow denaturation, but above this level, little further effect was observed. The rate of denaturation of both proteins was slower at pH 4 than at pH 9 (Hillier *et al.*, 1979). Other comparisons have shown that the method used for concentration may significantly affect the nature and detailed conformation of whey proteins, with membrane processing having much less severe effects than conventional evaporation.

Recently, Pearce (1983) has described a novel method for the manufacture of enriched α-lactalbumin and β-lactoglobulin fractions from whey, relying on previously unknown solubility characteristics of these proteins, and using temperatures below those at which denaturation is generally considered to start. The changes occurring in the proteins at these temperatures may well explain, in part, the improvement in flux commonly obtained by holding whey at 55°C before ultrafiltration (UF). Clearly, there is still much work to be done to understand the effect of heat, concentration and aqueous environment on the behaviour of whey products.

Nutritional aspects

Whey proteins, with their high cysteine content, are considered good quality foodstuffs for human consumption. For example, studies have shown the benefits of including whey proteins in foodstuffs for low birthweight babies (Berger *et al.*, 1979), while other reports indicate that ingesting high levels of dried whey has little effect on plasma lipids. The main nutritional problem of many whey-based products is the lactose intolerance of certain sectors of the population, particularly non-Caucasian races; the extent of this problem has been reviewed by Garfield (1980). In Japan, for instance, digestive problems from drinking milk affect more than 20–25 per cent of the population, and clinical trials indicate that over 50 per cent of Japanese children have low lactase activity.

One option available to whey processors supplying these markets is to hydrolyse the lactose in their products. However, trials on rats in the early 1970s suggested that a high intake of galactose might result in high blood galactose levels which, in turn, might be responsible for cataract formation. However, a number of studies have now indicated that, when galactose is ingested with equi-molar quantities of glucose (as in hydrolysed lactose products), blood galactose levels rise very little in comparison to the increase occurring on ingestion of similar amounts of galactose alone (Coton, 1980; Williams and MacDonald, 1982). It should be remembered that the incorporation of whey products into domestic animal foods — a growth market in the past decade — is also restricted to some extent by lactose intolerance in many animals; lactose hydrolysed products will also have a large potential application in this area.

A reduction of the serum cholesterol and high density, lipoprotein cholesterol levels in swine has been reported (Ritzel *et al.,* 1979) following whey feeds.

PROCESSING OF WHEY

Clarification

For many purposes, clarification is an essential part of whey processing. Clarification is particularly important for the manufacture of whey protein concentrates by ultrafiltration, where the presence of even small amounts of lipid can have an adverse influence on the functional properties of the finished powder. Recently described means for clarification include the modification of whey composition through demineralisation to specific levels, followed by sedimentation of the aggregated material (Wit and Klarenbeek, 1978), the use of anionic precipitating agents (Best *et al.,* 1982) and microfiltration (Merin *et al.,* 1983). In this latter process, whey was filtered through a membrane with a pore size of 1·2 nm, producing a permeate free of fat, but containing all the other constituents of whey. This product had a 30 per cent higher ultrafiltration flux rate than unfiltered whey. Small fat globules were, therefore, suggested to be major contributors to membrane fouling during the ultrafiltration of cheese whey.

Storage of Whey

A number of reagents have been suggested as additives to extend the useful storage life of whey through inhibition of bacterial action. The processes reported include the addition of benzoic acid (with pH adjustment to 3·0–4·2), centrifugal clarification followed by the addition of hydrogen peroxide, the addition of 0·3 per cent propionic acid (delaying spoilage of whey at 20 °C by about 2 days), the addition of orthophosphoric acid and an organic carboxylic acid, and the addition of sodium sorbate. It should be noted that some of these approaches may not be legally acceptable treatments for whey in many countries.

Evaporation and Concentration

A number of important developments are occurring in the industry in the field of evaporation technology. Alternatives to conventional, multi-stage evaporators are now becoming available, and include mechanical vapour recompression evaporators (MVR), submerged combustion techniques (Lovell-Smith, 1982), and membrane processing using reverse osmosis. Current evidence suggests that for large-scale applications, MVR and RO are closely competitive. For many applications, the most economic mix appears to be the use of RO to concentrate the whey to about 25 per cent solids, with further concentration by conventional methods. For small-scale operations in particular, the RO/evaporator combination appears to have economic advantages.

Dehydration

One problem in drying whey-based products with high lactose contents is caused by the hygroscopic nature of α-lactose. α-lactose, one of the two isomeric forms of lactose, is comparatively insoluble (7 g 100 ml^{-1} at 15 °C) compared with the other isomer, β-lactose (50 g 100 ml^{-1}). In aqueous equilibrium, a lactose solution contains about 63 per cent of the β form, due to mutarotation. On crystallisation, however, the more insoluble α form will initially precipitate, and some of the β-lactose will be converted to the α form; ultimately, crystallisation yields mostly alpha hydrate crystals in the product. Alpha-lactose exists as a monohydrate, whereas solid β-lactose contains no water of crystallisation. When solutions containing lactose are dried rapidly, there may be insufficient time for the α-lactose to form as the monohydrate, and the dried lactose is then, essentially, in the same form as the liquid. Neither α-lactose hydrate nor β-lactose are hygroscopic, but anhydrous α-lactose is strongly hygroscopic, and can absorb moisture from the air forming the hydrate, which occupies a greater volume than the anhydrous form. This action is responsible for the caking and lumping observed in many whey-based products.

Manufacturing procedures need to be modified to overcome this problem, and such processes involve the conversion of much of the lactose into crystalline α-lactose monohydrate before drying. This change may be effected either by holding the concentrate for sufficient time for crystallisation of the α-hydrate to occur, or by using techniques similar to instantising. Thus, if the surfaces of the particles emerging from the spray-drier are partially humidified, or the product is only partly dried, conversion of the α-lactose to the hydrated form can occur before final drying. Recent studies on the crystallisation process have been reported by Hynd (1980), and on the instantising process by Lewicki *et al.* (1981). In these latter studies, the best results were obtained by six treatments (30 s each) involving wetting of the product by air with a moisture content of 3·4 per cent, before final drying.

Storage Induced Changes in Powder

The major changes occurring during storage of whey-based products involve changes in the physical state of the lactose, and browning reactions. The rate of moisture uptake by whey-based products depends on a number of variables, particularly protein content, relative humidity, temperature and pH (Johns, 1982). Other studies (Saltmarch, 1980) have shown that moisture sorption isotherms for hygroscopic (but not non-hygroscopic) dried wheys stored at 25, 35 and 45 °C show discontinuities between water activities of 0·33 and 0·44, coinciding with the initiation of the transition from the amorphous to the crystalline form of α-lactose. Lactose crystallised at water activity 0·40 after about one week at 25 °C, and at water activity 0·33 after 1 week at 35 or 45 °C (Saltmarch and Labuza, 1980). The maximum rate of browning and loss of protein quality occurred at water activity 0·44. Loss of protein quality at constant water activity, but under fluctuating storage temperatures, was greater than at constant temperature (Labuza and Saltmarch, 1982). This has particular implications for countries exporting whey powders to tropical areas.

UNIT PROCESSES AND PRODUCTS

Demineralisation

The two main methods used commercially for the demineralisation of whey — ion exchange and electrodialysis — result in products of somewhat different composition. Ion exchange is relatively non-selective and removes both monovalent and polyvalent ions, whereas electrodialysis is more dependent on ionic mobility and tends preferentially to remove monovalent ions. The relative economics of each process have been examined by Marshall (1979). Economically, only about 90 per cent demineralisation is possible with electrodialysis, with 50 per cent being much more viable. By comparison, ion exchange processes can readily achieve close to 100 per cent demineralisation. The generally lower capital cost of ion exchange gives it a potential advantage in small systems, particularly where high degrees of demineralisation are required. Electrodialysis is likely to have advantages for plants with high hourly utilisation rates, for the manufacture of products which do not require high levels of demineralisation, and where low-cost electricity is available. Both systems produce substantial quantities of effluent, generally at least equal in volume to the quantity of whey processed.

The influence of demineralisation on overall whey processing costs is considerable, and may be comparable with the cost of spray-drying per tonne of solids. Such a high cost factor means that demineralisation is only employed where a high return for the product is guaranteed. Furthermore, the need to handle large quantities of difficult effluents from these processes has restricted their acceptance by industry.

A novel approach to demineralisation by ion exchange is the SMR process developed by the Central Laboratory of the Swedish Dairies Association (SMR) in conjunction with Wedholms AB. The cations in whey are exchanged for annonium ions, and the anions exchanged for bicarbonate ions. The resultant ammonium bicarbonate is then evaporated from the whey in the form of ammonia, carbon dioxide and water, and is recovered for regeneration of the ion exchange resins. High degrees of demineralisation are possible, and the product retains less than 0·25 per cent free ammonia. More than 80 per cent of the ammonium carbonate is recovered for use as regenerant, facilitating effluent treatment, and furthermore, only one solution needs to be passed through the two ion exchange systems. Total costs are estimated to be substantially less than for conventional ion exchange systems, and the mineral content of the effluent from the process is approximately half that of conventional systems (Jonsson and Forsman, 1978). A full-scale plant, with a capacity of more than 4 tonnes day^{-1}, has been installed in Sweden.

Manczak (1979) has indicated that improvements in the efficiency of demineralisation by electrodialysis may be achieved by processing at 50 °C rather than at 20 °C; less electrical energy is required. Replacement of the anionic membranes with neutral membranes resulted in a threefold increase in the electrical energy required for operation, and a less effective system for mineral removal. Nevertheless, this approach may be justified, in some cases, as anionic membranes are very expensive and wear out rapidly.

An alternative approach to improve the efficiency of electrodialysis has been reported by Ennis and Higgins (1981). In these studies, the calcium in permeate was replaced with sodium, by ion exchange, prior to electrodialysis. Permeate conductivity, the current carrying capacity of the plant, and the rate of ash removal all increased as the extent of replacement increased. In permeate where nominally 100 per cent of the calcium had been replaced, there was an initial rapid decline in ash removal rate, after which no further decline occurred. In the control samples, the ash removal rate followed an exponential decay pattern.

Two alternative developments in demineralisation technology have been described recently. 'Sirotherm', thermally regenerable ion exchange resins, may have potential in this field. These resins may be regenerated by treatment with warm water, rather than through the use of chemical reagents (Jackson, 1980). However, at present, some resins in the Sirotherm range are limited in their ability to process streams with the levels of calcium commonly found in whey, and have a low ion exchange capacity per kilogram of resin. Also of interest are the continuing developments in the field of 'open' RO membranes. These membranes are designed to allow a proportion of the ash content of whey to pass through into the permeate, whilst retaining most of the lactose and protein in the retentate. Recent reports (Wechsler, 1983) have indicated that progress has been made in this field, but lactose rejections are still a little too low. However, with the interest of membrane manufacturers throughout the world focussing on this problem, it is likely that such membranes will be commercially developed to a satisfactory performance level in the near future. In this event, it is very likely that open RO membranes for partial demineralisation, possibly in combination with ion exchange, may result in the ability to produce a wider range of demineralised products more economically.

Recently, Hayes (1982b) has described a novel technique for the demineralisation of permeate from the ultrafiltration of whey. The method involves the addition of calcium to the permeate to adjust the Ca/P ratio in such a way as to precipitate calcium phosphate in the apatite form. The precipitate can be removed by mild centrifugation, effectively reducing the calcium content of the permeate by more than 60 per cent, and the phosphorus content by more than 50 per cent. A similar approach has been described by Brothersen *et al.* (1982). This approach is simple and inexpensive, and offers advantages where only partial demineralisation is required, or as a preliminary treatment before conventional processing. The recovered calcium may have pharmaceutical applications.

Whey Protein Recovery and Modification

Lactalbumin and other protein fractions

Traditionally, lactalbumin is made by heat precipitation of proteins from whey at an acidic pH (Robinson *et al.,* 1976). The product is comparatively insoluble and, therefore, limited in its functional properties. However, as an ingredient of high nutritional quality, it still commands an important sector of the market place. Recent modifications suggested for the manufacture of lactalbumin include cation exchange of part of the whey, employing a novel approach to reduce the cost of pH adjustment prior to heat treatment. A portion of the whey is mixed with a cation exchange resin, resulting both in partial demineralisation and a drop in the pH of this fraction. This portion of the whey is then mixed with the remainder of the whey to obtain the desired pH for lactalbumin production by heat treatment. Process improvements are also claimed by coupling the traditional process with ultrafiltration to improve efficiency and yield (Buhler *et al.,* 1981).

Recently, two techniques have been described for the manufacture of whey fractions rich in α-lactalbumin and β-lactoglobulin (Amundson *et al.,* 1982; Pearce, 1983). Essentially the processes rely on modifying the pH, ionic composition and temperature of the whey to selectively precipitate one of the proteins. These processes may have important implications in the development of products from whey for pharmaceutical applications.

Gel filtration; co-precipitation with soluble polymers or phosphates

Each of these processes is capable of producing an essentially undenatured, whey protein fraction, although some end-products, for example those derived using soluble carboxymethyl cellulose, iron complexing and phosphate precipitation, would also contain some precipitant. It appears that there would be considerable difficulty in scaling-up many of these processes from laboratory and pilot-scale applications to commercial reality, and in spite of the high functionality of many of these fractions, markets for the products have not yet been developed. It is unlikely, therefore, that these techniques will be of commercial value in the near future.

However, there has been considerable research interest over the past decade in the development of these techniques for the manufacture of protein concentrates (Marshall, 1979). Recently, Humbert and Alais (1982) have reported on the effectiveness of some of these processes. Using carboxymethyl cellulose as the precipitant, protein recovery was

69 per cent, with the product containing about 60 per cent protein and 1·7 per cent ash; with ferric chloride, the corresponding values were 89 per cent, 79 per cent and 18 per cent; with ferric polyphosphate, the values were 92 per cent, 70 per cent and 28 per cent; for polyacrylic acid, the values were 63 per cent, 82 per cent and 13 per cent; and with sodium hexametaphosphate, the values were 76 per cent, 84 per cent and 13 per cent. All of the protein concentrates had excellent functional properties, and were considered to have considerable potential as food ingredients.

Adsorption onto insoluble supports — the 'Spherosil' and 'Bi-pro' processes

Recently, Rhone-Poulenc Chemie Fine have announced the development of a novel system for whey protein recovery, based on passing whey through a column containing porous, silica microbeads which specifically absorb protein (Mirabel, 1978). The protein is recovered by washing the column with a solution which has a slightly different pH to that of the whey. The process offers substantial advantages in protein recovery from whey, including production of a pure protein product with high functionality, low operational costs, and a protein-free waste material, similar in composition to UF permeate. The product has superior functional properties to 80 per cent whey protein concentrates prepared by UF, reflecting, perhaps, both the higher protein content of the Spherosil product, and the effect of energy input during ultrafiltration. Typically, whey at pH 6·5 is passed initially through a column of Spherosil QMA anion exchange resin, and then through Spherosil XOB cation exchange resin. The columns are washed with water, and eluted with 0·1M hydrochloric acid and ammonium hydroxide respectively. The columns may be reused after further washing. Spherosil QMA has an adsorption capacity of about 100 mg protein g^{-1}, and adsorbs mostly β-lactoglobulin, α-lactalbumin, serum albumin and some immunoglobulins, whereas the Spherosil XOB adsorbs mainly immunoglobulins. The recovered protein fractions are undenatured, and contain less than 3–4 per cent ash. At least two large commercial enterprises in France are now employing this process for the manufacture of high quality protein fractions from whey.

An alternative approach has been developed by Bio-Isolates, who have used an ion exchange process based on carboxymethyl cellulose for the manufacture of a 97 per cent protein powder (Bi-pro Dairy Albumin) from whey. The product is virtually pure protein, substantially free of fat and lactose, and with a low ash content. It has a digestibility of 99 per cent, a biological value of 94 per cent, net protein utilisation of 93 per cent and a protein efficiency ratio of 3·2. The functional properties of the product are also excellent, being similar to egg white in many respects. Current production is 30–40 tonnes per year, with a larger scale plant being planned (Hannigan, 1982).

Processes for the modification of whey proteins

Currently, there appears to be little commercial interest in the modification of whey proteins by enzymatic means, but this situation may well change in the future. Zadow (1979) has reviewed the status of existing techniques for modifying whey proteins. There has been recent interest in improving the solubility of lactalbumin by treatment with protease, so as to extend its applications (O'Keeffe and Kelly, 1981), and in chemically modifying the functionality of whey proteins by processes such as succinylation to improve solubility (Thompson and Reyes, 1980). Succinylation of heat-coagulated, cheese whey protein concentrate resulted in improved functionality, allowing it to absorb more than eleven times its weight of water. Possible applications for this product include baked goods and dairy analogues where high water and fat absorption are required.

Membrane Processes: Ultrafiltration and Reverse Osmosis

Processes

The application of membrane processes to the dairy industry has been the subject of several reviews, e.g. Matthews (1979). Over the past decade, there have been continuing developments and improvements in membrane composition and equipment design to increase flux and membrane life, and to reduce cost and improve ease of cleaning and sanitation. Membrane systems now include tubular, plate and frame and spiral wrap designs, and these developments have made membrane processing more economical for the industry. Although ultrafiltration has been most widely applied, it now appears that RO is coming into 'its own' as a means of concentration.

Of particular importance to the industry are the sanitation problems posed by membrane processing, particularly with the advent of the spiral wrap system. Beaton (1979) has discussed the hygienic design and operation of membrane systems, particularly considering cleanability and sanitation. Sanitation problems in UF systems have been virtually overcome, but some difficulties remain in ensuring sanitary operations of RO plants, particularly in milk processing.

An important factor in achieving commercial acceptance of UF for use on whey was the development of techniques which reduced flux decline during processing. Muller *et al.* (1973) and Hayes *et al.* (1974) showed that, by a combination of pH adjustment and heat treatment before processing, the extent of flux decline could be substantially reduced. The effect of such pretreatments on the processing characteristics of whey was reviewed by Muller and Harper (1979), and on whey processing in general by Zadow (1982).

The physical and chemical characteristics of the various types of whey markedly influence their performance during UF on different plants. Whey treatments, which have resulted in improved performance, include clarification, centrifugation (sometimes preceded by calcium addition), heating under conditions determined by the type of whey and pH, demineralisation, pH control and concentration; for RO, only demineralisation and pH adjustment seem to be effective. The factors controlling the performance of whey in membrane processing are, however, not well understood.

Recent studies have focussed on the components in whey involved in membrane fouling, and hence flux decline. Hickey and Hill (1980) have reported that the κ-casein macro-peptide present in Cheddar cheese whey contributes to fouling during ultrafiltration. Furthermore, the formation of calcium phosphate gels reduced permeation rates when circumstances favoured formation of the gel in membrane pores. The calcium ion concentration and the ionic strength of the whey directly influence permeation rates. Patel and Merson (1978) have examined the fouling characteristics of Cottage cheese whey fractions. Their results suggested that flux decline may be attributed to membrane compaction and fouling by proteinaceous materials which accumulate on the membrane surface; other studies have emphasised the role of whey salts in membrane fouling. Cheryan and Merin (1979) reported from model system studies that membranes may be negatively charged due to salt binding, and such adsorbed salts could act as bridges between the membranes and protein, thus increasing the fouling layer. It has been reported that salts have a detrimental effect on the rate of flux decline (Cheryan and Merin, 1980). Alpha-lactalbumin has the strongest gel forming tendencies, and β-lactoglobulin the worst long-term fouling effect. Whey showed greater fouling tendencies than did the model systems examined, due to the presence of both salts and protein. Removal of the salts appears, therefore, to be beneficial in purely practical terms, but the cost/benefit ratio may not be satisfactory overall.

It should be emphasised that whilst flux decline is still a major problem to the industry, existing methods of pretreating whey are adequate to make the process commercially viable, although further improvements are highly desirable.

The problem of membrane fouling is of less significance in RO processing of whey than in UF. Cheddar cheese whey causes much less fouling on RO than does casein whey, as also is the case with UF. However, the pretreatments described by Hayes *et al.* (1974) for fouling reduction in UF lead to an increase in fouling with RO. This may be caused by different ionic compositions of the membrane surface layer (Smith and MacBean, 1978).

Membrane fouling is not only dependent on the nature of the fluid being processed, it is also dependent on the membrane material itself. Currently, virtually all UF membranes are synthetic materials, mostly based on polysulphone. RO membranes have, until recently, been based on cellulose acetate, but now a range of membranes based on synthetic materials is becoming available, with resultant advantages to the dairy processor. A major problem with cellulose acetate membranes is that they are limited in their range of operation to temperatures below about 30 °C, and a pH range close to neutral. The synthetic membranes are more resistant to extremes of pH and heat, and cleaning processes may be more severe.

Recently, a new type of membrane has been reported (Maubois *et al.*, 1981) formed from inorganic, zirconium oxide supported on a graphite base. Unlike the best UF membranes, based on polysulphone and limited to a maximum operating temperature of about 75 °C within a pH range from 2 to 12, these zirconium membranes are resistant to temperatures up to 400 °C over the full pH range. Such membranes offer advantages in terms of sanitation, although there is not yet much information concerning their use in commercial operation.

There is little information available on the effect of membrane processing on protein structure. It is well recognised, however, that ultrafiltration of milk may result in significant changes in product functionality if the energy input during ultrafiltration is excessive; it is likely that this will also occur in UF of whey. Dziuba and Chojnowski (1982) have reported that the UF of whey, to give a 10–20-fold increase in protein concentration, results in a small increase in protein solubility. Changes in the UV spectra indicated an increase in hydrophobic groups, but this did not result in the association of whey proteins at the pH of the concentrates (5·8–6·0).

Of particular importance in ultrafiltration processes is the quality of the water used during diafiltration and washing, and very low levels of silica and/or iron will result in severe and permanent loss in flux. These materials appear to be irreversibly adsorbed onto the surface of the membranes. In most parts of the world, it is essential, therefore, to provide either facilities for the deionisation of wash and diafiltration water, or a supply of RO permeate. Care should also be taken in the selection of cleaning reagents employed in UF plants to ensure that they have very low levels of iron and silica. Problems due to the contamination of RO membranes by iron and silica are much less severe. However, most RO membranes are extremely sensitive to low levels of chlorine, and action must be taken to ensure that wash waters used in RO operations are chlorine-free.

Functionality

Whey protein concentrates (WPC) may be prepared with a very wide range of functional properties. It is this characteristic that can increase the attractiveness of WPC as an ingredient, as functionality may be tailor-made to specific requirements. Functionality is affected by a number of factors, such as source of whey, protein, mineral and lipid composition of the WPC, heat treatment given to the whey or retentate during manufacture, and the pH of the system in which the WPC is employed.

Although there has been much work done to develop an understanding of the relationship between compositional characteristics, manufacturing variables and the functional properties of WPC, it has generally been unsuccessful, even in model systems. However, as more detailed information becomes available from model food system studies, a greater understanding of the influence of molecular structure and interaction on the functional properties of WPC should be achieved. This area is particularly complex, and such an understanding is unlikely to be reached in the immediate future.

One complication is the lack of standard methods for the testing of even such straight-forward functional properties as gelation strength, emulsification capacity and foaming characteristics. This lack of standardisation is partly the result of the need to assess such properties in widely differing food systems. Standardisation of functionality tests is being considered currently by the International Dairy Federation.

Recent reviews covering the functionality of whey proteins include those of Morr (1979), Kamiya and Kaminogawa (1980) and Craig (1979). In detailed studies, Johns and Ennis (1981) have reported that 33, 67 or

100 per cent replacement of calcium ions with sodium ions in acid-casein whey, before UF treatment, substantially altered the functional properties of the resultant WPC. Protein solubility increased (especially at pH 4·9), and gel texture parameters improved when at least 67 per cent of the calcium was replaced, with gelation time being directly related to the level of replacement; whipping properties decreased with increasing replacement of calcium ions. Such modified WPC were, therefore, more suitable for applications requiring good gelling characteristics rather than for those requiring good whipping properties.

Dunkerley and Zadow (1981) have reported that the gelation strength of WPC is decreased by preheating the whey, and, in the main, by demineralisation. Gel strengths comparable to those of egg white were obtained under some circumstances. Other studies have compared the effects of calcium, magnesium and sodium ions on the heat aggregation of WPC, and in general, it has been found that aggregation was increased by addition of these salts, with calcium having the greatest effect. Denaturation temperatures of the proteins decreased on addition of calcium or magnesium, but increased in the presence of sodium (Varunsatian *et al.,* 1983).

Sulphydryl interactions are also involved in the formation of heat gels from whey proteins, and gel formation involves polypeptide chains cross-linked by disulphide bonds. The gels dissolve if a sulphydryl agent is added after heating, and their formation is retarded if compounds that react with sulphydryl groups are added before heating. Gels form more slowly at alkaline pH, perhaps because of the increased electrostatic repulsion between proteins (Hillier *et al.,* 1980). Proteose peptones do not appear to be involved in gelation characteristics (Hillier and Cheeseman, 1979).

Current studies support, therefore, the concept that gelation is controlled, to a large extent, by the mineral composition of the product and by sulphydryl interactions. Studies on these variables using multiple regression and response surface analysis have been reported by Schmidt *et al.* (1979).

Reports on factors affecting the emulsifying properties of WPC are conflicting, and often reflect the differing systems used for examination of this characteristic. In general, conditions affecting emulsifying properties are similar to those controlling foaming and whipping, and are affected by the system used for emulsification, the energy input, the ionic strength of the environment and the type of WPC employed. Functionality of proteins in foams depends, theoretically, on their ability to reduce interfacial tension, increase liquid phase viscosity and form strong films. These changes may be facilitated by alterations to the tertiary and quaternary structure of the proteins, the extent of which will depend on conditions in the system. Foaming performance of whey protein products has been found to be improved by the addition of calcium ions, hydrogen peroxide and hexametaphosphate, and impaired by the addition of sucrose, lipids and surfactants; interaction effects have also been observed. Reversible improvement of foaming by heat treatment seems to be related to protein/lipid interactions (Richert, 1979).

Many studies on functionality have shown that only minor changes in techniques of manufacture of WPC may have a very considerable impact on functionality. There are very significant functional differences, for example, in WPC prepared from whey preheated at 72 °C for 15 s, or 85 °C for 15 s. The degree of difference in denaturation indices in such WPC is small, indicating that this parameter is of little value in predicting the impact of processing on functionality.

Because of the sensitivity of functionality to minor changes in processing, there appears to be a strong case for functionality studies to use only samples prepared under strictly controlled conditions. As a minimum, it is essential that data on whey source, composition and processing history be included in all reports on WPC functionality.

Fermentation

Fermentation of whey and whey by-products may be used for the production of a wide range of products. A number of such processes are in commercial operation, with, for example, large-scale equipment for the manufacture of alcohol being installed in Eire and New Zealand. A company based in Wisconsin, USA, has been manufacturing single-cell protein for some years, and more recently has commenced production of potable alcohol. Friend and Shahani (1979) have reviewed the fermentation of whey and its constituents, with emphasis on factors influencing the production of alcohol, lactic acid, citric acid and single-cell protein.

Single-cell protein

France manufactures more than 600 tonnes per annum of single-cell protein for animal and food use. In one French plant, the process is based on fermentation of deproteinised whey with strains of *Kluyveromyces lactis* and *Kluyveromyces fragilis,* followed by centrifugation of the liquor to produce a 10–11 per cent concentrate, which is then further concentrated by filtration to 18–22 per cent solids, heated and dried. The yield corresponds to about 53 per cent of the weight of lactose in the initial whey, and the product contains about 50 per cent protein, 30 per cent carbohydrate, 6 per cent lipid, 8 per cent mineral, and moisture content of about 4·5 per cent (La Gueriviere, 1981).

Permeate from ultrafiltration of whey is a useful substrate for manufacture of single-cell protein. It has been reported that, of 10 strains examined, *Kluyveromyces fragilis* ATCC8582 was the most productive over a wide range of pH and lactose concentrations. The most economical permeate concentration from batch fermentation was 5 per cent (Gawel and Kosikowski, 1978). Mahmoud and Kosikowski (1978) examined the use of concentrated whey permeate for single-cell protein and alcohol manufacture. It was found that fermentation of a concentrated permeate (27 per cent solids) with five adapted strains of lactose fermenting *Kluyveromyces sp.* resulted in single-cell protein yields of 2–8 g dry matter per litre, with about 12 per cent ethanol produced from the process. The use of a two-stage fermentation with the initial stage using *Kluyveromyces fragilis* and the second *Propionibacterium sp.* gave a higher yield than did the *Propionibacterium sp.* alone (Giec *et al.,* 1978). In recent studies, Mahmoud and Kosikowski (1982) have reported that *Kluyveromyces fragilis* NRRL Y 2415 produced the highest yield of alcohol (9·1 per cent from concentrated whey permeate) and *Kluyveromyces bulgaricus* ATCC1605 gave the highest yield of biomass (13·5 mg ml^{-1}). High ash contents inhibited both biomass and alcohol production.

The use of hydrolysed whey as a substrate has also been examined, as with this material, non-lactose fermenting organisms, such as *Saccharomyces cerevisiae,* could be used for alcohol production. The metabolism of *Kluyveromyces fragilis* was slower with hydrolysed whey than with unhydrolysed whey, but alcohol yields were improved on fermentation with a high alcohol yielding strain of *Saccharomyces cerevisiae;* the galactose present was not utilised.

A review of the production of biomass from whey has been prepared by Meyrath and Bayer (1979).

Alcohol

The Carbery process of Express Dairies is the best known commercial operation for the production of alcohol from whey streams (Sandbach, 1981a,b). In the batch process, whey permeate (about 4·5 per cent lactose) is fed from a balance tank into either a 110 000 litre fermenter or a 26 000 litre yeast propagation tank. Six fermenters are used at Carbery, Eire, each approximately 15 m in diameter with a conical base. After 20 h fermentation, the yeast cream is recovered by centrifugation, and the clear liquid, containing about 2·8 per cent alcohol, is passed to a balance tank for distillation. The resultant ethanol (96·5 per cent by volume) is suitable for gin or vodka production. Particular care must be taken during production to avoid yeast and mould contamination from nearby cheese plants. The Carbery plant processes about 600 000 litres of whey per day yielding about 22 000 litres of alcohol. Efficiency of conversion is about 86 per cent. Waste products from fermentation and distillation have necessitated the installation of a purification plant for the daily treatment of about 900 000 litres of waste, with a BOD of 4 000 kg. During winter, when supplies of permeate are low, molasses may be used as a make up for the substrate. If it is assumed that the permeate has no value (alternative whey treatments may, in fact, often operate at a loss), the estimated annual return is about 28 per cent on the five million pound (UK) investment. Experiments have shown that concentration of the permeate to 10 per cent solids has little effect on process efficiency.

Plants using this system are considered economic provided that more than 500 000 litres of whey per day are available for processing. In Eire, potable alcohol is the most profitable outlet, but in other countries, anhydrous alcohol for industrial or power use may also be attractive. For example, in New Zealand there is little indigenous alcohol production, and until recently it was necessary for that country to import the majority of its alcohol. Under these circumstances, local manufacture of alcohol from permeate is likely to be attractive, and recent years have seen the installation of a number of such plants in New Zealand. Alcohol also has a potential use in 'gasohol', and aspects of utilising permeate in this product are described by Lyons and Cunningham (1980).

Some recent research interest in fermentation technology has centred on means of immobilising the organisms involved so as to develop continuous systems. *Kluyveromyces fragilis* immobilised in polyacrylamide gel has been used to treat whey permeate for alcohol production, with a column half-life of 50 days being achieved. With a column volume of 0·04 litres and a flow rate of 0·34 column volumes per hour, 80 per cent of the lactose was utilised and ethanol production was 45 mM h^{-1}. Ethanol production costs were estimated at NZ$0·59 per litre (Bartosh and Manderson, 1982). Other approaches have employed the immobilisation of cells, with and without β-galactosidase, and again, considerable success has been achieved using this approach (Linko *et al.,* 1981; Hartmeier, 1982). Developments of such systems to commercially viable processes may offer worthwhile economic gains to the industry.

Fuel gas

There is considerable potential for the use of anaerobic fermentation of the permeate for the production of fuel gas. Recent studies (Lang, 1980) have indicated that, on average, biogas production by the anaerobic fermentation of 1 litre of cheese whey at 55 °C yields 36·48 litres of biogas, with about 50 per cent methane and an energy content of 960 kJ litre^{-1}. A study on the application of biogas to a dairy complex of 284 farms and one central processing plant, in the USA, estimated that 75 per cent of the cheese plant's liquid fuel and propane requirements could be directly supplied by utilisation of whey permeate for fuel gas manufacture. In practice, however, direct applications of whey fermentation for biogas appear to be rare. Given the increasing cost of energy and the problems of whey disposal, utilisation of whey through such systems is likely to become more attractive in the near future. Research needs to place emphasis on the development of systems that are economic for small-scale manufacturers.

Ammonium lactate/lactic acid

Techniques for the fermentation of whey to lactic acid have been well known for some years, and there appear to have been few recent developments. However, there has been recent interest in the production of ammonium lactate by continuous fermentation of deproteinised whey to lactic acid, using ammonia to maintain constant pH. In studies on this process using a strain of *Lactobacillus bulgaricus* at pH 5·5 and 44 °C, lactate concentration was a maximum (58·7 g litre^{-1}) at a cell retention time of 27 h, and cell mass was at a maximum (1·9 g litre^{-1}) at 9·3 h (Stieber and Gerhardt, 1979). A mathematical model for the process has been developed, simulating the effect of cell retention time on substrate, product and cell mass concentration in a single-stage system.

Oil production

The production of oil by the fermentation of whey and permeate using *Candida curvata* and *Trichosporon cutaneum* has been examined by

Moon and Hammond (1978). Fermentation for 72 h reduced the COD of permeate by 95 per cent, and produced 4·0–5·6 g $litre^{-1}$ of oil, 19·6–26·8 g $litre^{-1}$ biomass and 2·2–2·3 g $litre^{-1}$ protein. *Candida curvata* was more efficient in producing oil and reducing COD. The oil contained approximately 50 per cent oleic, 30 per cent palmitic, 15 per cent stearic and 8 per cent linoleic acids. Extraction of the oil was best done by sequential treatment with methanol and benzene, then ethanol and hexane. The oil mass was a useful base for animal feed.

Citric acid production

Much of the world's supply of citric acid comes from fermentation of molasses, a commodity which may be variable in composition, and is often limited in supply. As there are few manufacturers of citric acid in the world, most countries import their requirements, but for some dairy-developed countries, there may be potential in employing local supplies of permeate as a feedstock for manufacture of this product. Australian research (Tonks and Hawley, personal communication, 1982) has indicated that permeate has potential in this area, although difficulties may be encountered using permeates with a high ash content. Other studies (Somkuti and Bencivengo, 1981) have reported that acid whey is a suitable substrate for production of citric acid by fermentation with *Aspergillus niger*. Optimal production of citric acid was reached after 8–12 days at 30 °C, with a maximum yield of 10 g $litre^{-1}$ being obtained with a permeate solids content of 15 per cent; throughout the fermentation, galactose was apparently co-metabolised with glucose. Aspects of the process have also been discussed by Hossain *et al.*, (1983).

Butanol

Production of the important solvent, n-butanol, by fermentation of permeate has been described by Maddox (1980). Whilst the traditional fermentation of corn mash results in a mixed end-product with a ratio of butanol: acetone: ethanol of 6:3:1, fermentation of permeate results in a product with the favourable ratio of 10:1:1, and a butanol yield of 1·3 per cent. Commercial development of this process is being considered in Europe.

Polysaccharides

The production of extracellular, microbial polysaccharides by the fermentation of whey, or hydrolysed whey, has also been reported. Test organisms were selected from strains known to produce gum when using glucose as a substrate, such as *Alcaligenes, Xanthomonas,*

Arthrobacter and *Zooglea spp.* All the organisms studied produced polymers from hydrolysed whey or glucose/galactose, but only *Alcaligenes viscosus* and *Zooglea ramigera* produced a gum from whey under the conditions employed (Stauffer and Leeder, 1978).

Other products

Whey-based products have been suggested as substrates for production of a number of other products, including 2, 3-butylene glycol, lysine, threonine, β-galactosidase, vitamin B_{12} and β-carotene, and for the growth of mushroom mycelium.

General

Of the various processes and products described above, only those for the manufacture of single-cell protein/biomass, alcohol and fuel oil appear to be economically viable at present, and of the remainder, only the use of permeate for citric acid production is likely to be of commercial importance in the near future. Many of the remaining processes have potential, however, and, with further development, may become industrial realities.

Lactose hydrolysis

In the USA, it appears unlikely that a market for hydrolysed products will develop to any considerable extent. The competition from corn syrup sweeteners, an established and highly competitive product, is likely to make the economics of lactose hydrolysis unattractive. In parts of the world such as Australia, however, which have high local sugar prices and no major corn syrup industry, lactose hydrolysis may have potential. In countries such as New Zealand, with no indigenous sugar industry, the potential could be even greater.

Lactose hydrolysis may be carried out either by treatment with acid at high temperatures, or through the use of β-galactosidase. The enzymes used for the latter process are extracted from yeasts, such as *Kluyveromyces lactis* (optimum pH of operation 6–7, optimum temperature about 35 °C), or *Kluyveromyces fragilis* (optimum pH about 6·5, optimum temperature about 40 °C), or from fungi such as *Aspergillus niger* (optimum pH about 4·8, optimum temperature about 50 °C). More recently, the development of a new lactase from a *Bacillus sp.* has been announced with a temperature optimum of 65 °C and a pH optimum of 6–8. Comparative studies on catalytic and enzymatic methods have indicated that catalysis produces a higher quality product of better keeping quality than does enzymatic hydrolysis. However, such results should be interpreted with care, as the quality of enzymes available for hydrolysis varies widely, and this may significantly affect the quality and storage characteristics of products produced by enzymatic processes. Aspects of lactose hydrolysis have been discussed by Zadow (1984).

Acid hydrolysis/cation exchange

This method is only viable for treatment of protein-free streams, and 80 per cent hydrolysis may be readily obtained. Adjustment of the pH may be made by direct addition of acid to the system, or by treatment of the permeate with an ion exchange resin in the cation form. Typically, the pH of the permeate is adjusted to 1·2 (perhaps by cation exchange), and the product heated at temperatures up to 150 °C for hydrolysis. During hydrolysis, a brown colour is formed, the extent of which is dependent on the non-protein nitrogen content of the sample. The product may be neutralised by addition of alkali, or by treatment with an anion exchange resin; a final clarification with charcoal may be employed, if desired. A number of modifications to the basic process have been suggested, particularly involving a combination of the steps of pH adjustment and hydrolysis. In this procedure, the permeate is passed directly onto the ion exchange resin at temperatures generally slightly below 100 °C, so that both processes proceed virtually simultaneously (MacBean *et al.,* 1979; Haggett, 1976; Demaimay *et al.,* 1978).

The costs of regeneration are significant in the economics of this process. As yet, there is apparently no commercial application of this procedure, although, in terms of simplicity and capital cost, it appears attractive.

Enzymatic hydrolysis

Three techniques may be employed for the enzymatic hydrolysis of lactose streams. These are the use of 'single-use' or 'throw-away' systems, lactase recovery systems (generally based on membrane recovery of β-galactosidase to allow its re-use) and immobilised enzymes. The relative costs of each of these processes are changing rapidly with developments in technology. The cost of single-use enzymes per tonne of hydrolysed lactose has dropped considerably in the past few years, with similar drops in the cost of immobilised enzyme systems. The cost of soluble enzymes is closely related to their purity, but the use of some of the cheaper soluble enzymes can result in the development of off-flavours in the product, probably from the activity of protease impurities.

Single-use enzymes

There is comparatively little detailed information on single-use enzymes for the hydrolysis of whey. Optimal conditions vary widely depending on the source of the enzyme and its purity. In selection of the enzyme, factors to be considered include the pH of the substrate, maximum temperature of hydrolysis possible, enzyme activity and cost. If economic concentrations of single-use enzymes are used, then several hours may be required for hydrolysis, and there are microbiological benefits in the use of enzymes which operate efficiently at temperatures where microbial growth is retarded.

Systems using ultrafiltration

Ultrafiltration equipment can be used in three ways in conjunction with enzymes for lactose hydrolysis. In the simplest approach, hydrolysis of protein-free whey is carried out in a batch process, and ultrafiltration is applied to recover the enzyme for further use (Norman *et al.,* 1978). A second approach involves immobilising the enzyme adjacent to the ultrafiltration membrane by a pressure-induced, flow regime (Maculan *et al.,* 1978). It has also been suggested by ultrafiltration equipment manufacturers that this equipment can be used in the feed/bleed mode as an enzyme reactor. In studies of this approach (Hayes, personal communication, 1982), a high concentration of enzyme was added to permeate being recycled in an ultrafiltration plant. When the desired level of hydrolysis was reached, the system was operated in the feed/bleed mode at the level of 75 per cent hydrolysis; continuous operation at 55 °C and pH 4·6 for 30 h resulted in only a small loss of enzyme activity.

These approaches can be used only for the treatment of protein-free, whey streams. With whole whey, the process involves initial ultrafiltration of the whey, treatment of the permeate by one of the techniques, and blending of the hydrolysed permeate with the UF retentate. The complexity of such a system can make it unattractive for commercial applications.

Immobilised systems

The use of immobilised enzyme systems for lactose hydrolysis appears to have the greatest potential for wide-scale application for the treatment of whey or permeate. A number of supports for the immobilisation have been suggested, including silica, ion exchange resins, Duolite, cellulose derivatives, phenol–formaldehyde, titanium dioxide, aluminium oxide, chitosan, diazotised glass, silica gel and silochrome. The most frequently used enzyme in immobilisation studies is that obtained from *Aspergillus niger*. Techniques employed for the utilisation of the immobilised systems involve both fixed-bed and partially, or wholly, fluidised-bed systems.

Pastore and Park (1981) have described immobilisation of lactase from *Scopulariopsis sp.* on Duolite. The enzyme did not lose activity after 60 h operation at 55 °C, but lost 45 per cent after 52 h at 60 °C. After 1284 h of suitable operation, the enzyme retained 94 per cent of its activity overall. Marconi *et al.,* (1980) employed a radial reactor for hydrolysis, comprising modular units in the form of bobbins prepared by winding fibres containing the immobilised lactase around perforated tubes. In a pilot plant containing 38 modules, 85–90 per cent hydrolysis was achieved in 130 litres of whey over 30 min. In a three month period, 182 000 litres were treated with little loss of activity.

A number of studies using immobilised enzymes in fluidised-bed reactors have reported reduction in activity with time, generally suggested to be caused by absorption of solid matter. In some cases, sonic cleaning of the enzyme was found to be beneficial; other aspects of sanitation and cleaning of immobilised systems are discussed later.

The Corning process is best known commercially for lactose hydrolysis using immobilised enzyme technology. The process employs lactase from *Aspergillus niger* covalently bound to a controlled-pore-size silica carrier. Its particle size is 0·4–0·8 mm, wet bulk density 0·6, activity about 500 U g^{-1} at 50 °C, optimal pH of operation between 3·2 and 4·3, and the estimated laboratory life of the resin is 2 years. A concentrated product, from a demineralised permeate (4 per cent solids), contained 61 per cent solids, 0·8 per cent ash, 12 per cent lactose, 23 per cent glucose and 25 per cent galactose. The process has been discussed in detail by Dohan *et al.,* (1980), and by Ennis (1984). Two plants are in commercial operation, one in the UK, and the other in the USA.

A novel approach has been directed by Griffiths and Muir (1980), who immobilised whole cells of *Bacillus stearothermophilus* by covalent linkage onto DEAE-cellulose treated with glutaraldehyde. There was no apparent loss of lactase activity as a result of immobilisation, and the half-life of enzyme activity was about 15 days at 60 °C and pH 7. Lactase activity increased with pH, and decreased with temperature and ionic strength. The bound cells were equally effective for hydrolysis of lactose in permeate, but were four times less efficient with skim-milk as the substrate.

Sanitation. Immobilised systems are designed for long-term operation, and sanitation is, therefore, a prime consideration. Three facets must be considered, these being maintenance of sanitary conditions during hydroloysis, development of adequate cleaning regimes on completion of hydrolysis, and maintenance of sanitary conditions whilst the plant is in storage between operations. The severity of sanitation and cleaning solutions which may be employed is restricted by the sensitivity of most immobilised enzyme systems to pH and temperature — the majority of common dairy detergents and cleaners cannot be safely employed. Current recommendations for sanitation include treatment of the resins with water, a range of acids, mild alkali and mild detergents with some bacteriological activity. Some resins are sensitive to chlorine, and care must be taken to avoid its contact with the resin. The chlorine may either inactivate the resin, or lead to problems with off-flavours in the product.

Oligosaccharide formation

The formation of oligosaccharides during lactose hydrolysis may have nutritional implications, as well as reducing monosaccharide yields. Galactose appears to be more involved in oligosaccharide formation than does glucose. Studies on lactose hydrolysis with a soluble lactase from *Saccharomyces lactis* have indicated that there is a linear rise in the formation of oligosaccharides with initial lactose concentration, up to 13 per cent by weight of total sugar, when 65–70 per cent of the lactose was hydrolysed to monosaccharides; of the six oligosaccharides identified, all were linear. It was concluded that the enzyme had a high transglycolation activity, with specificity for formation of β-(1–6) galactosidic bonds (Burvall *et al.*, 1979; Asp *et al.*, 1980). Other evidence suggests that the extent of oligosaccharide formation is much more important in systems employing soluble enzymes with long contact times, than in systems using immobilised enzymes with brief contact times. There appears to be little information available on the nutritional or toxicological characteristics of such oligosaccharides; there is a need for research in this area.

Determination of degree of hydrolysis

Cryoscopic methods have been recommended for determining the degree of lactose hydrolysis (Zarb and Hourigan, 1979). The method is based on the fact that the freezing point of the sample will be depressed by an amount depending on the molal degree of hydrolysis. Hayes (1982a) compared polarimetric and cryoscopic methods for the assessment of degree of hydrolysis, and between lactose concentrations of 1·15 and 10·29 per cent, the correlation coefficients for each method were better than 0·99; however, where a high level of precision was required, the polarimetric method was preferred. Chen *et al.* (1981) have reported that oligosaccharides formed during hydrolysis, and the presence of proteinases in the lactase may cause inconsistencies in results. Other techniques for determination of the degree of hydrolysis have been based on high performance liquid chromatography, chemiluminescence and spectrophotometry.

Sweetness of hydrolysed syrups

Sweetness and solubility data for sucrose, lactose, glucose and galactose are shown in Table IV. Studies on the relative sweetness of hydrolysed syrups have been reported by Shah and Nickerson (1978), using a model system based on unflavoured ice cream mix. The simulated hydrolysed syrups were evaluated at 25 and 50 per cent levels of replacement of sucrose in the control mix, with the results showing a synergistic effect of lactose. For example, less sucrose was needed for equi-sweetness when 25 per cent of the lactose was replaced with a glucose/galactose syrup (70 per cent hydrolysed) than when it was replaced with syrup showing total hydrolysis (100 per cent). Poutanen *et al.* (1978) have examined the conversion of glucose to fructose in glucose/galactose syrups, and found that the relative sweetness of the product was similar to that of sucrose. This approach may offer potential to the dairy industry, if the process is developed to commercial reality. Of even greater potential would be the development of a system which could

TABLE IV
Relative solubility and sweetness of saccharides

Saccharide	*Relative sweetness*	*Solubility (g/100 g solution)*		
		10°C	*30°C*	*50°C*
Sucrose	100	66	69	73
Lactose	16	13	20	30
Galactose	32	28	36	47
Glucose	74	40	54	70
Fructose	174	-	82	87

Data from Shah and Nickerson (1978) and Pazur (1970).

convert galactose, with its comparatively low sweetness, to a sweeter monosaccharide.

Drying of hydrolysed products

Drying lactose-hydrolysed products is extremely difficult using conventional spray-driers; in particular, spray-drying of hydrolysed low-protein or protein-free products is virtually impossible in conventional systems, because of the formation of sugar glasses on the surface of the spray-drier. However, hydrolysed products with protein contents similar to that of skim-milk may be dried, with adequate care. For drying low-protein hydrolysed products, the use of a Filtermat drier has been suggested. It is claimed that this unit is particularly suitable for such products, as the concurrent flow of product and air reduces the risk of deposits in the drier (Rheinlander, 1982).

Storage and properties of hydrolysed syrups

Storage of hydrolysed syrups poses a number of problems. In general, they cannot be spray-dried without the use of special equipment and, therefore, must be stored as concentrated liquids. However, problems may be encountered during storage, either due to the growth of micro-organisms, particularly osmophilic yeasts, or to crystallisation of lactose or galactose. Lactose will be crystallised if the degree of hydrolysis is low, and galactose will crystallise if the degree of hydrolysis and the total solids of the system is high. Recently, studies on factors affecting the storage characteristics of hydrolysed Cheddar cheese whey permeate syrups have been carried out in Australia (Hayes and Mitchell, 1983). It was found that both crystallisation and microbiological difficulties could be controlled by storage of the syrup at −10 to −20 °C, provided the total solids of the syrups were about 70 per cent. Lack of crystallisation at low storage temperatures was probably caused by high viscosity under these conditions. Storage of syrups at 50 °C to assist in prevention of bacterial growth resulted in severe browning of hydrolysed syrups in a few weeks.

In comparison with these results, Shah and Nickerson (1978) studied the crystallisation characteristics of simulated hydrolysed permeates (based on mixtures of lactose, glucose and galactose) containing 40–80 per cent solids. They reported that at 25 °C, crystallisation did not become a problem until the total solids exceeded 70 per cent. In view of the differing results reported above using permeate, it seems that it is not always possible to extrapolate from model systems to predict the behaviour of hydrolysed products.

Regulatory aspects

With the continuing development of a range of hydrolysed products in the dairy industry, regulatory aspects are being considered by a number of countries. French regulations provide for the use of enzymatically hydrolysed products in fine bakery goods, biscuits, desserts and confectionery. In Canada, hydrolysed products are authorised, with no restrictions on quantity; in the Netherlands, they are permitted in frozen confectionery but not yogurt; in Italy, hydrolysed market milk is permitted, and in Great Britain, a declaration is sufficient (Luquet, 1980). Doubtless, these regulations will be broadened as applications increase, and as further information on toxicology becomes available.

LACTOSE PRODUCTION

For many years, the production of lactose from whey typically involved protein removal (perhaps by lime treatment, or by heat and filtration), concentration of whey, refiltration, further concentration, induction of crystallisation and centrifugation to separate the crystals (Nickerson, 1974). In general, about 50 per cent of the lactose was recovered, and the mother liquor was sold as delactosed whey powder. Such processes have generally proved economic on a large scale, and there appears to be little potential in the market for substantial increases in lactose supply. The use of magnesium chloride to assist in the recovery of lactose has been suggested (Kwon and Nickerson, 1978). More recently, permeate has become a useful raw material for production of lactose, and aspects of its utilisation have been described by Madrid Vicente (1979). Other studies have shown improvements in yield of lactose by demineralisation of the permeate before crystallisation. Visser (1980) has recently reported that pharmaceutical grade lactose contains a substance with an acidic character which crystallises with the lactose and retards the rate of crystal growth. Ion exchange of the lactose removes this contaminant, yielding a non-ionic lactose which crystallises much more rapidly than does the pharmaceutical grade. It is likely that similar growth retardant materials will be present in whey and permeate, and contribute to the practical problems of lactose crystallisation. The value of demineralisation may be in the removal of such compounds.

Of the two isomers of lactose, β-lactose is slightly sweeter and more soluble. However, lactose produced by conventional processes contains mainly α-lactose. There is only a small demand for β-lactose as such, despite higher solubility and greater sweetness. Manufacture of this isomer involves crystallisation at temperatures above 90 °C, or treatment of α-lactose with alkaline methanol.

LACTOSE UTILISATION AND CHEMICAL DERIVATIVES

Lactosyl Urea

Although ruminants can use urea as a source of non-protein nitrogen for protein synthesis, it is not an ideal material for this purpose. The non-protein source should be palatable, have controlled nitrogen release, and be of low toxicity. Lactosyl urea, prepared by condensation of lactose and urea, appears to meet these requirements. The product may be prepared from whey or whey permeate, and yields in excess of 80 per cent have been claimed. The product is less toxic than a mixture of lactose and urea, and the linkage of lactose and urea results in slower degradation of the lactose, and consequently better utilisation, as well as removing any problems of crystallisation during storage (Widell, 1979). However, in spite of considerable initial interest in this product in both Europe and the USA, commercial development appears to have been slow.

Lactitol

Lactitol, which may be considered as the lactose equivalent of sorbitol, has been suggested for use as a non-nutritional sweetener. Details of lactitol manufacture have been outlined by Saijonmaa *et al.* (1978), covering hydrogenation of lactose by either Raney nickel or borohydride. Conversion rates of 97 per cent were achieved. Lactitol is hydrolysed as readily as lactose by the lactase from *Aspergillus niger*, but less rapidly by the enzyme from *Kluyveromyces fragilis*.

The use of lactitol as a base for lactose-derived surfactants has been suggested by Velthuijsen (1979).

Lactulose

Lactulose, an isomer of lactose, may be formed by molecular rearrangement of lactose under alkaline conditions. It may also be formed as a by-product of the action of some lactase enzymes on lactose. It has received much attention in the field of infant nutrition, and as a remedy for infant constipation. Matvievskii *et al.*, (1978) have reported that the best yield of lactulose, using calcium hydroxide as catalyst, was obtained from 15 per cent solutions of lactose held at 70 °C for 15–20 min at pH 11. More recently, a technique based on the use of boric acid, with either triethylamine or sodium hydroxide as catalyst, has been claimed to be a more economical and efficient process (Hicks *et al.*, 1983). Techniques for conversion of hygroscopic lactulose powder to the non-hygroscopic form by treatment with ethanol have also been developed. A number of methods for determining lactulose in dairy products are available based on techniques, such as spectrophotometry, enzymatic analysis, HPLC and thin-layer chromatography.

Lactulose is about half as sweet as sucrose at concentrations up to 15 per cent, the difference decreasing at higher concentrations. The water activity of lactulose solutions is significantly lower than that of sucrose syrups of equal concentration, and lactulose may, therefore, be useful as a humectant for intermediate moisture foods.

An extensive review of lactulose in infant nutrition has been prepared by Mendez and Olano (1979). The presence of lactulose in infant foods encourages the development of *Bifidobacterium bifidum* in the intestinal flora, similar to the flora found in breast fed infants. Lactulose also has medicinal uses in the treatment of portal systemic encephalopathy and chronic constipation. There has been some concern expressed, however, concerning possible laxative effects of lactulose, particularly with regard to consumption by infants. It has been suggested that such problems might be due to a low colonic pH.

Currently, lactulose commands a high price because of its unique properties of high solubility and non-digestibility. It has considerable potential for further market development.

Lactobionic Acid

There appears to be little commercial interest in production of this material from lactose. It may have some use as a food acidulant, and might be considered to have interesting chelating properties (Wright and Rand, 1973). Commercial development seems unlikely at present.

Gluconic Acid

A process for the conversion of lactose into gluconic acid and galactose has been described by Dahlgren (1978). In the process, lactose is treated with bromine under acidic conditions.

Binders

An interesting non-foodstuff application for lactose is its utilisation as a key ingredient of a binder system in processing iron fines captured in pollution control equipment. Commercial application of this process should result in large savings in energy expended in the manufacture of iron ore pellets, and permit recycling of iron/steel dust currently discarded (Ferretti and Chambers, 1979). Recently there has been much interest in the use of lactose as a basis for surfactants (Parrish, 1976). In 1959, the industrial manufacture of sucrose-based surfactants was commenced in Japan, with current production about 3000 tonnes per year. The method employed for sucrose-based surfactants is, however, not satisfactory for lactose, as it is necessary to use acid chlorides rather than esters for esterification. The properties of lactose-based surfactants are similar to those based on sucrose. Development of the process has been slow because of the requirement for expensive solvents. Work is continuing, however, on the development of solvent-free systems.

CONCLUSIONS

Of the various options discussed above for whey processing, many are unlikely to become commercially viable in the near future as their development is still at an academic level. Currently, therefore, the whey processor remains restricted to existing commercial operations for whey utilisation. However, the existing mix of processes employed will change significantly, particularly under the impact of new membrane technology and developments in lactose hydrolysis. The development of a wider range of whey protein concentrates, with specific functional properties for specific end uses, will result in expansion of this market. Such trends are already evident. These developments, together with developments in the manufacture of cheese using ultrafiltration of milk, will result in large quantities of permeate for processing or disposal.

For many countries, lactose hydrolysis, possibly in combination with the use of open RO membranes for demineralisation, is likely to become an economic alternative. In other countries, particularly the United States, disposal of lactose is likely to pose considerable difficulties in the near future. For large-scale manufacturers, options such as fermentation may be attractive for production of end-products with a defined market niche. However, given the capital-intensive nature of fermentation, it is probable that small producers will not find this option particularly attractive. In some cases, production of syrups for in-house use in dairy products may provide an alternative solution. The best option in areas where whey or permeate manufacture is concentrated may be for the dairy factories to supply permeate and/or whey to companies specifically organised to process these products. This approach has been shown to be particularly effective in New Zealand and may be a model for the rest of the world. Under these conditions, whey and permeate will be seen not as waste products, but rather as valuable raw materials.

REFERENCES

Allum, D. (1980). *Journal of the Society of Dairy Technology,* **33,** 59.

Amundson, C. H., Watanawanichakorn, S. and Hill, C. G. (1982). *Journal of Food Processing and Preservation,* **6,** 55.

Asp, N. G., Burvall, A., Dahlqvist, A., Hallgren, P. and Lundblad, A. (1980). *Food Chemistry,* **5,** 147.

Australian Dairy Corporation (1983). *Australian Dairy Corporation,* Planning Division, Australian Dairy Corporation, Melbourne.

Bartosh, J. A. and Manderson, G. J. (1982). *Brief Communications, XXI International Dairy Congress,* Vol. 1 (2), p. 526.

Beaton, N. C. (1979). *Journal of Food Protection,* **42,** 584.

Berger, H. M., Scott, P. H., Kenward, C., Scott, P. and Wharton, B. A. (1979). *Archives of Disease in Childhood,* **54,** 98.

Best, E., Plainer, H. and Sprossler, B. (1982). European Patent Application EP 0 057 273 A2.

Brothersen, C. F., Olson, N. F. and Richardson, T. (1982). *Journal of Dairy Science,* **65,** 17.

Buhler, M., Olofsson, M. and Fosseux, P. Y. (1981). United States Patent 4 265 924.

Burvall, A., Asp, N.-G. and Dahlqvist, A. (1979). *Food Chemistry,* **4,** 243.

Chebotarev, E. A., Nesterenko, P. G., Davydyants, L. E., Mikhailova, N. I. and Chebotareva, N. G. (1983). *Molochnaya Promyshlennost,* (2), 26.

Chen, S.-L. Y., Frank, J. F. and Loewenstein, M. (1981). *Journal of the Association of Official Analytical Chemists,* **64,** 1414.

Cheryan, M. and Merin, U. (1979). Abstracts of Papers, National Meeting, American Chemical Society, **178** (1), COLL 128.

Cheryan, M. and Merin, U. (1980). *Polymer Science and Technology,* **13,** 619.

Clark, W. S. (1979). *Journal of Dairy Science,* **62,** 96.

Coton, S. G. (1980). *Journal of the Society of Dairy Technology,* **33,** 89.

Craig, T. W. (1979). *Journal of Dairy Science,* **62,** 1695.

Dahlgren, S. A. (1978). United Kingdom Patent 1 526 903.

Delaney, R. A. M. (1979). *Proceedings, Whey Products Conference, Minneapolis,* 1978, p. 111.

Demaimay, M., Le Henaff, Y. and Printemps, P. (1978). *Process Biochemistry,* **13**(4), 3.

Dohan, L. A., Baret, J. L., Pain, S. and Delalande, P. (1980). *Enzyme Engineering,* **5,** 279.

Dunkerley, J. A. and Zadow, J. G. (1981). *New Zealand Journal of Dairy Science and Technology,* **16,** 243.

Dziuba, J. and Chojnowski, W. (1982). *Brief Communications, XXI International Dairy Congress,* Vol. 1 (2), p. 173.

Ennis, B. M. (1984). *New Zealand Journal of Dairy Science and Technology* (in press).

Ennis, B. M. and Higgins, J. J. (1981). *New Zealand Journal of Dairy Science and Technology,* **16,** 167,

Euber, J. R. and Brunner, J. R. (1979). *Journal of Dairy Science,* **62,** 685.

Ferretti, A. and Chambers, J. V. (1979). *Journal of Agricultural and Food Chemistry,* **27,** 687.

Food and Agriculture Organisation (1981). *FAO Production Yearbook,* Vol. 34, p. 234.

Friend, B. A. and Shahani, K. M. (1979). *New Zealand Journal of Dairy Science and Technology,* **14,** 143.

Garfield, E. (1980). *Current Contents: Life Sciences,* **23,** 5.

Gawel, J. and Kosikowski, F. V. (1978). *Journal of Food Science,* **43,** 1717.

Giec, J., Czarnecka, I., Baraniecka, B. and Skupin, J. (1978). *Acta Alimentaria Polonica,* **4,** 247.

Griffiths, M. W. and Muir, D. D. (1980). *Journal of the Science of Food and Agriculture,* **31,** 397.

Haggett, T. O. R. (1976). *New Zealand Journal of Dairy Science and Technology,* **11, 176.**

Hannigan, K. J. (1982). *Food Engineering,* **54** (3), 96.

Harper, W. J. (1979). *New Zealand Journal of Dairy Science and Technology,* **14,** 156.

Hartmeier, W. (1982). In: *Use of Enzymes in Food Technology,* International Symposium, Versailles, 1982 (ed. P. Dupuy), Technique et Documentation Lavoisier, Paris, p. 205.

Hayes, J. F. (1982a). *Working Papers, 2nd Dairy Technology Review Conference, Glemormiston, Victoria,* 1982, p. 211.

Hayes, J. F. (1982b). *Working Papers, 2nd Dairy Technology Review Conference, Glemormiston, Victoria,* 1982, p. 213.

Hayes, J. F. and Mitchell, I. R. (1983). *Proceedings, 2nd Whey Protein Collaborative Research Group Conference,* Highett, 1983, p. 379.

Hayes, J. F., Dunkerley, J. A., Muller, L. L. and Griffin, A. T. (1974). *Australian Journal of Dairy Technology,* **29,** 132.

Hickey, M. E. and Hill, R. D. (1980). *New Zealand Journal of Dairy Science and Technology,* **15,** 123.

Hicks, K. B., Raupp, D. L. and Smith, P. W. (1983). Abstracts of Papers, National Meeting, American Chemical Society, 186, AGFD 159.

Hillier, R. M. and Cheeseman, G. C. (1979). *Journal of Dairy Research,* **46,** 113.

Hillier, R. M., Lyster, R. L. J. and Cheeseman, G. C. (1979). *Journal of Dairy Research,* **46,** 103.

Hillier, R. M., Lyster, R. L. J. and Cheeseman, G. C. (1980). *Journal of the Science of Food and Agriculture,* **31,** 1152.
Hossain, M., Brooks, J. D. and Maddox, I. S. (1983). *New Zealand Journal of Dairy Science and Technology,* **18,** 161.
Humbert, G. and Alais, C. (1982). *La Technique Laitiere,* **952,** 41.
Hynd, J. (1980). *Journal of the Society of Dairy Technology,* **33,** 52.
Jackson, M. B. (1980). *Research Review,* CSIRO Division of Chemical Technology, 1980, 45.
Jenness, R. and Patton, S. (1959). *Principles of Dairy Chemistry,* Wiley, New York, p. 89.
Johns, J. E. M. (1982). *Brief Communications, XXI International Dairy Congress,* Vol. 1 (2), p. 196.
Johns, J. E. M. and Ennis, B. M. (1981). *New Zealand Journal of Dairy Science and Technology,* **16,** 79.
Jonsson, H. and Forsman, B. (1978). *Nordisk Mejeriindustri,* **5,** 636.
Kamiya, T. and Kaminogawa, S. (1980). *Japanese Journal of Dairy and Food Science,* **29,** A-15, A-27.
Kosikowski, F. V. (1977). *Cheese and Fermented Milk Foods,* 2nd edn, Edward Bros, Ann Arbor, Mich., p. 448.
Kwon, S. Y. and Nickerson, T. A. (1978). *Journal of Dairy Science,* **61** (suppl. 1), 112.
La Gueriviere, J.-F. de. (1981). *La Technique Laitiere,* **952,** 89.
Labuza, T. P. and Saltmarch, M. (1982). *Journal of Food Science,* **47,** 92.
Lang, F. (1980). *Milk Industry,* **82** (2), 30.
Lewicki, P. P., Galoch, I. and Slesinka, K. (1981). *Przemysl Spozywczy,* **35,** 102.
Linko, Y. Y., Jalanka, H. and Linko, P. (1981). *Biotechnology Letters,* **3,** 263.
Lipatov, N. N. and Chebotarev, E. A. (1981). *Izvestiya Vysshikh Uchehbnykh Zavedenii, Pischchevaya Tekhnologiya,* (2), 41.
Lovell-Smith, J. E. R. (1982). *New Zealand Journal of Dairy Science and Technology,* **17,** 161.
Luquet, F.-M. (1980). *La Technique Laitiere,* **940,** 23.
Lyons, T. P. and Cunningham, J. D. (1980). *American Dairy Review,* **42** (11), 42A.
MacBean, R. D., Hall, R. J. and Willman, N. J. (1979). *Australian Journal of Dairy Technology,* **34,** 53.
McGugan, W. A., Larmond, E. and Emmons, D. B. (1979). *Canadian Institute of Food Science and Technology Journal,* **12,** 32.
Maculan, T. P., Hourigan, J. A. and Rand, A. G. (1978). *Journal of Dairy Science,* **61** (suppl. 1), 114.
Maddox, I. S. (1980). *Biotechnology Letters,* **2,** 493.
Madrid Vicente, A. (1979). *Industrie Alimentari,* **18,** 811.
Mahmoud, M. M. and Kosikowski, F. V. (1978). *Journal of Dairy Science,* **61** (suppl. 1), 114.
Mahmoud, M. M. and Kosikowski, F. V. (1982). *Journal of Dairy Science,* **65,** 2082.
Manczak, M. (1979). *Przemsyl Spozywczy,* **33,** 382.
Marconi, W., Bartoli, F., Morisi, F. and Marani, A. (1980). *Enzyme Engineering,* **5,** 269.
Marshall, S. C. (1979). *New Zealand Journal of Dairy Science and Technology,* **14,** 103.

Matthews, M. E. (1978). *New Zealand Journal of Dairy Science and Technology*, **13**, 149.
Matthews, M. E. (1979). *New Zealand Journal of Dairy Science and Technology*, **14**, 86.
Matvievskii, V. Ya., Kravchenko, E. F. and Khramtsov, A. G. (1978). *Trudy, Vsesoyuznyi Nauchno-issledovatel'skii Institut Maslodel'noi i Syrodel'noi Promyshlennosti Nauchno-proizvodstvennogo Ob''edinenya 'Uglich'*, **26**, 13.
Maubois, J. L., Brule, G. and Gourdon, P. (1981). *Technique Laitiere*, **14**, 86.
Mendez, A. and Olano, A. (1979). *Dairy Science Abstracts*, **41**, 531.
Merin, U., Gordin, S. and Tanny, G. B. (1983). *New Zealand Journal of Dairy Science and Technology*, **18**, 153.
Meyrath, J. and Bayer, K. (1979). *Economic Microbiology*, **4**, (Microbial Biomass), 207.
Mirabel, B. (1978). *Informations Chimie*, **175**, 105.
Moon, N. J. and Hammond, E. G. (1978). *Journal of the American Oil Chemists Society*, **55**, 683.
Morr, C. V. (1979). *New Zealand Journal of Dairy Science and Technology*, **14**, 185.
Muller, L. L. (1979). *New Zealand Journal of Dairy Science and Technology*, **14**, 121.
Muller, L. L. and Harper, W. J. (1979). *Journal of Agricultural and Food Chemistry*, **27**, 662.
Muller, L. L., Hayes, J. F. and Griffin, A. T. (1973). *Australian Journal of Dairy Technology*, **28**, 70.
Nickerson, T. A. (1974). In: *Fundamentals of Dairy Chemistry* (ed. B. H. Webb, A. H. Johnson and J. A. Alford), AVI Publishing Co., Westport, Conn., p. 273.
Norman, B. E., Severinsen, S. G., Nielsen, T. and Wagner, J. (1978). *Nordeuropaeisk Mejeri Tidsskrift*, **44**, 129.
O'Keeffe, A. M. and Kelly, J. (1981). *Netherlands Milk and Dairy Journal*, **35**, 292.
Parrish, F. W. (1976). *Proceedings, Whey Products Conference, Atlantic City*, 1976, p. 104.
Pastore, G. M. and Park, Y. K. (1981). *Ciencia e Tecnologia de Alimentos*, **1**, 65.
Patel, P. C. and Merson, R. L. (1978). *Journal of Food Science and Technology, India*, **15**, 56.
Pazur, J. H. (1970). In: *Carbohydrates: Chemistry and Biochemistry* (ed. W. Pigman, D. Horton and A. Herp), 2nd edn., Academic Press, New York, Vol. IIA, p. 69.
Pearce, R. J. (1983). *Australian Journal of Dairy Technology*, **38**, 144.
Pepper, D. (1981). *Dairy Industries International*, **46**, 24.
Posati, L. P. and Orr, M. L. (1976). *Agriculture Handbook, United States Department of Agriculture*, no. 8–1, 144 pp.
Poutanen, K., Linko, Y.-Y. and Linko, P. (1978). *Nordeuropaeisk Mejeri Tidsskrift*, **44**, 90.
Rheinlander, P. M. (1982). *North European Dairy Journal*, **48**, 121.
Richert, S. H. (1979). *Journal of Agricultural and Food Chemistry*, **27**, 665.
Ritzel, G., Stahelin, H. B., Schneeberger, H., Wanner, M. and Jost, M. (1979). *International Journal for Vitamin and Nutrition Research*, **49**, 419.
Robinson, B. P., Short, J. L. and Marshall, K. R. (1976). *New Zealand Journal of Dairy Science and Technology*, **11**, 114.
Roeper, J. (1971). *New Zealand Journal of Dairy Science and Technology*, **6**, 112.

Ryder, D. N. (1980). *Journal of the Society of Dairy Technology,* **33,** 73.
Saijonmaa, T., Heikonen, M., Kreula, M. and Linko, P. (1978). *Milchwissenschaft,* **33,** 733.
Saltmarch, M. (1980). Dissertation, *Abstracts International,* B, **41,** p. 879.
Saltmarch, M. and Labuza, T. P. (1980). *Scanning Electron Microscopy,* **3,** 659.
Sandbach, D. M. L. (1981a). *Cultured Dairy Products Journal,* **16** (4), 17.
Sandbach, D. M. L. (1981b). *Proceedings, Second Bi-ennial Marschall International Cheese Conference,* Madison, 1981, p. 17.
Schmidt, R. H., Illingworth, B. L., Deng, J. C. and Cornell, J. A. (1979). *Journal of Agricultural and Food Chemistry,* **27,** 529.
Shah, N. O. and Nickerson, T. A. (1978). *Journal of Food Science,* **43,** 1575.
Smith, B. R. and MacBean, R. D. (1978). *Australian Journal of Dairy Technology,* **33,** 57.
Somkuti, G. A. and Bencivengo, M. M. (1981). *Developments in Industrial Microbiology,* **22,** 557.
Stauffer, K. R. and Leeder, J. G. (1978). *Journal of Food Science,* **43,** 756.
Stieber, R. W. and Gerhardt, P. (1979). *Journal of Dairy Science,* **62,** 1558
Strobel, D. R. (1972). *Foreign Agriculture,* **10** (46), 9.
Thompson, L. U. and Reyes, E. S. (1980). *Journal of Dairy Science,* **63,** 715.
Varunsatian, S., Watanabe, K., Hayakawa, S. and Nakamura, R. (1983). *Journal of Food Science,* **48,** 42.
Velthuijsen, J. A. van (1979). *Journal of Agricultural and Food Chemistry,* **27,** 680.
Visser, R. A. (1980). *Netherlands Milk and Dairy Journal,* **34,** 255.
Wechsler, R. M. (1983). *Proceedings, International Membrane Technology Conference, Sydney,* 1983, p. 189.
Widell, S. (1979). *Proceedings, Whey Products Conference, Minneapolis,* 1978, p. 53.
Williams, C. A. and MacDonald, I. (1982). *Annals of Nutrition and Metabolism,* **26,** 374.
Wit, J. N. de and Klarenbeek, G. (1978). British Patent 1 519 897.
Wright, D. G. and Rand, A. G. (1973). *Journal of Food Science,* **38,** 1132.
Zadow, J. G. (1979). *New Zealand Journal of Dairy Science and Technology,* **14,** 131.
Zadow, J. G. (1982). *Conference Proceedings, Second Australian Dairy Technology Review Conference, Glenormiston, Vic.,* 1982, p. 276.
Zadow, J. G. (1984). *Journal of Dairy Science* (in press).
Zaetz, N. E., Kubanskaja, D. M. and Kocherov, N. I. (1982). *Proceedings, XXI International Dairy Congress,* Vol. 1 (2), p. 657.
Zarb, J. M. and Hourigan, J. A. (1979). *Australian Journal of Dairy Technology,* **34,** 184.

Chapter 7

Utilisation of Milk Components: Casein

C. R. Southward
New Zealand Dairy Research Institute, Palmerston North, New Zealand

Casein, the principal protein in cow's milk, has been extracted commercially for most of the 20th century. The casein content of whole milk varies according to the breed of cow, stage of lactation and time of milking, but is generally in the range 24–29 g litre^{-1} (Jenness and Patton, 1959b). The skim-milk produced from the separation of whole milk to a 40 per cent fat cream (the first stage in the manufacture of casein) consequently contains from 25 to 31 g litre^{-1} of casein protein (Jenness and Patton, 1959a; McDowall, 1971), which accounts for 75–80 per cent of the nitrogen in skim-milk (Dolby *et al.*, 1969; Gordon and Kalan, 1974). With a content of 0·7–0·9 per cent phosphorus, covalently bound to the casein by a serine ester linkage, casein as a phosphoprotein is a member of a relatively rare class of proteins (Gordon and Kalan, 1974).

The amino acid content of casein has been well established, and includes high proportions of all amino acids essential to man (Gordon and Kalan, 1974) with the possible exception of cysteine.

Casein exists in milk in complex micelles which consist of casein molecules, calcium, inorganic phosphate and citrate (Farrell and Thompson, 1974). It has been established that whole casein is a heterogeneous mixture of several individual casein components (α_s, β, κ, etc), each of which has slightly different properties (Gordon and Kalan, 1974; Farrell and Thompson, 1974). For the purpose of commercial production, however, it is only necessary to consider whole casein, which contains all of these components.

EXTRACTION OF CASEIN FROM MILK

Casein is extracted from milk by the process outlined in Figs 1 and 2. After separation of the whole milk to produce cream and skim-milk, the casein is precipitated, washed and dried (Spellacy, 1953). The precipitation stage is outlined in Fig. 1. When the casein curd has been precipitated and the mixture heated, the curd is separated from the whey (Figs 2 and 3) and is subsequently washed several times with water in vats (Jordan, 1983; Fig. 4), prior to mechanical dewatering by pressing (Munro and Vu, 1983) or centrifuging (Munro *et al.,* 1983) and drying. In New Zealand, fluid-bed driers are typically used in this operation (Bates, 1960) (Fig. 5). The warm, granular casein is then cooled in a 'tempering' process (Fig. 2) in which casein is continuously circulated through several bins

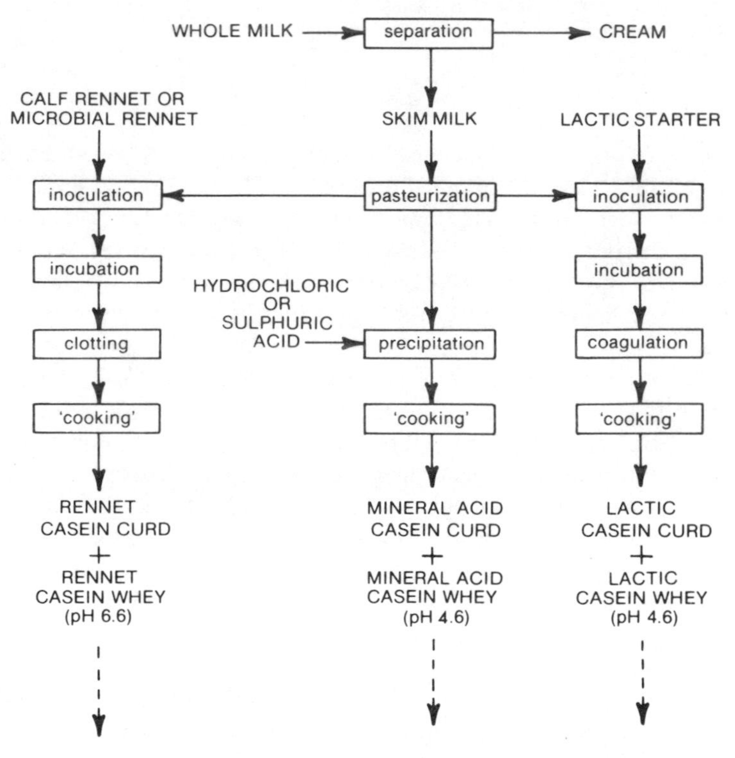

Fig. 1. Processing steps involved in the precipitation of acid and rennet caseins from milk. (Reproduced from Southward and Walker (1980) by permission of the publishers.)

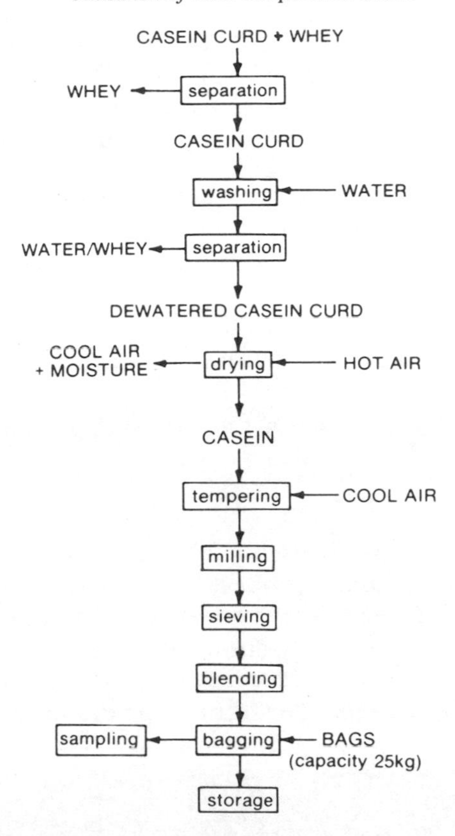

Fig. 2. Processing operations in the washing and drying of caseins. (Reproduced from Southward and Walker (1980) by permission of the publishers.)

to attain equilibration of moisture throughout the different sized particles. The fully tempered casein is passed over a magnetic separator prior to grinding in roller mills. The milled casein is then sieved to various particle sizes, depending upon market requirements, blended (King and Jebson, 1970) and finally bagged, ready for storage or shipment.

Two basic types of casein are produced, and these are named in accordance with the coagulating agent employed. They are acid casein and rennet casein. Three types of acid casein are produced commercially: lactic, hydrochloric and sulphuric casein. In New Zealand, lactic casein is by far the most common product, although a small quantity of sulphuric casein is also produced for particular requirements. In Australia and Europe, however, hydrochloric acid is commonly employed as a precipitant in casein manufacture, since this acid is available as a relatively inexpensive by-product of the chemical industry.

The other type of casein manufactured commercially is rennet casein, which is produced from skim-milk that has been clotted by the action of calf chymosin (rennin).

Manufacturing processes for casein have been reviewed by Muller (1971, 1982a, b) and Gwozdz (1978).

Acid Casein Manufacture

In the manufacture of acid casein, the following processing operations occur (Fig. 1).

Mineral acid casein

Skim-milk (pH 6·6) is pasteurised (72 °C/15 s) and mixed with dilute (0·5N) acid at a temperature of about 20 °C to a pH of approximately 4·6. The mixture is heated, for instance by injection of steam, to a temperature of 50–55 °C in order to aid agglomeration of the casein particles. Following a short period of residence in a 'cooking' line and 'acidulation' vat, the resultant curd is separated from the whey, washed and dried.

Fig. 3. Separation of casein curd and whey by inclined stationary screens.

Fig. 4. Washing vats for casein curd.

Fig. 5. Vibrating tray-drier for casein.

Lactic acid casein

Pasteurised skim-milk for lactic casein manufacture is inoculated with strains of lactic acid-producing bacteria, known as 'starter' (e.g. *Streptococcus cremoris*, 0·1–0·5 per cent of milk volume) at a temperature of 22–26 °C (Thomas and Lowrie, 1975a, b; Heap and Lawrence, 1984). The milk is incubated in several silos, each with a capacity of up to 250 000 litres, for 14–16 h. During this period, some of the lactose in the milk is fermented to lactic acid by the starter, and the pH of the milk is reduced to about 4·6 causing coagulation of the casein. The lactic coagulum is then 'cooked' and processed further in a manner similar to that described for mineral acid casein.

Process schematics for lactic, hydrochloric and acid casein manufacture in New Zealand, Australia and Europe, respectively, have been published (King, 1970; Muller, 1971, 1982a). Throughputs in commercial plants in New Zealand range from 20 000 to 90 000 litre h^{-1} of skim-milk, representing production of 0·6–2·7 tonne h^{-1} of casein. During the seasonal peak (October/November), some units produce 35–45 tonne casein/day.

Rennet Casein Manufacture

In the manufacture of rennet casein (Fig. 1), calf rennet or microbial rennet (Southward and Elston, 1976) is added to skim-milk (ratio 1:7000) at a temperature of about 29 °C (Weal and Southward, 1974). Where rennet casein is intended for industrial purposes, it is important that the skim-milk should not be pasteurised, for it has been observed (Munro *et al.*, 1980) that rennet casein plastic produced from pasteurised milk has a darker colour than casein plastic from unpasteurised milk.

During the renneting of the skim-milk, the enzyme specifically cleaves one of the bonds in κ-casein, releasing a glycomacropeptide (Ernstrom and Wong, 1974). This action destabilises the casein micelles and they form a three-dimensional clot with calcium ions. This process normally takes place in about 30 min at pH 6·6. The clotted milk may then be 'cooked', and the casein processed in a manner similar to that described for acid casein.

Yield

The yield of commercial casein from 100 kg skim-milk is approximately 3 kg.

Co-precipitates

The action of heat on skim-milk helps to induce complex interactions between casein and the whey proteins (mainly β-lactoglobulin). This phenomenon forms a basis for the commercial preparation of a third general group of casein products, namely the milk protein co-precipitates. These products are manufactured by heating skim-milk (generally to between 85 °C and 95 °C) to induce casein–whey protein interactions, and subsequently causing co-precipitation and coagulation of protein either by reduction of the pH of heated milk, or by addition of inorganic calcium salts to the hot skim-milk (Muller *et al.,* 1967).

All casein and co-precipitate products produced by the three processes described above are insoluble in water. However, addition of alkali (such as sodium, potassium, calcium or ammonium hydroxide) will cause the undried curd or the rehydrated casein product to dissolve. The soluble proteins formed by reaction of acid casein with alkali are known as caseinates.

MANUFACTURE OF CASEINATES

Caseinates may be produced from acid casein curd or from dry acid casein by reaction under aqueous conditions with any one of several different dilute alkalis, as outlined in Fig. 6. The resulting homogeneous caseinate solution may be dried by the spray or roller process to produce a caseinate powder having a moisture content of 3–8 per cent, depending on the manufacturer and customer requirements.

Spray-dried Sodium Caseinate

Aspects of the manufacture of spray-dried sodium caseinate have been reviewed by Fox (1968) and Muller (1971, 1982a), and described by Burston *et al.* (1967), Australian Society of Dairy Technology (1972), Bergmann (1972), Towler (1976a) and Segalen (1982).

Raw Materials

In casein manufacturing countries, it is common practice to prepare sodium caseinate from fresh, acid casein curd, since it is considered to be blander in flavour than sodium caseinate made from dried casein (Cayen and Baker, 1963; Burston *et al.,* 1967). Sodium caseinate

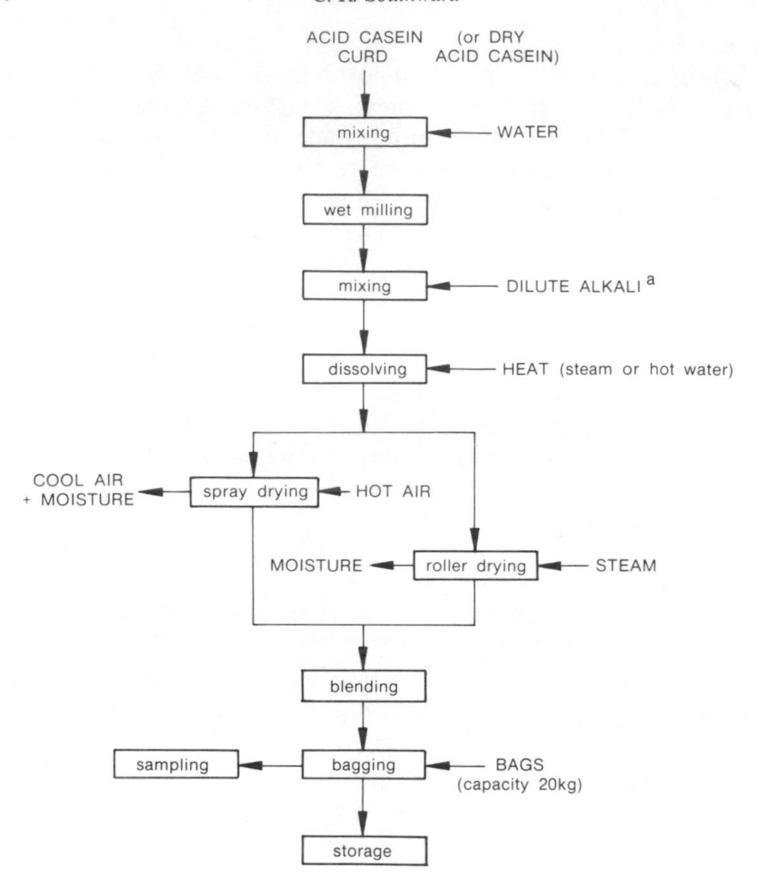

Fig. 6. Manufacture of spray-dried or roller-dried caseinate from acid casein curd or dry, acid casein.

prepared from dry casein will also incur the additional manufacturing costs associated with drying, dry processing, bagging and storage of the casein prior to its conversion to sodium caseinate.

The most common alkali used in the production of sodium caseinate is sodium hydroxide solution, with a strength of 2·5M (Australian Society of Dairy Technology, 1972; Towler, 1976a). The amount of sodium hydroxide required is generally 1·7–2·2 per cent by weight of the casein solids (Australian Society of Dairy Technology, 1972). Other alkalis may be used, such as sodium bicarbonate or sodium phosphates

(Segalen, 1982), but the amount required and their cost are greater than those of sodium hydroxide.

Equipment

Milling

One or more colloid mills are required to reduce the particle size of curd for rapid dissolving. When dry casein is used, 30 mesh casein is preferred to the finer 80–90 mesh product which tends to form lumps when added to water (Burston *et al.*, 1967; Fox, 1968). The use of a funnel in a recirculation line (as used in reconstituting milk powder) may help to overcome this problem.

Alkali dosing

Suitable equipment for metering the alkali into the casein slurry is necessary.

Dissolving vats

One or more dissolving vats, equipped with powerful agitators and recirculating pumps, are needed. The capacity and number of vats may vary, depending on the throughput of the drier, and whether the dissolving process is batch or continuous. It is preferable to keep the size of any dissolving vat below 5000 litres, since it is difficult and expensive to achieve good mixing of the viscous caseinate solution in larger vessels.

Heating

The caseinate solution must be heated during the dissolving process. Indirect heating may be carried out by means of a tubular heat exchanger in the recirculation line, or through the walls of a jacketed dissolving vat. Alternatively, it is possible to inject steam directly into the caseinate solution, provided that the steam is of good quality. However, this results in dilution of the caseinate, and has the potential for producing scorched particles.

pH and viscosity measurement

It is essential for pH and viscosity of the caseinate solution to be controlled, and suitable instruments are necessary for the measurement of these properties. It is possible to use a measurement of refractive index of a solution of caseinate to estimate its solids content, and hence use this information to assist in the control of the dissolving operation (Towler, 1976c).

Spray-drying

Sodium caseinate can be dried in spray-driers equipped with either disc or pressure nozzle atomisation (Segalen, 1982) from solutions with a concentration of up to 20 per cent solids.

Roller-dried Sodium Caseinate

By comparison with the literature on spray-dried caseinates, very little information has been published on roller-drying these products, though it is known that the technique is used quite extensively in Europe (Towler, 1977). Roller-dried caseinate is used mainly in the meat industry (Gwozdz, 1978). Solutions of sodium caseinate for roller-drying may be prepared in a similar manner to those intended for spray-drying (Fox, 1968; Muller, 1971). A recent patent (Lippe *et al.,* 1983) describes using a drum-drier for drying sodium caseinate solutions with a solids content of 44–47 per cent. Alternatively, casein curd with a moisture content of 50–65 per cent may be mixed with sodium carbonate or bicarbonate, and fed on to the hot drum of a roller-drier to produce a completely reacted sodium caseinate product before drying is completed (Towler, 1976b).

Granular Sodium Caseinate

The high costs of spray-drying solutions of sodium caseinate, together with the problems associated with the highly-viscous nature of the solutions during dissolving and the costs of packaging and transport of the relatively low bulk-density product, have prompted investigations into the production of granular sodium caseinate (Towler, 1977, 1978; Gwozdz, 1978, 1983; Roeper, 1982; Segalen, 1982). The manufacture of this product, which may be either partly or fully reacted, may include the partial drying of casein curd, mixing with an alkali (such as sodium carbonate) and, finally, drying of the caseinate (Towler, 1977, 1978). Where partial drying of the curd is not carried out prior to addition of alkali, it is possible to produce an extremely sticky and rubbery curd-like mass which is difficult to subsequently process. Methods of overcoming this have included low initial drying temperatures for the caseinate, mixing of curd with the sodium salt of an organic acid (e.g. sodium citrate) prior to addition of the alkali (Wakodo Co., 1969), and cooling either the reaction/mixing vessel (Towler, 1978) or the curd itself

(Gwozdz, 1983). Alternatively, the solids content of the curd can be increased by addition of dry casein (Gwozdz, 1983).

Preparation of Concentrated Caseinate Solutions

One further technique for the manufacture of sodium caseinate has been reported recently (Lippe *et al.*, 1983). A mixture of casein (curd or powder), water and a solubiliser, such as NaOH, is pumped into a steam-heated reaction chamber. This concentrated solution (typically 33–47 per cent solids) is prepared in two steps, first to pH 5·0–5·3 (where the viscosity is at a minimum) and then to pH 6·5–7·0. The solution may then be spray- (at up to 33 per cent solids) or roller-dried. Alternatively, the solution may be forced from the chamber through orifices to produce fine jets of caseinate which can be dried in hot air to form threads of sodium caseinate.

Extruded Sodium Caseinate

As noted by Towler (1977), it is possible to produce sodium caseinate from casein in the presence of a limited amount of water by use of extrusion techniques. In most of the published information, dry casein is used as the starting material, and water and alkali are added to form a mixture for extrusion. The moisture content (10–30 per cent) is generally lower than that used for the granular caseinates discussed above, and fairly high pressures and temperatures may be used during extrusion to facilitate the formation of sodium caseinate.

Ammonium Caseinate/Potassium Caseinate

As observed by Gwozdz (1978), ammonium caseinate, potassium caseinate and sodium caseinate have similar physical properties. Thus, it is possible to add ammonium hydroxide to a slurry of casein and water to produce a solution of ammonium caseinate which may be subsequently spray-dried in the same manner as sodium caseinate. Alternatively, a granular ammonium caseinate may be produced by exposing either acid casein curd (Krause, 1953) or dry acid casein (Girdhar and Hansen, 1974) to gaseous ammonia. Excess ammonia is removed during drying of the converted curd, or on air flushing of the converted casein.

It has also been reported that small amounts of ammonium hydroxide may be used to increase the pH of calcium caseinate solutions either when there is an upper limit placed on the calcium content of the dried product (Muller, 1971), or to decrease the amount of sedimentable matter (Roeper, 1977b; Lippe *et al.,* 1983). Most of the added ammonia is evaporated during the subsequent drying process.

Potassium caseinate may be prepared in a similar manner to sodium caseinate by substituting potassium hydroxide, for example, in place of sodium hydroxide (e.g. Towler, 1977; Lippe *et al.,* 1983).

Calcium Caseinate

In contrast to the translucent, viscous, straw-coloured sodium, potassium and ammonium caseinates, calcium caseinate forms micelles in water, producing an intensely white, opaque, 'milky' solution of relatively low viscosity (Roeper, 1977a, b). Its preparation follows the same general process as that used for sodium caseinate with one or two important exceptions. Calcium caseinate solutions may be destabilised on heating (Roeper, 1977a, b), especially at pH values below 6 (Hayes *et al.,* 1968). This sensitivity decreases with increase in pH and decrease in concentration, and is manifested as a reversible heat gelation, first reported by Zittle *et al.* (1956).

During the dissolving process, it has been found that the reaction between acid casein curd and calcium hydroxide (the alkali most commonly used in the production of calcium caseinate) proceeds at a much slower rate than that between curd and sodium hydroxide. The temperature of conversion is a particularly important factor in determining the completeness of solubility (amount of sedimentable matter) of the calcium caseinate (Roeper, 1977a, b). Therefore, the dissolving process must be closely monitored to ensure production of calcium caseinate of good solubility.

Since calcium hydroxide itself has very limited solubility in water, it is common practice to add it to the mixture of casein in water as an aqueous slurry.

Other Caseinates

Magnesium caseinate has also been briefly mentioned in the literature (Towler, 1977; Muller, 1982a), being prepared from casein and a magnesium base or basic salt such as magnesium oxide (Zavagli and

Kasik, 1978), magnesium hydroxide, carbonate or phosphate (Segalen, 1982) or by ion exchange (Hidalgo *et al.,* 1981).

Compounds of casein with aluminium have been prepared for medicinal use (Schuette, 1939; Paterson, 1955), or for use as an emulsifier in meat products (Barrat, 1978).

Heavy metal derivatives of casein which have been used principally for therapeutic purposes (Schuette, 1939), include those containing silver, mercury, iron and bismuth. Iron and copper caseinates have also been prepared by ion exchange for use in infant and dietetic products (Hidalgo *et al.,* 1981).

MANUFACTURE OF CO-PRECIPITATES

Following addition of small quantities of calcium chloride or acid to skim-milk, the mixture is heated to 85–95°C and held at that temperature for a period of 1–20 min, to allow interaction between the caseins and the whey proteins. Precipitation of the proteins from the heated milk is then effected by controlled addition of either calcium chloride solution (to produce high-calcium co-precipitate) or dilute acid (to produce medium-calcium or low-calcium co-precipitate, depending upon the amount of acid added and the pH of the resultant whey). The curd is subsequently washed, and either dried to produce granular, insoluble co-precipitates, or dissolved in alkali as described in the methods for manufacture of caseinates to produce soluble or 'dispersible' co-precipitates. While co-precipitate manufacture was first reported in the USSR in the early 1950s (e.g. D'yachenko *et al.,* 1953), the commercial potential of co-precipitates was developed in the 1960s and 1970s by the Australian Commonwealth Scientific and Industrial Research Organization (CSIRO). This, and subsequent work, has been extensively reviewed (Muller, 1971, 1982a, b; Southward and Goldman, 1975).

The processes which are described above cause heat denaturation of the whey proteins. As a consequence, the solubility of the resultant co-precipitates in alkali at pH 6·6–7·2 is not complete (Southward and Aird, 1978). By application of a lower heat treatment to skim-milk under alkaline conditions, however, it has been possible to alter the whey proteins so that they precipitate with the casein at pH 4·6, but dissolve completely with the casein when alkali is added to pH 6·6–7·0 (Connolly, 1983). The product is referred to as 'Total Milk Protein'.

Another, similar, product has also recently been developed in the Netherlands (Stichting Nederlands Instituut voor Zuivelonderzoek, 1982).

COMPOSITION OF CASEIN

The typical composition of adequately washed acid casein is shown in Table I. When the same manufacturing operations are employed the caseins produced from lactic, sulphuric or hydrochloric acid precipitation are indistinguishable from one another (Roeper, 1974). As indicated in Table I, rennet casein differs from acid casein particularly in ash content, and in the pH of a water extract. During the acidification process in the manufacture of acid casein, the calcium and inorganic phosphate, which were associated with the casein micelle in milk, are dissolved and leached from the curd leaving only the organic phosphorus and a small residue of calcium. Rennet casein contains about 3 per cent calcium and approximately 1·4 per cent phosphorus (McDowell, 1968).

PROPERTIES OF CASEIN

Acid and rennet caseins are not soluble in water. When the dry powders are wetted, however, the particles will absorb water and swell. Both types of casein may be dissolved in alkali and, for virtually all industrial applications, acid casein is dissolved prior to use. Acid casein is soluble at pH 7 or above in most alkalis, such as sodium and potassium hydroxides, carbonates and bicarbonates, trisodium phosphate, borax and lime. Rennet casein may be rendered soluble in solutions of complex phosphates, such as sodium tripolyphosphate, at a pH of 7–8, or may be dissolved in sodium hydroxide at a pH of 9·5 or higher.

Solutions of all caseins (except rennet casein in sodium tripolyphosphate) are very sticky and make very good adhesives and films. The solutions are also very viscous, especially at concentrations above 150 g $litre^{-1}$.

COMPOSITION OF CASEINATES

The typical composition of sodium and calcium caseinate, produced from well-washed acid casein, is shown in Table II. The ash content of

TABLE I
Composition of casein and granular co-precipitates

	Acid casein[a]	*Rennet casein*[a]	*Co-precipitate*[b] *High calcium*	*Medium calcium*	*Acid*
Moisture (%)	11·4	11·4	9·5	9·5	9·5
Protein (%)	85·4	79·9	81·7	85·6	86·7
Ash (%)	1·8	7·8	7·7	3·7	2·4
Lactose (%)	0·1	0·1	0·5	0·5	0·5
Fat (%)	1·3	0·8	0·6	0·7	0·9
pH	4·6–5·4	7·3–7·7	6·5–7·2	5·6–6·2	5·4–5·8
pH of whey after separation of curd	4·3–4·6	6·5–6·7	5·8–5·9	5·1–5·3	4·9–5·1

[a]Caseins dried to a maximum moisture content of 12 per cent.
[b]Co-precipitates washed twice in water and dried to a nominal maximum moisture content of 10 per cent (data of Southward and Aird, 1978).

TABLE II
Composition and physical properties of spray-dried soluble co-precipitates and caseinates

	Sodium caseinate	*Calcium caseinate*	*Soluble co-precipitate*[a] *High calcium*	*Medium calcium*	*Acid*
Moisture (%)	3·8	3·8	(4·0)[b]	(4·0)	(4·0)
Protein (N × 6·38) (%)	91·4	91·2	81·4	88·1	90·5
Ash (%)	3·6	3·8	13·5	6·7	4·1
Lactose (%)	0·1	0·1	(0·5)	(0·5)	(0·5)
Fat (%)	1·1	1·1	0·6	0·7	0·9
Sodium (%)	1·2–1·4	<0·1	1·9	–	1·1
Calcium (%)	0·1	1·3–1·6	2·9	1·2	0·5
Iron (mg kg^{-1})	3–20	10–40			
Copper (mg kg^{-1})	1–2	1–2			
Lead (mg kg^{-1})	<1	<1			
pH	6·5–6·9	6·8–7·0	7·1–7·2	6·6–7·2	6·6–7·2
Solubility (%)	100	90–98	92	95–98	97–98
Farinograph water absorption[c] (%)	271	143	278	282	292

[a]Data of Southward and Aird (1978).
[b]Assumed values in parentheses.
[c]Data of Knightbridge and Goldman (1975).

the caseinates includes approximately 1·8 per cent derived from the organic phosphorus which forms an integral part of the casein molecule. With a pH generally in the range of 6·5–7·0, sodium caseinate will usually contain 1·2–1·4 per cent sodium, while the calcium content of calcium caseinate is generally in the range 1·3–1·6 per cent.

COMPOSITION OF CO-PRECIPITATES

The proximate compositional analysis of granular (insoluble) co-precipitates, prepared from curd which was washed twice in water, is shown in Table I. The data show that the calcium content of co-precipitates decreases steadily with a reduction in the ash content, both of which are caused by a decrease in the pH of precipitation. The fat content of the co-precipitates increases from 0·6 to 0·9 per cent as the pH of precipitation is reduced from 5·9 to 4·9. The high ash content of high calcium co-precipitate leads to a consequential reduction of 3–5 per cent in the protein content of the product relative to lactic and acid caseins and acid co-precipitate.

Proximate compositional analysis and physical properties of soluble co-precipitates are presented in Table II. Soluble high calcium co-precipitate has a very high ash content (13·5 per cent) due in part to the presence of sodium tripolyphosphate and contains almost 2 per cent sodium. Soluble acid co-precipitate has a composition similar to that of sodium caseinate (Table II). In order to render medium calcium co-precipitate substantially (> 90 per cent) soluble at pH 7·5, it is necessary to use both sodium hydroxide and sodium tripolyphosphate (or other complex phosphate).

PROPERTIES OF CASEINATES AND CO-PRECIPITATES

Solubility in Water

The granular co-precipitates shown in Table I are, like acid and rennet caseins, insoluble in water. Of the 'soluble' casein products shown in Table II, sodium caseinate is the only one which is completely soluble, i.e. can be centrifuged without depositing any insoluble particles, and produces a pale straw-coloured translucent solution. Calcium caseinate and the various soluble co-precipitates are not completely soluble in water, but vary in solubility from just over 90 per cent to about 98 per cent in solubility. Calcium caseinate produces a very white colloidal dispersion. The heat denaturation of the whey proteins during the manufacture of co-precipitates causes these products to exhibit some measure of insolubility, though the amount of insoluble material is only a fraction of the whey protein present in the co-precipitate. Variables which affect the appearance of solutions of co-precipitates and their solubility have been discussed by Smith and Snow (1968) and Hayes *et al.* (1969).

Functional Properties

The terms 'functional property' and 'functionality' are generally used to describe the physical effects of (in this case) casein products on foods. 'Functionality' may be used to include almost all properties (e.g. colour, flavour, nutrition and solubility as well as the functional properties listed below) (American Chemical Society, 1981). However, the functional properties of more specific interest which are provided by casein products include water absorption, fat emulsification, viscosity and gelation, whipping and foaming, and texturisation (Morr, 1982). A number of reviews of milk protein functionality have recently been published (Morr, 1979, 1981, 1982; Fox and Mulvihill, 1983; Kinsella, 1983).

Farinograph Water Absorption

The water absorption characteristics of the dispersible and soluble casein products are shown in Table II. These values vary from about 130 to approximately 290 per cent (i.e. 1·3–2·9 g water absorbed/g product). By contrast, acid and rennet caseins and the insoluble co-precipitates have much lower absorptions (80–110 per cent), while calcium caseinate shows an intermediate absorption (143 per cent) (Knightbridge and Goldman, 1975). The water absorption properties give an indication of the suitability of casein products for different bakery applications; products with high water absorption (>200 per cent), such as sodium caseinate and the soluble co-precipitates, function as water binders in intermediate and high moisture foods, while products of medium (100–200 per cent) (e.g. calcium caseinate, dispersible and insoluble co-precipitates) and low water absorption (< 100 per cent) (e.g. casein and insoluble co-precipitates) find a variety of applications ranging from use in bread and baked goods to confections and frozen desserts (Knightbridge and Goldman, 1975).

Viscosity and Gelation

The viscosity characteristics of solutions of sodium caseinate, rennet casein (dissolved in sodium tripolyphosphate) and soluble co-precipitates have been studied (Towler, 1974; Southward and Goldman, 1978; Roeper and Winter, 1982) over a range of concentration and temperature. The products with the highest viscosity, soluble rennet casein and soluble high calcium co-precipitate, also have the highest calcium content. Calcium caseinate has a relatively high calcium content but a comparatively low viscosity, since it exists in water as a white, colloidal dispersion (Roeper, 1977b).

The viscosity of casein solutions has an important bearing on their manufacturing costs and in their applications. Thus, a number of investigations have been carried out to decrease the viscosity of casein solutions by enzymatic hydrolysis either during manufacture (Hooker *et al.*, 1982), or for paper coating (Muller and Hayes, 1963; Salzberg and Simonds, 1965). On the other hand, an increase in viscosity, leading to the formation of casein gels, can also be of value in various food applications. Thus, acidified casein gels may have application in the preparation of milk protein-fruit juice drinks and jellies (Korolczuk, 1982).

Fat Emulsification

Sodium caseinate has been used for many years as an emulsifier of oils and fats in such food products as comminuted meats, whipped toppings and coffee whiteners. Many studies have been carried out to determine the relative emulsifying power and emulsion stability of milk proteins (Tornberg, 1980; Reimerdes and Lorenzen, 1983; Fox and Mulvihill, 1983), but it is apparent that a standardised procedure is needed to enable true comparisons to be made among the different groups of research workers.

Whipping and Foaming

The surfactant properties of milk proteins make them particularly useful in whipping and foaming applications, such as in whipped toppings. A brief discussion on foam formation and stability is given by Kinsella (1983). Measurement of whipping and foaming properties of casein products has been made by Southward and Goldman (1978), Goldman and Toepfer (1978) and Kitabatake and Doi (1982).

Texturisation

The ability of caseins to form or alter the texture of various foods has been used to produce meat-like products either by spinning from a caseinate solution (Amiot *et al.,* 1979; Kelly *et al.,* 1983) using techniques similar to those used in the late 1930s for producing textile fibres (Ferretti, 1938), or by production of a 'curd' (Goldman, 1974). Casein has also been used extensively as the basic texture-forming matrix in imitation cheese products (Vernon, 1972).

Flavour of Casein Products

The flavour of casein products has assumed a much greater importance during the past 10 years as their use in foods has increased (Southward *et al.,* 1978). Various workers have studied the effects of manufacturing technique and treatment of the final casein on its flavour (Cayen and Baker, 1963; Hansen *et al.,* 1970), and considerable research has been undertaken in an attempt to isolate the gluey (Ramshaw and Dunstone, 1969a, b), or musty (Walker, 1972, 1973; Walker and Manning, 1976) storage off-flavours in acid caseins and their derivatives. It has generally been found, however, that fresh curd caseinates (Cayen and Baker, 1963) and rennet casein (Walker, 1972) have the blandest and most stable flavour of the casein products. The flavour of co-precipitates tends to follow a similar pattern to that of the caseins, high-calcium co-precipitates being more stable than low-calcium (acid) co-precipitates, and fresh curd, soluble co-precipitates being better than those reconstituted from dry, granular, insoluble co-precipitates (Southward and Goldman, 1978). The co-precipitates, as a class, may also tend to exhibit 'cooked flavour' overtones as a result of the high heat treatment given to the milk during their manufacture. Particle size of granular caseins and co-precipitates also appears to affect their flavour; finely-ground products tend to exhibit stronger off-flavours than coarser fractions.

Nutritional Properties

Amino acid analysis

The nutritive quality of a protein is determined primarily by its essential amino acid composition. For adult man, eight amino acids are essential, i.e. they must be supplied in the diet. These are isoleucine, leucine, lysine, methionine, phenylalanine, threonine, tryptophan, and valine. The infant requires histidine as well. In 1973, an 'ideal' protein composition was published (Food and Agriculture Organization, 1973) which was derived from the probable pattern of amino acid requirements in man. By comparing the essential amino acid composition of dietary proteins with that of the reference protein, a preliminary evaluation of their nutritive quality can be made. In comparison with the reference protein, caseins are only deficient (or limiting) in the sulphur-containing amino acids methionine and cystine.

Biological evaluation

At present, chemical analysis of the amino acids present in proteins is not adequate as the sole measure of their quality and nutritive value. Biological availability of the amino acids cannot always be determined by chemical methods, and for this and other reasons, biological tests of protein quality continue to be necessary. Protein Efficiency Ratio (PER), Biological Value (BV) and Net Protein Utilization (NPU) are all used to assess protein quality. Rats are the most widely used test animals and, of the methods based on the growth of rats, the PER is the most widely used method. A standardised version of this test, which is the officially recommended method for the biological estimation of protein quality (Association of Official Analytical Chemists, 1970), uses ANRC casein as the reference protein, with a PER of 2·5. PER is defined as the ratio of weight gained to protein eaten. The nutritional aspects of caseins and other milk proteins are more fully described by Hambraeus (1982).

Casein Production and Trade

From the data in Table III, it can be seen that the mean annual production of casein during 1977–83 has been approximately 225 000 tonnes, substantially higher than earlier estimates of 150–170 000 tonnes (Southward and Walker, 1980, 1982). This higher figure is also substantiated by Morr (1984), who estimates recent (1982–3) production of casein at 220–226 000 tonnes. Major producers are New Zealand (mean production 60 000 tonnes), France (31 000 tonnes) and Poland (29 000 tonnes). The USSR also appears to produce some 25–30 000 tonnes of casein each year, but complete statistical data for this period are not available. World production in 1984 is expected to decline from 1983 levels (United States Department of Agriculture, 1984).

According to the figures for exports and imports of casein, shown in Tables IV and V, respectively, approximately 150 000 tonnes of casein

TABLE III
Production of casein 1977–83 (tonnes)[a]

	1977	*1978*	*1979*	*1980*	*1981*	*1982*	*1983*
Austria	700	700	600	1 800	1 700	1 800	666[b]
Denmark	700	1 200	2 600	4 500	4 400	6 800	8 900
France	17 800	21 900	31 700	38 700	29 785[c]	39 982[c]	38 798[c]
Irish Republic	7 600	11 200	12 900	17 000	14 483[c]	19 400[c]	24 400[c]
Netherlands	13 000[d]	16 000[d]	18 000[d]	15 000[e]	19 000[f]	19 000[f]	20 000
Norway	3 500	3 600	3 500	3 400	4 322[c]	4 212[c]	4 270[c]
Poland	29 000[g]	30 000[f]	30 000[e]	30 000[e]	15 556[b]	25 475[b]	40 645[b]
Spain	1 287[b]	1 063[b]	2 351[b]	2 151[b]	1 892[b]	1 850[b]	1 856[b]
USSR	26 300[h]		27 400[i]			24 400[i]	28 000[i]
United Kingdom	1 000[d]	1 000[d]	1 000[d]	3 600	2 300	4 100	3 250[b]
West Germany	12 841[b]	14 800	16 300	19 200	15 707[c]	17 010[c]	21 105[c]
Argentina	4 355[b]	3 300	2 700	2 700	2 800	2 200	2 000[b]
Uruguay	818[b]	900[b]	1 192[b]	1 940[b]	1 840[b]	2 331[b]	4 182[b]
Australia[j]	18 484[k]	17 200[k]	15 100[k]	15 120[k]	8 437[k]	11 465[k]	13 273[l]
New Zealand[m]	56 724[n]	63 326[n]	66 162[n]	59 590[n]	47 384[n]	65 151[n]	63 065[l]

[a]Source: Commonwealth Secretariat (1982–84) unless shown otherwise.
[b]Source: Food and Agriculture Organization (1980–84).
[c]Source: Agra Europe (London) Ltd (1984c).
[d]Source: Milk Marketing Board (1979–83).
[e]Source: Agra Europe (London) Ltd (1981).
[f]Source: United States Department of Agriculture (1979–84).
[g]Source: Anon. (1978).
[h]Source: Agra Europe (London) Ltd (1978).
[i]Source: Agra Europe (London) Ltd (1984a).
[j]Twelve months ending 30 June in following year.
[k]Source: Australian Dairy Corporation (1978–83).
[l]Source: Agra Europe (London) Ltd (1984d).
[m]Twelve months ending 31 May in following year.
[n]New Zealand Dairy Board (1982–84).

TABLE IV
World trade in casein. Exports 1977–83 (tonnes)[a]

	1977	*1978*	*1979*	*1980*	*1981*	*1982*	*1983*
Belgium/Luxembourg	232	184	348	596	499	157	
Denmark	1 005	1 230	2 602	2 538	3 963	7 148	4 200[b]
France	16 612	16 395	26 948	32 715	24 812	31 349	30 000[c]
Irish Republic	5 963	9 701	14 978	14 741	13 731	17 301	20 000[d]
Italy	132	242	313	241	361	415	
Netherlands	1 206	1 472	4 763	3 725	4 186	3 225	3 000[c]
Poland	24 000[b]	26 000[b]	18 000[b]	15 000[c]	8 000[c]	8 000[c]	15 000[d]
United Kingdom	1 784	1 245	2 027	3 632	2 991	4 175	5 000[d]
West Germany	9 637	10 467	12 287	12 702	12 311	14 671	15 300[c]
Argentina	4 000[d]	2 000[d]	1 000[d]	900[b]	1 100[b]	1 000[c]	1 000[d]
Canada	52[e]	127[f]	103[f]	78[f]	1[f]	34[f]	
Australia[g]	16 277[h]	16 560[h]	13 687[h]	10 364[h]	10 072[h]	9 929[h]	10 000[i]
New Zealand[g]	71 971[d]	67 351[j]	67 875[j]	49 859[j]	54 848[j]	54 989[j]	57 993[j]

[a]Source: Milk Marketing Board (1979–83) unless shown otherwise.
[b]Source: Commonwealth Secretariat (1982–84).
[c]Source: Agra Europe (London) Ltd (1984b).
[d]Source: United States Department of Agriculture (1979–84).
[e]Source: Agra Europe (London) Ltd (1979).
[f]Source: LaBrosse and McSorley (1981); Aube (1983).
[g]Twelve months ending 30 June of following year.
[h]Source: Australian Dairy Corporation (1978–83).
[i]Source: Agra Europe (London) Ltd (1984c).
[j]Source: New Zealand Department of Statistics (1979–84).

TABLE V
World trade in casein. Imports 1977–83 (tonnes)[a]

	1977	*1978*	*1979*	*1980*	*1981*	*1982*	*1983*
Belgium/Luxembourg	4 449	5 277	6 697	5 938	4 345	4 166	
Denmark	2 172	2 056	2 332	2 363	2 051	1 931	
France	7 871	8 264	7 301	6 962	3 436	6 100	5 000[b]
Greece				304	444	325	
Irish Republic	223	375	449	653	686	1 113	1 000[b]
Italy	10 347	10 125	11 759	11 060	15 764	9 157	10 000[c]
Netherlands	4 747	13 951	14 331	10 650	7 419	8 411	10 000[b]
United Kingdom	4 747	4 631	5 732	4 396	4 669	4 487	4 500
West Germany	17 644	19 145	20 141	19 352	14 053	14 962	14 600[c]
Canada	2 275[e]	1 195[f]	2 483[f]	2 330[f]	1 471[f]	1 198[f]	
United States	65 429[b]	62 204[b]	68 415[b]	69 019[b]	57 981[b]	80 196[b]	72 381[b]
Japan	17 884[g]	21 381[g]	23 484[g]	22 265[g]	18 690[h]	22 377[h]	23 153[h]

[a]Source: Milk Marketing Board (1979–83) unless shown otherwise.
[b]Source: United States Department of Agriculture (1979–84).
[c]Source: Agra Europe (London) Ltd (1984b).
[d]Source: Commonwealth Secretariat (1982–84).
[e]Source: Agra Europe (London) Ltd (1979).
[f]Source: LaBrosse and McSorley (1981); Aube (1983).
[g]Source: Japan Ministry of Agriculture, Forestry and Fisheries (1982).
[h]Source: Japan Finance Ministry (1981–83).

enters international trade each year. New Zealand contributed 60 000 tonnes of this total (40 per cent), while French and Polish exports amounted to an average of 25 000 and 16 000 tonnes, respectively, during this period. The major casein importing countries are United States (66 000 tonnes yr^{-1}), Japan (20 000 tonnes) and West Germany (17 000 tonnes). Data from 1920 to 1978 have been reported earlier (Southward and Walker, 1980, 1982).

INDUSTRIAL USES OF CASEIN

The principal industrial uses of acid casein are shown in Table VI.

Casein as an Adhesive

Wood glues

Casein glues have been marketed in two forms — prepared glues and wet-mix glues. Prepared glues are sold in powder form and contain all the necessary ingredients except water. The user simply mixes the powder with water (usually in the proportions of 1 of powder to 2 parts water by weight). Once mixed, the glue must generally be used within one day or less. Wet-mix glues are made as required from ground casein, water, and the additional chemicals required by the formula (Browne and Brouse, 1939).

Besides casein and water, an alkali must be used to dissolve the casein. This is commonly sodium hydroxide and provides the third ingredient in the production of a simple glue. Where a water-resistant

TABLE VI
Principal industrial uses of acid casein

Coatings for paper and board
Adhesive for wood, e.g. plywood
Paints
Joint cements
Textile sizing
Synthetic fibres
Horticultural spreaders
Leather tanning
Stock foods

glue is required, it is necessary to also add lime to this formulation. The lime promotes cross-linking of the casein molecules and, over a period of several hours, will cause a casein glue to irreversibly form a jelly which is insoluble in water. Many different additives may be used to promote various properties of casein glues. For instance, addition of sodium silicate may be employed to prolong the working life of a glue, while use of copper chloride will increase the glue's water resistance.

Casein glues are often employed in interior wood working, for example in the production of interior doors for houses. In 1967, approximately 5000 tonnes of casein were used for glues in the USA (Fox, 1970).

Paper coating

In a development parallel to that of adhesives for wood, casein has also been employed as an adhesive in coating of board and paper (Salzberg *et al.,* 1961), especially in the production of high quality art paper. During the period from 1940 to 1962, paper coating accounted for the major use of casein — generally more than 50 per cent (Robinson, 1964) as shown in Fig. 7. Even in 1967, paper coating accounted for about one-third of the approximately 50 000 tonnes of casein products imported into the USA (Fox, 1970).

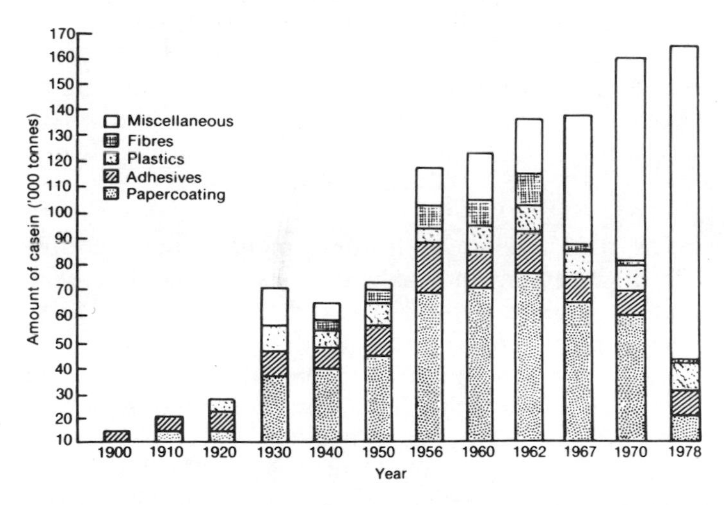

Fig. 7. Probable utilisation of casein 1900–1978. Sources: for period 1900–1962: Robinson (1964); for period 1967–78: Southward and Walker (1980). (Reproduced from Southward and Walker (1980) by permission of the publishers.)

When used in coatings with a high solids content, casein forms highly viscous solutions (Salzberg *et al.,* 1961) which limit the solids content of the coating mix and hence the speed of the coating process. The viscosity of casein solutions has been reduced in commercial applications by the addition of urea or dicyandiamide (Salzberg and Marino, 1975) which are considered to be flow modifiers. There has been a number of successful attempts to lower the viscosity of casein solutions by alkaline or enzymatic hydrolysis (Muller and Hayes, 1963; Salzberg and Simonds, 1965). Disulphide bond-reducing agents, such as mercaptoacetic acid and its ammonium and sodium salts, and 2-mercaptoethanol, have also been examined for this purpose (Towler *et al.,* 1981).

Sizing

The film-forming ability of casein is retained even when it is deposited from a very dilute solution. Thus, casein may be applied to wool to reduce its felting properties (Jackson and Backwell, 1955) and for textile sizing generally. Paper surfaces may be made repellent to solvents, oils and greases by applying a film of casein to them ('sizing'). This casein film may also be given a high degree of water resistance by the inclusion of 'hardeners' in the solution or by post application to the film. Casein then becomes a permanent finish which is applied to paper to enhance its lustre or stiffness (Salzberg, 1965).

Casein may also be employed as an adhesive for other paper bonds such as laminates, cigarette side-seaming, bottle and can-labelling, etc. (Salzberg, 1962; Salzberg *et al.,* 1974), and in bonding the striking heads to matches.

Casein Fibre

The process of producing casein fibres includes the following steps. Casein is dissolved in an alkali, such as sodium hydroxide, at a concentration of about 200 g litre^{-1}, and the solution is then forced through a spinneret into a coagulating bath. The bath usually contains acid, inorganic salts and often heavy metal salts. The fibres thus formed resemble wool, except that they have a lower tensile strength and do not 'felt' (i.e. shrink on washing) like wool.

Following commercial development of casein fibre by Ferretti (1938), considerable effort was expended in Europe and the USA in perfecting new techniques for hardening this regenerated casein fibre. Amongst the more successful hardening processes was acetylation (Brown *et al.,* 1944).

The principal proprietary casein fibres which were developed throughout the world in the decade from 1936 to 1945 include (Wormell, 1954): *Aralac* (National Dairy Products Corp., USA), *Casolana* (Co-op Condensfabriek Friesland, Netherlands), *Fibrolane* (Fig. 8) (Courtaulds Ltd, UK), *Lanital* and *Merinova* (Snia Viscosa, Italy). Of these, only *Fibrolane* and *Merinova* were still in production by 1971 (Roff and Scott, 1971).

Estimates of world production suggest that around 5000 tonnes per annum of casein fibre were made during the decade 1940–50. Manufacture increased during the 1950s to about 10 000 tonnes by 1960 (Robinson, 1964) (see also Fig. 7).

Casein fibres were employed during and after the war years, usually in combination with wool and other fibres, such as cotton, viscose, rayon, etc., in a variety of products, such as flannel, woollen spun cloth (overcoats, blankets), felt hats (up to 25 per cent casein fibre with wool), filling materials such as artificial horsehair, and in carpets and rugs (Wormell, 1954). Bristles were also produced from casein fibres (McMeekin *et al.*, 1945) for use in brushes of various types.

The importance of casein fibre in the USA and Europe has now declined in the face of competition from other fibres. However, copolymer fibres containing casein have been prepared in Japan as a

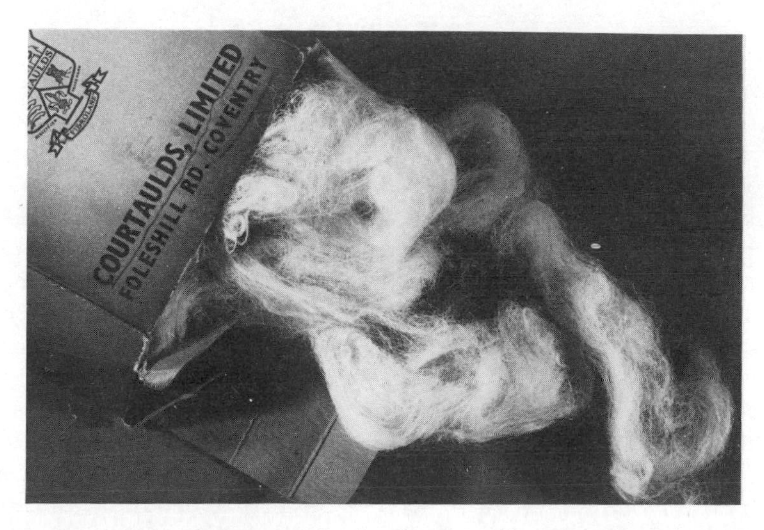

Fig. 8. Courtauld's regenerated casein fibre — *Fibrolane.* (Reproduced from Southward and Walker (1980) by permission of the publishers.)

substitute for silk (Fox, 1970), and the author has a number of ties, recently manufactured in Japan from casein and casein co-polymer fibres.

Casein in Paints

Casein has been used for many years as a binder and pigment vehicle in water-based paints (Scholz, 1953), and as a stabiliser/emulsifier for oil paints (Mummery, 1949). More recently, it has come to be used as a pigment dispersing agent, stabiliser and bodying agent in latex paints, (Melsheimer and Hoback, 1953; Salzberg, 1964), usually at a concentration of 1–2 per cent (w/w) of the paint.

Casein in the Leather Industry

The use of casein in the leather industry is confined almost entirely to the last of the finishing operations, which consist in coating leather with certain preparations, and then subjecting it to mechanical operations, such as glazing, plating, brushing and ironing. After finishing in this way, leather is said to have been seasoned (Cavett, 1939; Landmann, 1962). Casein is generally used in these operations in combination with other substances, such as acrylates.

Casein in Animal Feeds

Formaldehyde-treated casein for ruminants

During the past ten years or so, extensive nutritional studies have been carried out in the feeding of casein to ruminants. During normal digestion in the rumen, proteins, such as casein, are almost completely broken down by the rumen microflora to ammonia, and therefore become unavailable for nutrition of the animal. This problem may be overcome by 'protecting' the casein by first treating it with formaldehyde. Under such circumstances, the treated casein passes undigested through the rumen (neutral pH), and is digested under acid conditions in the abomasum to provide a nutritional supplement. Formaldehyde treated caseins have consequently been examined for their effectiveness in promoting wool growth in sheep (Reis and Tunks, 1969; Barry, 1972), and milk production in lactating cows (Wilson, 1970; Broderick and Lane, 1978). Studies have also been undertaken to determine the effect of feeding ruminants with polyunsaturated oils, encapsulated in formalin-treated caseins, on the amount of polyunsaturated fat in meat and milk (Scott *et al.,* 1970; McGilliard, 1972).

Pet foods

Casein, generally in the form of sodium caseinate, is also finding application as a nutritional supplement and binder in calf milk replacers (Battelle Memorial Institute, 1974) and various pet foods (Burkwall *et al.,* 1976), where it may comprise from 3 to 30 per cent of the food. Figure 9 shows an example of one type of pet food which contains casein.

Miscellaneous Uses of Acid Casein

Amongst the large number of other industrial applications in which casein has been used (or at least claimed to be used) are the production of concrete, particularly in Eastern Europe; in joint cement for plaster board (Robinson, 1964), as an emulsifying agent in asphalt and bitumen, in photolithography (Salzberg, 1965), as a reinforcing agent for rubber (Genin, 1961), in smoking mixtures, in soaps and dishwashing liquids and in cosmetics, such as cold-wave lotions (Genin, 1966), hair sprays and hand creams (Salzberg, 1967). Casein has also been used as a spreader in horticultural sprays and even as a fertiliser. In beverages, casein has been employed as a clarifying agent for wines and beer, and in removal of colour from apple juice (Lodge and Heatherbell, 1976), at

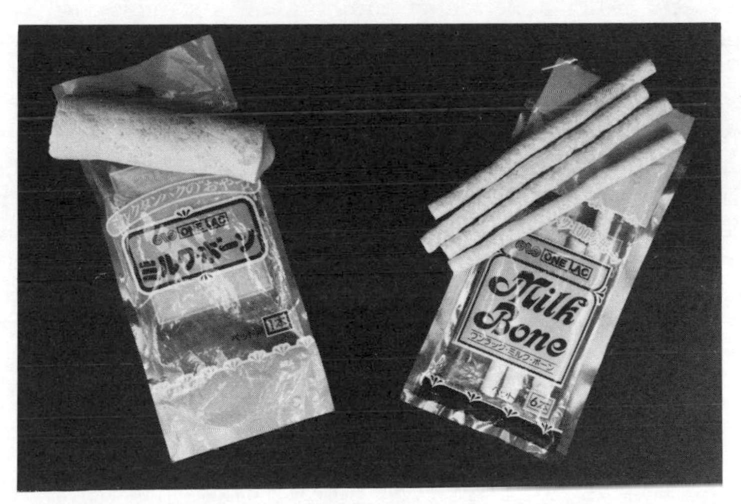

Fig. 9. Pet food — 'Milk Bone', made from casein.

a concentration usually less than 0·1 per cent. One of the most recent uses claimed for casein is for the extraction and recovery of heavy metals from mining, tanning and electro-plating wastes (Davey *et al.*, 1980).

USES OF RENNET CASEIN

The principal industrial uses of rennet casein are in the production of plastics, such as buttons, buckles and imitation ivory knife handles. Casein plastics were first produced before the turn of the century, and commercial production of Galalith (Greek: 'milk stone') was commenced in France and Germany by the International Galalith Gesellschaft Hoff & Co. in 1904. After the First World War, the manufacture of casein plastics increased all over the world, producing such trade names as *Erinoid* (UK), *Aladdinite* (USA), *Casolith* (Netherlands) and *Lactoloid* (Japan) (Brother, 1939). Since that time, it is estimated that about 10 000 tonnes of casein plastic have been made throughout the world each year (Southward, 1974).

Rennet casein produces a plastic which is far superior to that from acid casein. In the production of casein plastic, the casein (if unground) is milled to pass a 40 mesh (350 μm) sieve and mixed with dyes and fillers, as required, and water to a final moisture content between 20 and 35 per cent. After a period of equilibration, the moistened casein mixture is extruded through heated nozzles at relatively high pressures (15–75 MPa). The plastic rod which emerges is subsequently cooled in water, trimmed and cut into button blanks which are cured for a period of several days or weeks in a dilute solution of formaldehyde prior to machining and polishing (McDowell *et al.*, 1976; Munro *et al.*, 1980). Alternatively, if sheet plastic is required, the freshly extruded rods may be placed side by side and formed into sheets in a heated hydraulic press. Samples of casein plastics are shown in Fig. 10. Generally, casein plastics are fabricated today into buttons and buckles. Only a small proportion of this production is channelled into other items, such as imitation ivory knife handles.

EDIBLE USES OF CASEIN PRODUCTS

While casein and its derivatives were used in foods well before 1940, mainly for nutritional purposes (Schuette, 1939), it was not until the

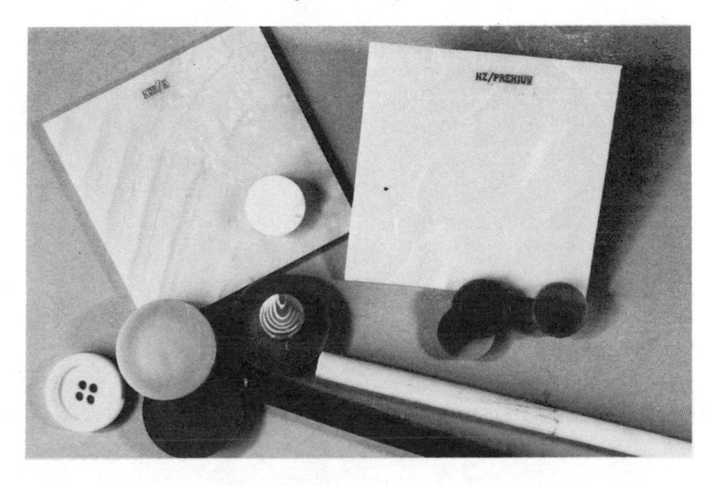

Fig. 10. Sheets, rods, buttons and button blanks made from rennet casein plastic. (Reproduced from Southward and Walker (1980) by permission of the publishers.)

1960s that commercially significant amounts of casein products were employed for this purpose (Poarch, 1967; Fox, 1970), particularly in the developed countries, such as the United States (Hammonds and Call, 1970; Centre National du Commerce Exterieur, 1970) and in Europe (Kreveld, 1969). The significant increase in the 'Miscellaneous' uses category for casein products in 1967 and the 1970s (Fig. 7) is due mainly to the use of casein and its derivatives in food and feed applications. The uses of milk proteins in formulated foods have been reviewed recently (Evans, 1982; Southward and Walker, 1982), and the principal uses of casein products in foods are shown in Table VII.

TABLE VII
Principal uses of casein products in foods

Meat products
Coffee whiteners
Whipped toppings
Bakery
Instant breakfasts and beverages
Pasta
Cheese products
Pharmaceuticals

Casein in Meat Products

The main properties imparted by casein products to comminuted meats are protein enrichment, fat emulsification, water-binding and general improvement of consistency (Visser, 1982). Usually, the amount of added casein is fairly low (less than 5 per cent of the weight of the meat product), though higher amounts of up to 20 per cent of the meat may be employed for specific purposes, such as lower cost or an increase in nutritional quality. Some countries have promulgated regulations which restrict the amount of non-meat proteins in meat products, however, thus highlighting some basic problems in this field (Pfaff, 1974; Fogh, 1975). Synthetic or artificial meats (meat analogues) which possessed the fibrous and chewy texture of meat muscle have also been prepared from casein using the following two basic techniques:

1. The spinning of protein fibre by methods developed in the 1930s for producing textile fibres, in which a protein solution is forced through spinnerets into a coagulating bath (Burgess and Coton, 1982).
2. The production of chewy meat-like gels either by precipitation from concentrated casein solutions, or from moistened casein products by such techniques as extrusion cooking (Skurray and Osborne, 1976).

One of the main problems encountered in the production of fibrous casein for foods has been its relative sensitivity to heat; when such fibres are cooked in a moist environment, they rapidly lose both their strength and fibrous nature. Similar problems were encountered in the manufacture and use of casein textile fibres. Future prospects for casein texturisation have been recently reviewed (English, 1981).

Casein in Baked Products and Cereals

Early uses of casein derivatives in baked products were in various dough and biscuit formulations for the principal purpose of nutritional fortification of the wheat flour. One of the most important functional characteristics of the casein derivatives in this application was found to be water binding.

While sodium caseinate has apparently found acceptance in controlling texture and uniformity of such baked products as doughnuts, biscuits, waffles and yeast raised doughs (Craig and Colmey, 1971), it has also been found to produce doughs with an unacceptable 'slimy' mouthfeel in some farinaceous products. However, when it is employed in a mixture with whey, its suitability for use in baked foods is greatly extended (Kirk, 1973).

Considerable interest in co-precipitates as milk protein ingredients for baked products was generated some 14 years ago (Craig and Colmey, 1971). Co-precipitates having a range of controlled water absorption characteristics were prepared and used in such foods as the Australian milk biscuit (Townsend and Buchanan, 1967), cake mixes for diabetics (Buchanan and Henderson, 1971), bread and in a pastry glaze (Bready, 1966).

Casein in Whipped Toppings

The main constituents of whipped toppings are water, vegetable fat, sugar(s), protein (such as sodium caseinate), emulsifiers and stabilisers. Since sodium caseinate is not defined as a dairy product in the United States, whipped toppings which incorporate this protein ingredient are considered to be 'non-dairy' substitutes for cream (Webb, 1970).

In the preparation of a whipped topping, the water-soluble ingredients (i.e. sugar, sodium caseinate, stabiliser) are usually blended together and then dissolved in the water (Knightly, 1968; Hedrick, 1969). The fat and emulsifiers are melted together, generally at a temperature of 38–46 °C, and added to the aqueous solution which is heated to the same temperature. Following pasteurisation and two-stage homogenisation, generally at low pressures of about 7 and 3·5 MPa for first and second stage, respectively, the whipped topping mixture may be either cooled rapidly to below freezing point or spray-dried. The effect of processing variations on the properties of whipped toppings has been reviewed by Knightly (1968).

In whipped toppings, sodium caseinate functions as a film former to entrap aerating gases (Knightly, 1968), as a fat encapsulating agent (Cameron *et al.*, 1959), stabiliser (Pader and Gershon, 1965) and as a bodying agent (Clarke and Love, 1974). The amount of caseinate employed for this purpose is usually 5–10 per cent of the dry weight of ingredients (Centre National du Commerce Exterieur, 1970). Other casein derivatives, such as casein hydrolysates, and calcium and potassium caseinates are also occasionally employed in whipped topping formulations.

Casein in Ice Cream and Frozen Desserts

While the effect of sodium caseinate in products such as ice cream was studied as early as 1935 (Bird *et al.*, 1935), it has not had an extensive commercial use in this application because of various legal restrictions. For example, in the United States, standards of identity for ice cream do not permit the use of casein derivatives as part of the minimum requirement for 10 per cent milk solids-not-fat.

Ice cream substitutes have appeared, however, in the United States market (Hedrick, 1969; Centre National du Commerce Exterieur, 1970), and a number of them have contained sodium caseinate. The function of sodium caseinate in frozen desserts and ice cream is to impart body (Webb, 1970), and to act as a stabiliser (Little, 1966). It serves a similar function in milk shakes and instant puddings where foam stability is also important. The amount of caseinate commonly used in ice cream and other desserts varies from 1 to 10 per cent according to the manufacturer's formula (Centre National du Commerce Exterieur, 1970).

Casein in Coffee Whiteners

The greatest single use of sodium caseinate in food products in the United States during the 1960s was in coffee whiteners (Centre National du Commerce Exterieur, 1970). In 1968, approximately 4000 t of sodium caseinate, representing 22·7 per cent of the total food use of caseinates for that year, was employed in coffee whiteners. Data recently obtained suggest that approximately 6000 t of caseinate was used in coffee whiteners in 1978 (United States International Trade Commission, 1979).

Surprisingly little information has been published on the production of 'non-dairy' coffee whiteners, presumably because the major manufacturers have preferred to keep the details confidential. They are based, however, on vegetable fat and are formulated in a similar manner to whipped toppings. Three retail packs of coffee whiteners are shown in Fig. 11.

The function of sodium caseinate (or other protein) in coffee whiteners is multipurpose: it provides emulsification, some whitening, imparts body, improves flavour and promotes resistance to 'feathering' (Clarke and Love, 1974). It is usually incorporated into the formulation at a level of 7–10 per cent of the dry ingredients. Coffee whiteners may be used in the fluid or frozen state, or as a spray-dried powder (Knightly,

Fig. 11. Coffee creamers/whiteners from the United States, Korea and New Zealand.

1969). The importance of manufacturing variables is reviewed by Knightly (1969), who describes the various stages in the preparation of coffee whiteners. While the process is similar to that described for the production of whipped toppings, one important difference is in the first stage homogenisation pressure. This is recommended to be 14–17 MPa (Hedrick, 1969; Knightly, 1969) in order to reduce the diameter of the fat globules to 0·7–1 μm, whereas a lower first stage pressure of about 7 MPa is employed in the homogenisation of whipped toppings.

Casein in Instant Breakfast, Imitation Milk, and Other Beverages

Instant breakfast formulations are a feature of the United States domestic food scene (Centre National du Commerce Exterieur, 1970). They usually contain skim-milk powder, sucrose, sodium caseinate, vitamins and minerals, and are sold in different flavours. One sachet of an instant breakfast may be mixed with a glass of milk to provide adequate nutrition for an adult's breakfast. Instant breakfast formulations usually contain 2–4 per cent sodium caseinate (Centre National du Commerce Exterieur, 1970), though some variants may contain considerably more. One product, which is formulated specifically for nutritional purposes as a diet beverage (Sobotka, 1971), is believed to contain about 30 per cent sodium caseinate.

The incorporation of casein products (mostly in the form of sodium caseinate) in imitation milk beverages, which are based on vegetable fat and a carbohydrate source such as corn syrup solids, also appears to have originated mainly in the United States (Hedrick, 1969; Rusch, 1971; Filer, 1972), although it has been observed that some artificial milk has also been produced in the United Kingdom (Waite, 1972), and in the Soviet Union (Voropalva and Ionkina, 1979). Probably the main reasons for the appearance of imitation milks has been the lower ingredient cost and absence of lactose, for which some people show intolerance, compared with cow's milk.

Formulations of imitation milks generally show them to consist of 3–4 per cent vegetable fat, 1–5 per cent protein (usually 1–2 per cent of either sodium caseinate or soy protein), 6–10 per cent of carbohydrate (usually corn syrup solids and possibly sucrose), and various stabilisers, emulsifiers, etc. Vitamins and minerals are other optional ingredients (Hedrick, 1969; Rusch, 1971; Filer, 1972; Waite, 1972). Processing of these products is carried out in a manner similar to those methods used for coffee whiteners and whipped toppings (Rusch, 1971). Generally, imitation milks are nutritionally inferior to cow's milk and to filled milk, especially in terms of protein, vitamin and mineral content (National Dairy Council, 1968). The low protein content is mainly a consequence of problems encountered by the formulations in providing a beverage of acceptable flavour and consistency (Filer, 1972).

The nutritional properties of casein may be used to good advantage, however, in the fortification of various milk products, including fluid milk, as recent reports have indicated (Solms-Baruth, 1972; Downes and Merwe, 1982).

A recent new application for sodium caseinate is as a stabiliser in cream-based liqueurs which are one of the fastest-growing markets for cream. The main source of cream-based liqueurs has been the Irish Republic, although recent work on these products has been carried out in Scotland (Banks *et al.,* 1983).

Casein in Cultured Milk Products, Soups and Gravies

In the United States, the use of sodium caseinate in sour cream products based on vegetable fat, i.e. imitation sour cream, has been reported (Centre National du Commerce Exterieur, 1970). In such products, the sodium caseinate acts as a stabiliser, and as an emulsifier for the fat. The amount of sodium caseinate incorporated into sour cream has been estimated to be 2–3 per cent of the weight of all ingredients (Centre

National du Commerce Exterieur, 1970). Mayonnaise is another product in which sodium caseinate has been used, its function being to act as an emulsifier for the fat (Cameron *et al.,* 1959).

In cultured dairy foods, sodium caseinate has been used as a stabiliser in yoghurt in the United States (Centre National du Commerce Exterieur, 1970; Modler *et al.,* 1983), and in cultured cream products in the Soviet Union for the purpose of increasing the viscosity (Bogdanova *et al.,* 1978).

Casein products have also been used for nutritional fortification and thickening of soups and gravies (Powell, 1974), while flavour enhancement of these other foods has been achieved by using hydrolysates of casein (Connell, 1966). Several substitute foods based on caseins, such as vegetable cutlets, synthetic caviar (Nesmeyanov *et al.,* 1971) and a nut substitute have also been prepared in Poland and the USSR.

Casein in High-fat Powders, Shortenings and Spreads

The emulsifying properties of casein derivatives have been used to good effect in a number of high-fat foods, e.g. butter substitutes (Roberts, 1959) and shortening products for baking purposes (Cameron *et al.,* 1959). Casein, usually in the form of sodium caseinate, is generally incorporated at a level of up to 10 per cent by weight of the product, and is used in combination with other emulsifiers. It has consequently been possible to manufacture spray-dried butter powders with a fat content of up to 80 per cent (Hansen, 1963; Snow *et al.,* 1967).

A number of butter-like foods (Stavrova and Mochalova, 1978; Rasic *et al.,* 1978) and cheese spreads (Elenbogen and Baron, 1968) have been made by the incorporation of sodium and/or calcium caseinate or co-precipitate in emulsions of edible fats, usually based on vegetable oils. Whipping fat and whipping cream containing sodium caseinate have also been produced (Cooper and Peacock, 1979).

Casein in Infant Foods

A number of infant food formulations, which contain casein derivatives, have been reported in the literature. In some cases, the formulations have been prepared specifically for infants with particular dietary problems. For instance, Lofenalac®[a], a casein hydrolysate which has been specially treated to remove almost all the phenylalanine, is used

[a]*Lofenalac®* is a registered trademark of Mead Johnson & Co.

for feeding infants suffering from phenylketonuria (Owen, 1969). For those children who have problems in digesting lactose or other sugars, a carbohydrate-free infant food has been formulated (Henderson and Buchanan, 1973). Other infant formulas containing casein have been produced in the USA (Kennedy and Bernhart, 1953), Japan (Morinaga Milk Industry Co. Ltd, 1975), the USSR (Korobkina *et al.,* 1974) and Yugoslavia (Caric *et al.,* 1981).

Casein in Pasta Products and Snack Foods

The reason for using casein derivatives in pasta and snack foods is mainly to enhance their nutritional quality. Casein has been used to enrich macaroni (Tolstogusow *et al.,* 1975) and spaghetti (Durr and Neukom, 1972), co-precipitate has been formed into imitation rice (Markh *et al.,* 1982) and caseinates have been used in the fortification of rice, pasta and bread (Humphries and Roeper, 1974). Recent work suggests there is some interest in Japan in the fortification of noodle products with casein (Minami *et al.,* 1979). While the amount of casein used to fortify these products is usually between 5 and 20 per cent of the total weight, some applications suggest that the pasta is a synthetic or imitation product which contains casein as one of the major ingredients. In such cases, the casein obviously has a major effect on texture.

Casein products, such as casein, caseinates and co-precipitate curd, can be used to form relatively high-protein (30–75 per cent expressed on the basis of the dry food product) snack foods (Wong and Parks, 1970). While these milk proteins form a source of nutrition for the consumer, they are also formed, during the extrusion or other manufacturing process, into the basic matrix which provides the texture necessary for their appeal as a snack food. One such product, Pro-Teens®, has undergone market evaluation in the United States (McCormick, 1973).

Casein in Confectionery

Casein products are used in candy in order to develop a chewy, firm body which is neither sticky nor tough (Webb, 1970; Kinsella, 1970; Hugunin and Nishikawa, 1977). They also contribute to the colour and flavour of confectionery. During the cooking of products, such as caramel and fudge, the casein apparently coagulates (Webb, 1970), and it is considered that the fineness of the coagulum contributes towards the final texture of the candy. An extensive review on this topic has been published by Lim (1980).

Casein in Cheese Products

While casein products have been used in cheese-like foods such as cheese spreads (Elenbogen and Baron, 1968), it was not until the early 1970s that so-called 'non-dairy' or imitation cheese products, based entirely on vegetable oil and casein, were produced (Vernon, 1972). These were developed initially in the United States, and considerable research and development was applied in attempts to produce 'non-dairy' cheese with functional and nutritional properties similar to traditional cheese products, such as Mozzarella and processed cheese. Initially, at least, it appears that imitation cheese products were used for 'industrial' purposes by the manufacturers of foods such as pizza and lasagna, by institutional food service companies, and by suppliers to fast-food franchises for use in cheeseburgers and sandwiches.

The volume of cheese substitutes produced in 1978 was 43 000 tonnes, and represented 2·7 per cent of the natural cheese produced in the United States (Shaw, 1984). Within two years, the production of cheese substitutes had increased to more than twice the 1978 level and, at 90–95 000 tonnes, was equivalent to about 5 per cent of the 1·8 million tonnes of natural cheese produced in the United States in 1980 (Shaw, 1984).

The growth of imitation cheese (real and potential) in the USA has led to considerable study of the subject (Vakaleris, 1980), and an expression of concern that imports of casein should be curtailed (Graf, 1979). However, both Vakaleris (1980) and Shaw (1984) note that it is generally believed that the introduction of imitation cheese products has increased the total amount of 'cheese' consumed, rather than displaced any natural cheese.

Interest in imitation cheese has also been shown in other countries. In 1980, production of imitation cheese, based on casein and vegetable fat, commenced in the United Kingdom (Shaw, 1984) and in Japan (Akino and Yoshioka, 1982).

Casein derivatives are used in cheese products primarily for their functional properties, e.g. binding of fat and water, texture, melting properties, stringiness and shredding properties (Petka, 1976). While the list of casein derivatives which are used in cheese products includes mainly acid casein and caseinates, it is evidently a field where other types of casein, such as rennet casein and co-precipitates, may also find application (Roeper, 1976; Schreiber Cheese Company, Inc., 1978). Figure 12 shows a number of imitation cheese products derived from casein.

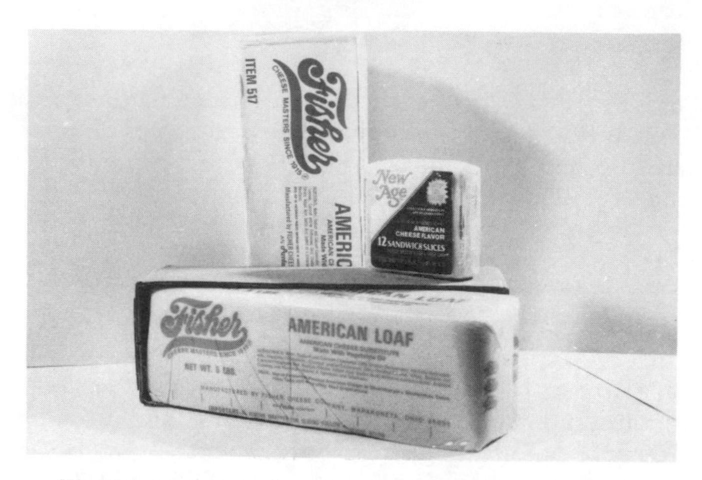

Fig. 12. Imitation cheese products derived from casein.

Use of Casein in Pharmaceutical Products

Casein and its derivatives have been employed for many years in a variety of pharmaceutical applications such as tonic foods, in the treatment of convalescent and undernourished patients, as a therapeutic agent (in combination with some heavy metals) in dressing wounds, and in ointments (Schuette, 1939).

During and just after World War II, a number of studies were undertaken to determine the nutritional effect of feeding various casein hydrolysate preparations to hospital patients following surgery (Robertson and Smaill, 1947; Bell, 1947), and a recent review describes the use of milk protein and casein hydrolysates in pharmaceutical applications (Manson, 1980).

Various casein products have been used in dietary foods and drinks (Fig. 13), for meal replacement, for weight reduction (Sobotka, 1971), in candy for space feeding (Dymsza *et al.,* 1966) and in high protein supplements (Fig. 14). Applications for casein products in the treatment of medical conditions such as anaemia, digestive problems, cancer and disorders of the pancreas have been claimed.

Other, pharmaceutical, applications include the use of casein derivatives in dentifrice and dental paste, as disintegrating agents in pharmaceutical tablets, in a protective handcream and in stabilisation

Fig. 13. Dietary drink containing casein.

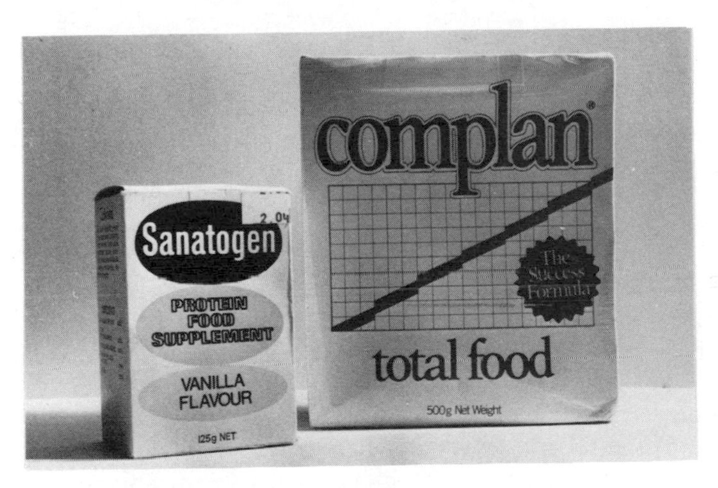

Fig.14. Pharmaceutical products containing casein.

of vitamins A, B_1 and C as listed by Southward and Walker (1982).

Arising from some of the studies noted above, numerous proprietary pharmaceutical compounds which contain casein have been marketed in recent years. A selection of these is shown in Table VIII.

TABLE VIII
Proprietary[a] pharmaceutical products containing casein

Casein derivative	*Pharmaceutical product*	*Manufacturer*	*Reference*
Sodium caseinate	Meritene	Doyle	Lagua *et al.* (1974)
	Nutrament	Drackett	Lagua *et al.* (1974)
	Portagen	Mead Johnson	Lagua *et al.* (1974)
	Sustacal	Mead Johnson	Lagua *et al.* (1974)
Calcium caseinate	Casec	Mead Johnson	Robertson and Smaill (1947)
	Casilan	Glaxo	Bender (1975)
	Complan	Glaxo	Bender (1968)
	Metrecal	Mead Johnson	Bender (1968)
	Sustagen	Mead Johnson	Lagua *et al.* (1974)
Casein hydrolysate (either acid or enzymatic)	Amigen	Mead Johnson	Bell (1947)
	Aminosol	Vitrum	Jorpes *et al.* (1946)
	Lofenalac	Mead Johnson	Lagua *et al.* (1974)
	Nutramigen	Mead Johnson	Lagua *et al.* (1974)
	Parenamine	Frederick Stearns	Bell (1947)
	Pronutrin	Herts Pharmaceuticals	Bell (1947)

[a]Proprietary product names are registered trademarks of the appropriate manufacturer.

CONCLUSIONS

Casein has had a long history of use in industrial applications. During the past twenty years, however, there has been a significant decline in its non-food uses in some countries because of lower cost alternatives. In the same period, however, edible applications for casein products have increased significantly, particularly in the United States, Europe and Japan. Although casein does have some application in nutritional foods as a high quality protein, its major value appears to be as a specialised food ingredient, included for its functional properties such as water binding, fat emulsification, whipping and foaming properties and texturisation characteristics, as discussed earlier. Future trends for its production and world trade are difficult to predict since government trade and dairy support policies will continue to influence the situation. The dairy support programmes in the EEC and the United States have, for instance, tended to favour the production of skim-milk powder rather than casein. On the other hand, major casein producers, such as New Zealand, are to some extent limited in the expansion of their production because of trade policies which restrict access to world markets for the remaining butterfat portion of the milk.

REFERENCES

Agra Europe (London) Ltd. (1978). *Agra Europe,* No. 776, May 26, 1978.

Agra Europe (London) Ltd. (1979). *Agra Europe,* No. 820, April 6, 1979.

Agra Europe (London) Ltd. (1981). *EEC Dairy Policy and Markets in the 1980s,* Agra Europe Special Report No. 8, p. 55, Agra Europe (London) Ltd.

Agra Europe (London) Ltd. (1984a). *Agra Europe East Europe Agriculture,* No. 16, January, 1984, p. 7.

Agra Europe (London) Ltd. (1984b). *Preserved Milk,* a monthly supplement to Agra Europe, No. 65, May 1984.

Agra Europe (London) Ltd. (1984c). *Preserved Milk,* a monthly supplement to Agra Europe, No. 65, June/July 1984; No. 67, August/September 1984.

Agra Europe (London) Ltd. (1984d). *Agra Europe,* No. 1094, August 10, 1984.

Akino, Y. and Yoshioka, S. (1982). *Food Industry (Japan),* **25** (14), 20.

American Chemical Society (1981). *Protein Functionality in Foods* (Ed. J. P. Cherry), ACS Symposium Series 147, American Chemical Society, Washington, D.C.

Amiot, J., Brisson, G. J., Castaigne, F., Goulet, G. and Boulet, M. (1979). *Canadian Institute of Food Science and Technology Journal,* **12**, 23.

Anon. (1978). *Zuivelzicht,* **70** (11), 255.

Association of Official Analytical Chemists (1970). *Official Methods of Analysis of the Association of Official Analytical Chemists* (Ed. W. Horwitz), 11th edn, A.O.A.C., Washington, D.C., p. 800.

Aube, D. L. (1983). *Canada's Trade in Agricultural Products 1980, 1981 and 1982* (Communications Branch Agriculture Canada, Ed.), Publication No. 83/4, Marketing and Economics Branch, Agriculture Canada, Ottawa, pp. 83, 110.

Australian Dairy Corporation (1978–83). Annual Reports for years ended 30 June 1978, 1980–81, 1981–2, 1982–3, Melbourne, Victoria.

Australian Society of Dairy Technology (1972). *Casein Manual,* Australian Society of Dairy Technology, Parkville, Victoria, Australia, pp. 25–27.

Banks, W., Muir, D. D. and Wilson, A. G. (1983). In: *Proceedings of IDF Symposium on Physico-Chemical Aspects of Dehydrated Protein-rich Milk Products,* 17–19 May, 1983, Statens Forsogsmejeri, Hillerod, Denmark, pp. 331–8.

Barrat, J. -F. (1978). French Patent 2 370 441.

Barry, T. N. (1972). *New Zealand Journal of Agricultural Research,* **15**, 107.

Bates, J. G. (1960). New Zealand Patent 122 331.

Battelle Memorial Institute (1974). British Patent 1 350 647.

Bell, M. E. (1947). *New Zealand Medical Journal,* **46**, 255.

Bender, A. E. (1968). *Dictionary of Nutrition and Food Technology,* 3rd edn, Butterworths, London.

Bender, A. E. (1975). *Dictionary of Nutrition and Food Technology,* 4th edn, Newnes-Butterworth, London, pp. 48, 198.
Bergmann, A. (1972). *Journal of the Society of Dairy Technology,* **25**, 89.
Bird, E. W., Sadler, H. W. and Iverson, C. A. (1935). *The Preparation of Non-Desiccated Sodium Caseinate Sol and its Use in Icecream,* Research Bulletin No. 187, Agricultural Experiment Station, Iowa State College of Agriculture and Mechanic Arts (Dairy Industry Section), USA.
Bogdanova, E. A., Padaryan, I. M., Lavrenova, G. S. and Inozemtseva, V. F. (1978). *XX International Dairy Congress, Brief Communications,* Paris, p. 845.
Bready, P. J. S. (1966). *Australian Journal of Dairy Technology,* **21**, 153.
Broderick, G. A. and Lane, G. T. (1978). *Journal of Dairy Science,* **61**, 932.
Brother, G. H. (1939). In: *Casein and Its Industrial Applications* (Ed. E. Sutermeister and F. L. Browne), 2nd edn, Reinhold Publishing Corporation, New York, pp. 181–232.
Brown, A. E., Gordon, W. G., Gall, E. C. and Jackson, R. W. (1944). *Industrial and Engineering Chemistry,* **36**, 1171.
Browne, F. L. and Brouse, D. (1939). In: *Casein and Its Industrial Applications* (Ed. E. Sutermeister and F. L. Browne), 2nd edn, Reinhold Publishing Corporation, New York, pp. 233–92.
Buchanan, R. A. and Henderson, J. O. (1971). *Journal of the Dietetic Association of Victoria,* **22**, 7 (*Dairy Science Abstracts* (1971), **33**, 5048).
Burgess, K. J. and Coton, G. (1982). In: *Food Proteins* (Ed. P. F. Fox and J. J. Condon), Applied Science Publishers, Ltd, London, pp. 211–24.
Burkwall, M. P., Jr., Leyh, J. C., Jr. and Reagen, J. G. (1976). United States Patent 3 984 576.
Burston, D. O., Muller, L. L. and Hayes, J. F. (1967). *International Dairy Federation Seminar on Caseins and Caseinates,* 31 May–2 June 1967, PARIS-SEM Subject 4, Paris, France.
Cameron, D. E., Chilson, W. H., Elsesser, C. C. and Windmuller, R. (1959). United States Patent 2 913 342, British Patent 822 614.
Caric, M., Gavaric, D., Milanovic, S., Jakimov, N., Karic, A. and Markovic, D. (1981). *Mljekarstvo,* **31**, 79 (*Dairy Science Abstracts* (1982), **44**, 2732).
Cavett, E. S. (1939). In: *Casein and Its Industrial Applications* (Ed. E. Sutermeister and F. L. Browne), 2nd edn, Reinhold Publishing Corporation, New York, pp. 354–65.
Cayen, M. N. and Baker, B. E. (1963). *Journal of Agricultural and Food Chemistry,* **11**, 12.
Centre National du Commerce Exterieur (1970). *The United States Market for Edible Caseins and Caseinates,* Centre National du Commerce Exterieur, Paris.
Clarke, R. J. and Love, G. (1974). *Chemistry and Industry,* No. 4, 16 Feb 1974, 151.
Commonwealth Secretariat (1982–84). *Meat and Dairy Products,* pp. 8, 29 (May 1982); p. 9 (Nov. 1982); pp. 8, 17 (May 1983); p. 9 (Nov. 1983); pp. 10, 20 (May 1984). Commonwealth Secretariat Publications, London.
Connell, J. E. (1966). *Canadian Food Industries,* **37** (2), 23.
Connolly, P. B. (1983). United States Patent 4 376 072.
Cooper, H. R. and Peacock, I. C. (1979). *New Zealand Journal of Dairy Science and Technology,* **14**, 291.

Craig, T. W. and Colmey, J. C. (1971). *Bakers Digest,* **45** (1), 36.
Davey, P. T., Williams, D. R. and Winter, G. (1980). *Journal of Applied Biochemistry,* **2**, 60.
Dolby, R. M., Creamer, L. K. and Elley, E. R. (1969). *New Zealand Journal of Dairy Science and Technology,* **4**, 46.
Downes, T. E. H. and van der Merwe, N. L. (1982). *South African Journal of Dairy Technology,* **14**, 59.
Durr, P. and Neukom, H. (1972). *Lebensmittel-Wissenschaft und-Technologie,* **5**, 132.
D'yachenko, P., Vlodavets, I. and Bogomolova, E. (1953). *Molochnaya Promyshlennost',* **14** (6), 33 (*Dairy Science Abstracts* (1953). **15**, 962).
Dymsza, H. A., Stoewsand, G. S., Donovan, P., Barrett, F. F. and Lachance, P. A. (1966). *Food Technology, Chicago,* **20**, 1349.
Elenbogen, G. D. and Baron, M. (1968). United States Patent 3 397 994.
English, A. (1981). *Journal of the Society of Dairy Technology,* **34**, 70.
Ernstrom, C. A. and Wong, N. P. (1974). In: *Fundamentals of Dairy Chemistry* (Ed. B. H. Webb, A. H. Johnson and J. A. Alford), 2nd edn, AVI Publishing Co., Inc., Westport, Connecticut, pp. 662–771.
Evans, E. W. (1982). In: *Developments in Food Proteins — 1* (Ed. B. J. F. Hudson), Applied Science Publishers Ltd, London and New Jersey, pp. 131–69.
Farrell, H. M., Jr and Thompson, M. P. (1974). In: *Fundamentals of Dairy Chemistry* (Ed. B. H. Webb, A. H. Johnson and J. A. Alford), 2nd edn, AVI Publishing Co., Inc., Westport, Connecticut, pp. 442–73.
Ferretti, A. (1938). British Patent 483 731.
Filer, L. J., Jr (Chairman) (American Academy of Pediatrics, Committee on Nutrition) (1972). *Pediatrics,* **49**, 770.
Fogh, A. W. (1975). *Maelkeritidende,* **88**, 352.
Food and Agriculture Organization (1973). *Energy and Protein Requirements,* p. 63. FAO Nutrition Meetings Report Series No. 52. WHO Technical Report Series No. 522. FAO and WHO, Rome.
Food and Agriculture Organization (1980–84). *FAO Monthly Bulletin of Statistics,* **3** (12), 30 (1980); **6** (9), 26 (1983); **7** (3), 37 (1984); **7** (6), 27 (1984).
Fox, K. K. (1970). In: *Byproducts from milk* (Ed. B. H. Webb and E. O. Whittier), 2nd edn, AVI Publishing Co., Inc., Westport, Connecticut, pp. 331–55.
Fox, P. F. (1968). *Casein, Caseinates and Casein Co-precipitates,* Dairy Research Review Series No. 4, An Foras Taluntais, Fermoy, Co. Cork, Irish Republic.
Fox, P. F. and Mulvihill, D. M. (1983). In: *Proceedings of IDF Symposium on Physico-Chemical Aspects of Dehydrated Protein-rich Milk Products,* 17–19 May 1983, Statens Forsogsmejeri, Hillerod, Denmark, pp. 188–259.
Genin, G. (1961). *Le Lait,* **41**, 44.
Genin, G. (1966). *Le Lait,* **46**, 283.
Girdhar, B. K. and Hansen, P. M. T. (1974). *Journal of Food Science,* **39**, 1237.
Goldman, A. (1974). *Food Technology in New Zealand,* **9** (7), 25.
Goldman, A. and Toepfer, N. G. (1978). *XX International Dairy Congress, Brief Communications,* Paris, pp. 418–19.
Gordon, W. G. and Kalan, E. B. (1974). In: *Fundamentals of Dairy Chemistry* (Ed. B. H. Webb, A. H. Johnson and J. A. Alford), 2nd edn, AVI Publishing Co., Inc., Westport, Connecticut, pp. 87–124.

Graf, T. F. (1979). *Hoard's Dairyman,* **124**, 1531.
Gwozdz, E. (1978). *XX International Dairy Congress,* Conferences Science and Technique No. 51ST, Paris.
Gwozdz, E. (1983). *La Technique Laitiere,* **979**, 19.
Hambraeus, L. (1982). In: *Developments in Dairy Chemistry — 1. Proteins* (Ed. P. F. Fox), Applied Science Publishers Ltd, London and New York, pp. 289–313.
Hammonds, T. M. and Call, D. L. (1970). *Utilization of Protein Ingredients in the US Food Industry. Part I. The Current Market for Protein Ingredients. Part II. The Future Market for Protein Ingredients,* Cornell University Agricultural Experimental Station, New York State College of Agriculture, Ithaca, New York.
Hansen, P. M. T. (1963). *Australian Journal of Dairy Technology,* **18**, 79.
Hansen, P. M. T., Harper, W. J. and Sharma, K. K. (1970). *Journal of Food Science,* **35**, 598.
Hayes, J. F., Dunkerley, J. and Muller, L. L. (1969). *Australian Journal of Dairy Technology,* **24**, 69.
Hayes, J. F., Southby, P. M. and Muller, L. L. (1968). *Journal of Dairy Research,* **35**, 31.
Heap, H. A. and Lawrence, R. C. (1984). *New Zealand Journal of Dairy Science and Technology,* **19**, 119.
Hedrick, T. I. (1969). *Dairy Industries,* **34**, 127.
Henderson, J. O. and Buchanan, R. A. (1973). *Australian Journal of Dairy Technology,* **28**, 7.
Hidalgo, J., Wenner, V. and Forni, F. (1981). United States Patent 4 303 580.
Hooker, P. H., Munro, P. A. and O'Meara, G. M. (1982). *New Zealand Journal of Dairy Science and Technology,* **17**, 35.
Hugunin, A. G. and Nishikawa, R. K. (1977). *Milk Derived Ingredients for Confectionery Products,* Dairy Research Inc., Rosemont, Illinois, USA.
Humphries, M. A. and Roeper, J. (1974). *XIX International Dairy Congress, Brief Communications,* Volume 1E, India, pp. 778–9.
Jackson, D. L. C. and Backwell, A. R. A. (1955). *Australian Journal of Applied Science,* **6**, 244 (*Australian Journal of Dairy Technology* (1955). **10**, 142).
Japan Finance Ministry (1981–83). Statistics for Imports. Commodities 35·01-100 (Casein) and 35·01–200 (Caseinate) for 1981 (p. 0139), 1982 (p. 0137) and 1983 (p. 0136).
Japan Ministry of Agriculture, Forestry and Fisheries (1982). *The 57th Statistical Yearbook of Ministry of Agriculture, Forestry and Fisheries,* 1980–81, Association of Agriculture and Forestry Statistics, Tokyo, Japan, pp. 534–5.
Jenness, R. and Patton, S. (1959a). *Principles of Dairy Chemistry,* John Wiley, New York, pp. 1–29.
Jenness, R. and Patton, S. (1959b). *Principles of Dairy Chemistry,* John Wiley, New York, pp. 101–57.
Jordan, P. J. (1983). *New Zealand Journal of Dairy Science and Technology,* **18**, 27.
Jorpes, J. E., Magnusson, J. H. and Wretlind, A. (1946). *The Lancet,* **2**, 228.
Kelly, P. M., O'Keefe, A. M. and Phelan, J. A. (1983). In: *Proceedings of IDF*

Symposium on Physico-Chemical Aspects of Dehydrated Protein-rich Milk Products, 17–19 May 1983, Statens Forsogsmejeri, Hillerod, Denmark, pp. 260–75.
Kennedy, J. G. and Bernhart, F. W. (1953). United States Patent 2 639 235.
King, D. W. (1970). *New Zealand Journal of Dairy Science and Technology,* **5**, 100.
King, D. W. and Jebson, R. S. (1970). *XVIII International Dairy Congress,* Volume 1E, p. 420.
Kinsella, J. E. (1970). *Manufacturing Confectioner,* **50** (10), 45.
Kinsella, J. E. (1983). In: *Proceedings of IDF Symposium on Physico-Chemical Aspects of Dehydrated Protein-rich Milk Products,* 17–19 May 1983, Statens Forsogsmejeri, Hillerod, Denmark, pp. 12–32.
Kirk, D. J. (1973). *Bakers Digest,* **47** (5), 76.
Kitabatake, N. and Doi, E. (1982). *Journal of Food Science,* **47**, 1218.
Knightbridge, J. P. and Goldman, A. (1975). *New Zealand Journal of Dairy Science and Technology,* **10**, 152.
Knightly, W. H. (1968). *Food Technology, Chicago,* **22**, 731.
Knightly, W. H. (1969). *Food Technology, Chicago,* **23**, 171.
Korobkina, G. S., Danilova, E. N., Pokrovskii, A. A., Arbatskaya, N. I. and Agienko, K. S. (1974). *XIX International Dairy Congress, Brief Communications,* Volume 1E, India, p. 566.
Korolczuk, J. (1982). *New Zealand Journal of Dairy Science and Technology,* **17**, 135.
Krause, G. A. (1953). German Federal Republic Patent 896 449.
Kreveld, A. van (1969). *Voeding,* **30**, 231.
LaBrosse, G. and McSorley, E. (1981). *Canada's Trade in Agricultural Products 1978, 1979 and 1980* (Information Services Agriculture Canada, Ed.). Publication No. 81/4. Marketing and Economics Branch, Agriculture Canada, Ottawa, pp. 50, 65.
Lagua, R. T., Claudio, V. S. and Thiele, V. F. (1974). *Nutrition and Diet Therapy Reference Dictionary,* 2nd edn, C. V. Mosby Co., St Louis, USA, pp. 314–17.
Landmann, A. W. (1962). *Journal of the Society of Leather Trades' Chemists,* **46**, 97.
Lim, D. M. (1980). *Functional Properties of Milk Proteins with Particular Reference to Confectionery Products,* Scientific and Technical Survey No. RA 120, British Food Manufacturing Industries Research Association, Leatherhead, Surrey.
Lippe, F., Ottenhof, H. A. W. E. M. and de Boer, R. (1983). United States Patent 4 407 747.
Little, L. L. (1966). United States Patent 3 236 658.
Lodge, N. and Heatherbell, D. A. (1976). *New Zealand Journal of Dairy Science and Technology,* **11**, 263.
McCormick, R. D. (1973). *Food Product Development,* **7** (2), 16.
McDowall, F. H. (1971). *New Zealand Journal of Dairy Science and Technology,* **6**, 128.
McDowell, A. K. R. (1968). In: *Annual Report, New Zealand Dairy Research Institute,* New Zealand Dairy Research Institute, Palmerston North, New Zealand, p. 52.
McDowell, A. K. R., Southward, C. R. and Elston, P. D. (1976). *New Zealand Journal of Dairy Science and Technology,* **11**, 40.

McGilliard, A. D. (1972). *Journal of the American Oil Chemists' Society,* **49**, 57.
McMeekin, T. L., Reid, T. S., Warner, R. C. and Jackson, R. W. (1945). *Industrial and Engineering Chemistry,* **37**, 685.
Manson, W. (1980). *Bulletin, International Dairy Federation,* Document 125, 60.
Markh, A. T., Feldman, A. L., Ponomarenko, S. F. and Strashnenko, E. S. (1982). *XXI International Dairy Congress, Brief Communications,* Volume 1, Book 2, Moscow, p. 85.
Melsheimer, L. A. and Hoback, W. H. (1953). *Industrial and Engineering Chemistry,* **45**, 717.
Milk Marketing Board (1979–83). *EEC Dairy Facts and Figures* 1979, pp. 152, 153; 1980, pp. 152, 153; 1982, pp. 139, 162, 163; 1983, pp. 162, 163, Economics Division, Milk Marketing Board, Surrey, England.
Minami, J., Shigato, M. and Ishibashi, S. (1979). UK Patent Application 2 010 658A (*Dairy Science Abstracts* (1980). **42**, 1951).
Modler, H. W., Larmond, M. E., Lin, C. S., Froehlich, D. and Emmons, D. B. (1983). *Journal of Dairy Science,* **66**, 422.
Morinaga Milk Industry Co. Ltd. (1975). United States Patent 3 901 979.
Morr, C. V. (1979). In: *Functionality and Protein Structure* (Ed. A. Pour-El), ACS Symposium Series 92, American Chemical Society, Washington, D.C., pp. 65–79.
Morr, C. V. (1981). In: *Protein Functionality in Foods* (Ed. J. P. Cherry), ACS Symposium Series 147, American Chemical Society, Washington, D.C., pp. 201–15.
Morr, C. V. (1982). In: *Developments in Dairy Chemistry — 1. Proteins* (Ed. P. F. Fox), Applied Science Publishers Ltd, London and New York, pp. 375–99.
Morr, C. V. (1984). *Deutsche Molkerei-Zeitung,* **105**,1066.
Muller, L. L. (1971). *Dairy Science Abstracts,* **33**, 659–74.
Muller, L. L. (1982a). In: *Developments in Dairy Chemistry — 1. Proteins* (Ed. P. F. Fox), Applied Science Publishers Ltd, London and New York, pp. 315–37.
Muller, L. L. (1982b). In: *Food Proteins* (Ed. P. F. Fox and J. J. Condon), Applied Science Publishers Ltd, London, pp. 179–89.
Muller, L. L. and Hayes, J. F. (1963). *Australian Journal of Dairy Technology,* **18**, 184.
Muller, L. L., Hayes, J. F. and Snow, N. (1967). *Australian Journal of Dairy Technology,* **22**, 12.
Mummery, W. R. (1949). *New Zealand Journal of Science and Technology,* **B30**, 297.
Munro, P. A., Southward, C. R. and Elston, P. D. (1980). *New Zealand Journal of Dairy Science and Technology,* **15**, 177.
Munro, P. A., and Vu, J. T. (1983). *New Zealand Journal of Dairy Science and Technology,* **18**, 93.
Munro, P. A., Vu, J. T. and Mockett, R. B. (1983). *New Zealand Journal of Dairy Science and Technology,* **18**, 35.
National Dairy Council (1968). *US Dairy Council Digest,* **39**, 7.
Nesmeyanov, A. N., Rogozhin, S. V., Slonimsky, G. L., Tolstoguzov, V. B. and Ershova, V. A. (1971). United States Patent 3 589 910.
New Zealand Dairy Board (1982–84). Annual Reports for Years Ended 31 May 1982, 1983, 1984, New Zealand Dairy Board, Wellington.
New Zealand Department of Statistics (1979–84). Exports 1977–78. pp. 53–4

(June 1979); Exports 1978–79. p. 69 (June 1980); Exports 1979–80. p. 70 (Dec. 1980); Exports 1980–81. p. 66 (March 1982); Exports 1981–82. (Microfiche) Table EA71, pp. 367–8 (30 Nov. 1982); Exports 1982–83 (Microfiche) Table EA71, pp. 373–4 (6 Dec. 1983); Exports 1983–84 (Microfiche) Table EM23, pp. 481–5 (20 July 1984) (provisional), Wellington.

Owen, G. M. (1969). *American Journal of Clinical Nutrition,* **22**, 1150.

Pader, M. and Gershon, S. D. (1965). United States Patent 3 224 883.

Paterson, L. O. (1955). United States Patent 2 721 861.

Petka, T. E. (1976). *Food Product Development,* **10** (10), 26.

Pfaff, W. (1974). *Fleischwirtschaft,* **54**, 1740.

Poarch, E. A. (1967). *International Dairy Federation Seminar on Caseins and Caseinates,* 31 May–2 June, 1967, PARIS-SEM Subject 6, Paris, France.

Powell, L. A. (1974). United States Patent 3 843 805.

Ramshaw, E. H. and Dunstone, E. A. (1969a). *Journal of Dairy Research,* **36**, 203.

Ramshaw, E. H. and Dunstone, E. A. (1969b). *Journal of Dairy Research,* **36**, 215.

Rasic, J., Bosic, Z. and Vracar, L. (1978). *XX International Dairy Congress, Brief Communications,* Paris, pp. 987–8.

Reimerdes, E. H. and Lorenzen, P. C. (1983). In: *Proceedings of IDF Symposium on Physico-Chemical Aspects of Dehydrated Protein-rich Milk Products,* 17–19 May 1983, Statens Forsogsmejeri, Hillerod, Denmark, pp. 70–93.

Reis, P. J. and Tunks, D. A. (1969). *Australian Journal of Agricultural Research,* **20**, 775.

Roberts, J. G. (1959). United States Patent 2 878 126.

Robertson, H. R. and Smaill, D. W. (1947). *Canadian Medical Association Journal,* **56**, 59.

Robinson, G. H. (1964). Casein in the New Zealand Dairy Industry. Some Economic Implications and Some International Marketing Aspects. M. A. Thesis, Victoria University of Wellington.

Roeper, J. (1974). *New Zealand Journal of Dairy Science and Technology,* **9**, 128.

Roeper, J. (1976). *New Zealand Journal of Dairy Science and Technology,* **11**, 62.

Roeper, J. (1977a). In: *Proceedings of Jubilee Conference on Dairy Science and Technology,* 15–17 March 1977. New Zealand Dairy Research Institute, Palmerston North, pp. 81–3.

Roeper, J. (1977b). *New Zealand Journal of Dairy Science and Technology,* **12**, 182.

Roeper, J. (1982). *Bulletin, International Dairy Federation,* Document 147, 21.

Roeper, J. and Winter, G. J. (1982). *XXI International Dairy Congress, Brief Communications,* Volume 1, Book 2, Moscow, pp. 97–8.

Roff, W. J. and Scott, J. R. (1971). *Fibres, Films, Plastics and Rubbers — A Handbook of Common Polymers,* Butterworths, London, pp. 197–208.

Rusch, D. T. (1971). *Food Technology, Chicago,* **25**, 486.

Salzberg, H. K. (1962). In: *Handbook of Adhesives* (Ed. I. Skeist), van Nostrand Reinhold Company, New York, pp. 129–47.

Salzberg, H. K. (1964). *American Paint Journal,* **48** (42), 104; (43), 110; (45), 90.

Salzberg, H. K. (1965). In: *Encyclopedia of Polymer Science and Technology* (Ed. H. F. Mark, N. H. Gaylord and N. M. Bikales), Vol. 2, Interscience Publishers, New York, pp. 859–71.

Salzberg, H. K. (1967). *American Perfumer and Cosmetics,* **82** (Nov.), 41.

Salzberg, H. K., Britton, R. K. and Bye, C. N. (1974). In: *Testing of Adhesives* (Ed. R. G. Meese), Tappi Monograph Series No. 35, Technical Association of the Pulp and Paper Industry, Atlanta, Georgia, pp. 30–51.
Salzberg, H. K., Georgevits, L. E. and Karapetoff Cobb, R. M. (1961). In: *Synthetic and Protein Adhesives for Paper Coating* (Ed. L. H. Silvernail and M. W. Bain), Tappi Monograph Series No. 22, Technical Association of the Pulp and Paper Industry, New York, pp. 103–66.
Salzberg, H. K. and Marino, W. L. (1975). In: *Protein Binders in Paper and Paperboard Coating* (Ed. R. Strauss), Tappi Monograph Series No. 36, Technical Association of the Pulp and Paper Industry, Atlanta, Georgia, pp. 1–74.
Salzberg, H. K. and Simonds, M. R. (1965). United States Patent 3 186 918.
Scholz, H. A. (1953). *Industrial and Engineering Chemistry,* **45**, 70.
Schreiber Cheese Company, The L. D. (1978). French Patent 2 381 474.
Schuette, H. A. (1939). In: *Casein and Its Industrial Applications* (Ed. E. Sutermeister and F. L. Browne), 2nd edn, Reinhold Publishing Corporation, New York, pp. 366–90.
Scott, T. W., Cook, L. J., Fergusson, K. A., McDonald, I. W., Buchanan, R. A. and Loftus Hills, G. (1970). *Australian Journal of Science,* **32**, 291.
Segalen, P. (1982). In: *Proteines Animales. Extraits, Concentres et Isolats en Alimentation Humaine* (Ed. C. -M. Bourgeois and P. LeRoux), Technique & Documentation Lavoisier, Paris, pp. 151–71.
Shaw, M. (1984). *Journal of the Society of Dairy Technology,* **37**, 27.
Skurray, G. R. and Osborne, C. (1976). *Journal of the Science of Food and Agriculture,* **27**, 175.
Smith, D. R. and Snow, N. S. (1968). *Australian Journal of Dairy Technology,* **23**, 8.
Snow, N. S., Townsend, F. R., Bready, P. J. and Shimmin, P. D. (1967). *Australian Journal of Dairy Technology,* **22**, 125.
Sobotka, J. J. (1971). *Current Therapeutic Research,* **13**, 636.
Solms-Baruth, H. Graf zu (1972). *Deutsche Milchwirtschaft, Hildesheim,* **23** (48), 2057.
Southward, C. R. (1974). *Food Technology in New Zealand,* **9** (8), 11.
Southward, C. R. and Aird, R. M. (1978). *New Zealand Journal of Dairy Science and Technology,* **13**, 77.
Southward, C. R. and Elston, P. D. (1976). *New Zealand Journal of Dairy Science and Technology,* **11**, 144.
Southward, C. R. and Goldman, A. (1975). *New Zealand Journal of Dairy Science and Technology,* **10**, 101.
Southward, C. R. and Goldman, A. (1978). *New Zealand Journal of Dairy Science and Technology,* **13**, 97.
Southward, C. R., Humphries, M. A. and Creamer, L. K. (1978). *XX International Dairy Congress, Brief Communications,* Paris, p. 910.
Southward, C. R. and Walker, N. J. (1980). *New Zealand Journal of Dairy Science and Technology,* **15**, 201.
Southward, C. R. and Walker, N. J. (1982). In: *CRC Handbook of Processing and Utilization in Agriculture* (Ed. I. A. Wolff), Volume **1**, Chemical Rubber Company Press, Boca Raton, Florida, pp. 445–552.

Spellacy, J. R. (1953). *Casein, Dried and Condensed Whey,* Lithotype Process Co., San Francisco.

Stavrova, E. R. and Mochalova, K. V. (1978). *XX International Dairy Congress, Brief Communications,* Paris, pp. 981–2.

Stichting Nederlands Instituut voor Zuivelonderzoek (1982). Netherlands Patent Application 82.04923

Thomas, T. D. and Lowrie, R. J. (1975a). *Journal of Milk and Food Technology,* **38**, 269.

Thomas, T. D. and Lowrie, R. J. (1975b). *Journal of Milk and Food Technology,* **38**, 275.

Tolstogusow, W. B., Tschimirow, Ju. I., Braudo, E. E., Wajnermann, E. S. and Kosmina, E. P. (1975). *Nahrung,* **19**, 33.

Tornberg, E. (1980). *Journal of Food Science,* **45**, 1662.

Towler, C. (1974). *New Zealand Journal of Dairy Science and Technology,* **9**, 155.

Towler, C. (1976a). *New Zealand Journal of Dairy Science and Technology,* **11**, 24.

Towler, C. (1976b). *New Zealand Journal of Dairy Science and Technology,* **11**, 140.

Towler, C. (1976c). *New Zealand Journal of Dairy Science and Technology,* **11**, 285.

Towler, C. (1977). In: *Proceedings of Jubilee Conference on Dairy Science and Technology,* 15–17 March 1977, New Zealand Dairy Research Institute, Palmerston North, pp. 83–5.

Towler, C. (1978). *New Zealand Journal of Dairy Science and Technology,* **13**, 71.

Towler, C., Creamer, L. K. and Southward, C. R. (1981). *New Zealand Journal of Dairy Science and Technology,* **16**, 155.

Townsend, F. R. and Buchanan, R. A. (1967). *Australian Journal of Dairy Technology,* **24**, 113.

United States Department of Agriculture (1984). *Dairy Outlook and Situation, DS-396,* Economic Research Service, U.S. Department of Agriculture, Washington, D.C., USA.

United States Department of Agriculture (1979–84). Foreign Agriculture Circular FD1-79 (1979); FD1-80 (1980); FD2-81 (1981); FD1-82 (1982); FD 2-83 (1983); FD 1-84 (1984), Foreign Agricultural Service, U.S. Department of Agriculture, Washington, D.C., USA.

United States International Trade Commission (1979). Casein and Its Impact on the Domestic Dairy Industry, USITC Publication 1025, United States International Trade Commission, Washington, D.C., USA.

Vakaleris, D. G. (1980). In: *Proceedings from the First Biennial Marschall International Cheese Conference,* September 10–14, 1979, Marschall Dairy Ingredients Division, Miles Laboratories, Inc., Madison, Wisconsin, USA, pp. 261–71.

Vernon, H. R. (1972). *Food Product Development,* **6** (5), 22.

Visser, F. M. W. (1982). *Bulletin, International Dairy Federation,* Document 147, 42.

Voropalva, V. S. and Ionkina, A. A. (1979). *Molochnaya Promyshlennost',* No. 10, 37 (*Dairy Science Abstracts* (1980). **42**, 3319).

Waite, R. (1972). *Journal of the Society of Dairy Technology,* **25**, 92.

Wakodo Co. Ltd. (1969). New Zealand Patent 150 402.

Walker, N. J. (1972). *Journal of Dairy Research,* **39**, 231.

Walker, N. J. (1973). *Journal of Dairy Research,* **40**, 29.

Walker, N. J. and Manning, D. J. (1976). *New Zealand Journal of Dairy Science and Technology,* **11**, 1.
Weal, B. C. and Southward, C. R. (1974). *New Zealand Journal of Dairy Science and Technology,* **9**, 2.
Webb, B. H. (1970). In: *ByProducts from Milk* (Ed. B. H. Webb and E. O. Whittier), 2nd edn, AVI Publishing Company, Inc., Westport, Connecticut, pp. 285–330.
Wilson, G. F. (1970). *Proceedings of the New Zealand Society of Animal Production,* **30**, 123.
Wong, N. P. and Parks, O. W. (1970). *Journal of Dairy Science,* **53**, 978.
Wormell, R. L. (1954). *New Fibres from Proteins,* Butterworths, London.
Zavagli, S. B. and Kasik, R. L. (1978). United States Patent 4 115 376.
Zittle, C. A., Dellamonica, E. S. and Custer, J. H. (1956). *Journal of Dairy Science,* **39**, 1651.

Chapter 8

Automation in the Factory

L.-E. Nilsson
Alfa-Laval, Food & Dairy Engineering AB, Lund, Sweden

Over the past few decades, the dairy industry has, just like all other food manufacturing industries, lived through a period of rapid change. Manual production methods have now been replaced by modern industrial mass production, and the small, local dairy has been made obsolete by large, centrally located units.

Most dairy processes up to the 1950s used processing equipment and piping systems that had to be manually disassembled for cleaning every day. Setting up different configurations of process equipment for the manufacture of separate products was a very heavy and time-consuming business, but today, the demand for new products and more even product quality, better working environment, more capacity, better profitability, longer shelf-life and so on, have resulted in the development of automatic process control systems and process equipment suitable for automatic control.

One major development in the industry during the 1950s was the introduction of cleaning-in-place (CIP) systems. Machines no longer needed to be disassembled for cleaning; they were so designed that they could be cleaned by detergent solutions which were circulated through the product lines according to a fixed cleaning programme. Further developments during the 1950s and 1960s led to air operated automatic valves, automatic control of CIP and other limited processes and all-welded piping systems for both products and CIP solutions.

MECHANISATION

A far-reaching mechanisation of dairy operations gradually took place, with the result that more of the heavy manual labour was taken over by machines. Hoists being installed for heavy lifting, and trucks and trolleys being replaced by conveyor belts, are two examples. Mechanisation also led, together with the rapid expansion of production capacity, to a substantial increase in the number of operations. More and more valves had to be operated, more motors had to be started and stopped, and every operation had to be executed at a specific point in the process. The timing of individual operations also became more tightly constrained due to the increasing throughputs; operating a valve too soon or too late, for example, began to involve product losses of increasing magnitude. Every malfunction in the process and every wrong decision made by an operator could have serious economic consequences.

REMOTE CONTROL AND MONITORING

As time went on, even more remote control facilities were introduced to relieve the operator. Manually operated valves were replaced by electric and pneumatic ones. Control wiring for actuation and shut-off of valves, pumps, agitators and other motors was connected to strategically located control panels. Other wiring was installed to transmit process status readings (pressures, levels, temperatures, pH, flow rates, etc.) to instruments installed adjacent to the control panel.

To notify the operator that valves and motors had in fact obeyed the command signals sent to them (open/shut and start/stop), components were equipped with devices to transmit feedback signals to indicator lights on the panel. It thus gradually became possible for the operator at a control panel to monitor and control the entire process.

WHAT IS AUTOMATION?

'Automation' is probably the most improperly used word in technical history. Professional salesmen can, without any hesitation, tell an attentive audience, fantastic things like:

'If I push the button on the control panel, the pump over there will start automatically.' Such an operation is naturally not more 'automatic' than turning on the light switch at home.

Automation comes from the Greek word 'AUTO', that meaning 'of itself'. Unfortunately there is not much in the factory that can operate 'of itself', for there has to be a person, the operator, in command. He or she will do the thinking and take the necessary decisions, and then set up some parameters and initiate the operations. Then the process control system can take over the activities. The control system will operate the process with optimal efficiency on the basis of instructions that have been fed into the system in the form of a control program.

In an automated process, the control system must be in communication with every controlled component and every transmitter, and some examples of the kinds of signals passed between the control system and the process which it controls are:

- output (command) signals which actuate components in the process;
- input (feedback) signals from valves and motors which inform the control system that the component in question has been actuated;
- input (analog) signals from temperature, pressure and other transmitters which provide information on the momentary status of process variables;
- input signals from 'monitors' in the system, i.e. transmitters which report when a given condition has been attained. Examples of such conditions are maximum or minimum levels in a tank, or a signal to set minimum temperature etc.

Signals are processed by the logic unit of the control system, and before we go on, we shall study the meaning of the term 'logic'.

Logic

Logic is a fundamental concept in automation. It denotes the decision making mechanism which makes it possible to perform a given task according to a given pattern. The human mind is programmed by education and experience to perform a task in a certain way. Figure 1 illustrates how an operator uses logic to resolve a control problem which consists of supplying a process line with milk from a battery of tanks. From the process he receives a number of pieces of information, e.g. that tank T1 will soon be empty, that tank T2 is currently being cleaned, that tank T3 is full of product, etc. The operator processes this information

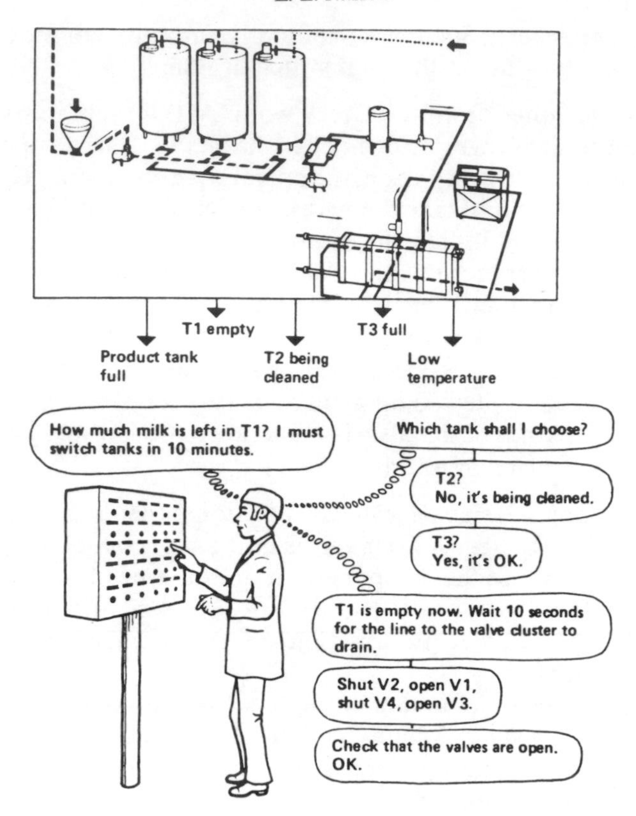

Fig. 1. What is logic?

logically; the figure illustrates his train of thought — the questions he puts and the decisions he makes. Finally he implements his decisions by pushing buttons on his panel to actuate the appropriate valves, pumps and other components.

The operator has no great difficulty in solving this control problem. Yet there is a potential for error. If, by mistake, the operator decides to draw product from tank T2, detergent and milk will be mixed. If he opens the line from tank T3 too late, the equipment in the process line will run dry, which is apt to result in charring on the heat transfer surfaces. If on the other hand he cuts in T3 too soon, the product remaining in T1 will be wasted when the tank is cleaned. The risk of such errors increases if the operator is responsible for several similar sections of the process at once, for he is then apt to be rushed and under stress, which increases the likelihood that he will make a mistake.

At first glance, it is easy to gain the impression that the operator is constantly faced with choices between a large number of alternative solutions to control problems, but closer study reveals that such is not by any means the case. In many hours of operation, the dairy has confirmed which control sequences result in optimum product quality, safety and economy. In other words, the operator has acquired a more or less permanent control logic; he selects tanks according to established routines, he uses a stopwatch to time drainage of milk from one tank so that he knows exactly when to switch to a full tank to minimise product losses and so on. Every process can be analysed in this way, and it is possible, on the basis of the analysis, to determine the control logic which produces optimum results.

WHY AUTOMATE?

The motives for introducing automatic control can be many and various, but generally speaking they can be summarised under the following:

— Limitation of the effects of hazardous malfunctions.
— Uniformly high product quality.
— High reliability of production.
— Economical operation.
— Good working conditions.

Hazardous malfunctions are faults which may cause injury to dairy personnel, or to consumers (via the product), or damage to equipment and the environment. Hazardous malfunctions can be avoided by efficient process equipment design, and correctly applied automation of crucial steps in the process.

Uniformly high product quality can be secured if the process is correctly automated. This can eliminate all the errors liable to occur in a manual process, e.g. the inadvertent mixing of products, or contamination of products with detergents. The equipment is also cleaned thoroughly according to a proven program which guarantees uniformly high product quality. Uniform treatment of the product and cleaning of the equipment offer the best guarantee of product quality, i.e. flavour, shelf life and appearance.

High reliability of production is an important economic requirement which can be satisfied by properly applied automation. In a manually controlled plant, the risk of stoppages is high because even skilled operators have difficulty in supervising large-scale processes adequately. Observation and manual logic processing are often so demanding that wrong decisions and wrong control operations can easily occur. Such risks can be easily eliminated by process adapted automation.

Economical operation is a major consideration when it comes to investing in automatic control systems. Properly designed process control can yield profit in more efficient utilisation of process equipment, smaller product losses and lower consumption of heating and cooling media.

One example is the automatic control of a milk pasteuriser (see Fig. 2). Here we can easily see how automatic control can result in better utilisation of the milk pasteuriser with one more hour production time per day, or one hour less working time per day with the same total production. No consideration is given to other effects.

Working conditions were formerly a neglected area, but today there is a social demand for a better environment, both at work and outside it. By introducing automation, it is possible to relieve operators of heavy, monotonous work, reduce exposure to stress, noise, and toxic and corrosive substances, and provide them instead with interesting work in more congenial conditions.

Important Factors for Automation

When designing a dairy, several factors must be taken into consideration. It follows that the final design is always a compromise between product needs, process needs and economic factors where the external requirements on the plant must be satisfied. The external requirements concern such things as labour, type and amount of products, product quality, hygiene and other legal considerations, production availability and flexibility and economic factors.

The product needs include raw materials, product treatment and quality of the end-product, while the process needs include choice of process equipment to satisfy external requirements. Even if the product processing lines in the plant are selected primarily to achieve the stated product quality, different compromises must be made, especially if many different products are to be manufactured. Such conditions apply

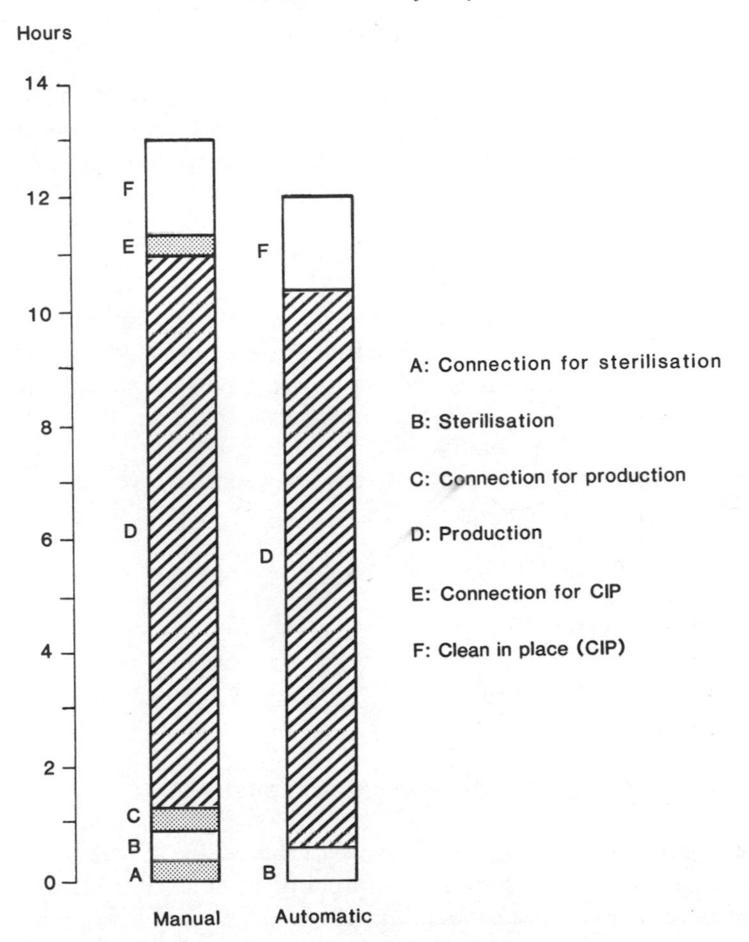

Fig. 2. Operating time for a manual and an automatically controlled milk pasteuriser.

to, for example, the cleaning requirements of the equipment and its suitability for connection to the cleaning system proposed. Other compromises must also be made, such as the consumption of energy and service media, and the suitability of equipment to be automated. In this context, it is important to state that when choosing process equipment it is extremely easy to get a bad solution for process automation.

AUTOMATION LEVELS

The automation level determines what task the control system shall perform, and how the responsibility to observe, memorise, decide and act is split between the control system and the process operator. We need a certain degree of mechanisation to enable automation to be implemented, for it is impossible to automate without any controlled items and sensor for process parameters. In a dairy where most processes include the transport of different liquids between parts of the plant, it is necessary to incorporate remote controlled valves as well as other flow equipment like pumps. If we install remote controlled valves in all connection points and cross points in the pipe system, we will have, together with sensors and other equipment, the base for the highest possible degree of mechanisation.

It must be clearly understood that the automation level depends on a number of factors such as:

— Mechanisation level
— Control level
— Information level
— Safety level

all of which affect the control system.

Mechanisation Level

The mechanisation level is a measure of the degree to which a plant is prepared for automation by introducing remote controlled objects instead of using manual operations.

Automation is possible only when the plant has a certain degree of mechanisation, i.e. there must be certain components controlling the flow of the product.

There may be three different levels of mechanisation.

Mechanisation level 1

All products routes and all connections of the CIP-system are done manually.

Necessary components are swing bends, key-pieces, pipe connection panels (swing bend panel) and hand-operated valves (see Fig. 3).

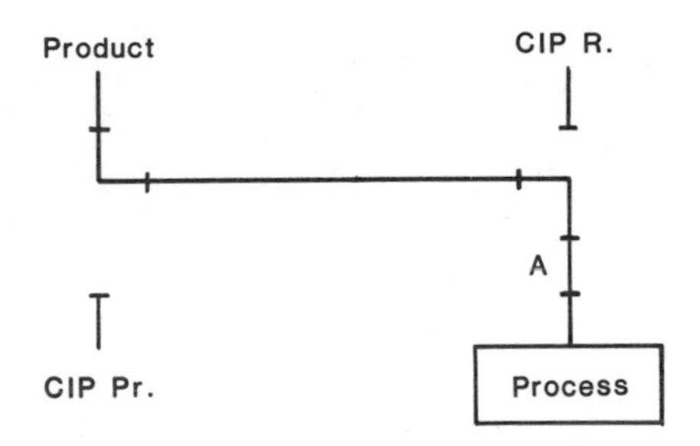

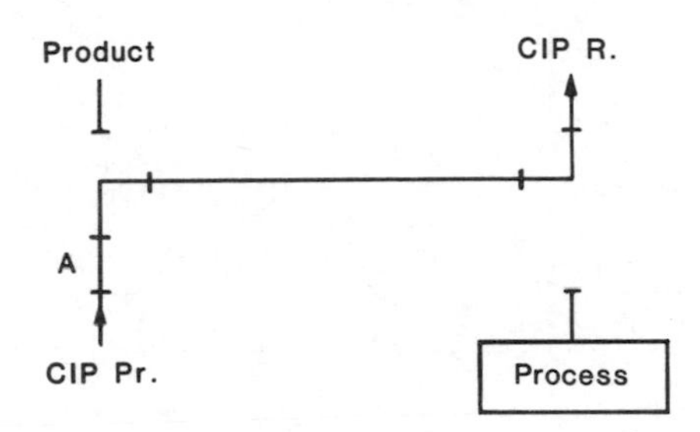

Fig. 3. Mechanisation level 1.

Mechanisation level 2

All product and CIP-routes are equipped with remote controllable valves.

CIP-routes are connected to product routes manually (key-pieces, swing bends, etc.) (see Fig. 4).

Mechanisation level 3

This level, the highest mechanisation level, implies that all product and CIP-routes are equipped with remote-controlled valves (see Fig. 5). It is obvious that the levels of mechanisation and automation are not directly dependent on each other. A mechanised plant does not necessarily have to be automated.

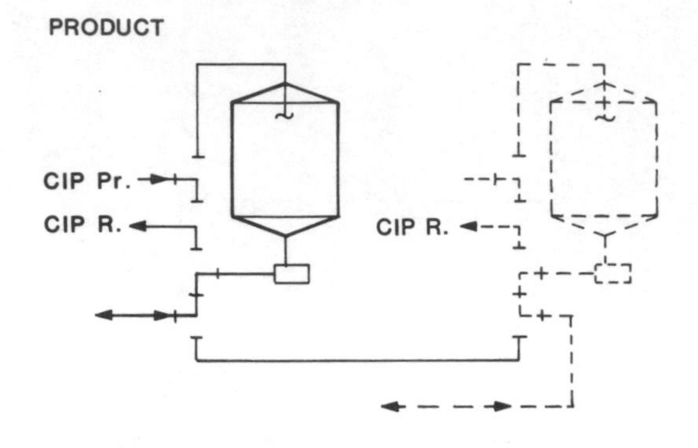

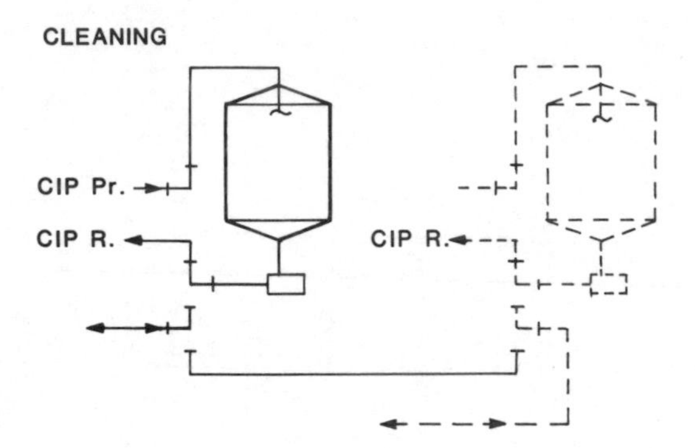

Fig. 4. Mechanisation level 2.

In reality, several levels of mechanisation are possible in one plant, all depending upon the line of process and plant. Rules and regulations must, of course be taken into consideration in the construction of the valve system.

Control Level

It is possible for a complex plant built according to mechanisation level 3, to be controlled by means of a very simple system. The control level defines to which degree the decisions necessary to control the process objects are delegated from the operator to the control system.

Control level 1

Every controlled object is operated by its own electrical or pneumatic switch (see Fig. 6).

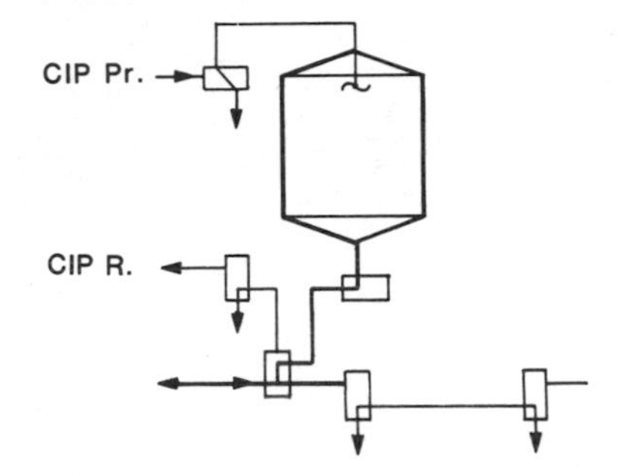

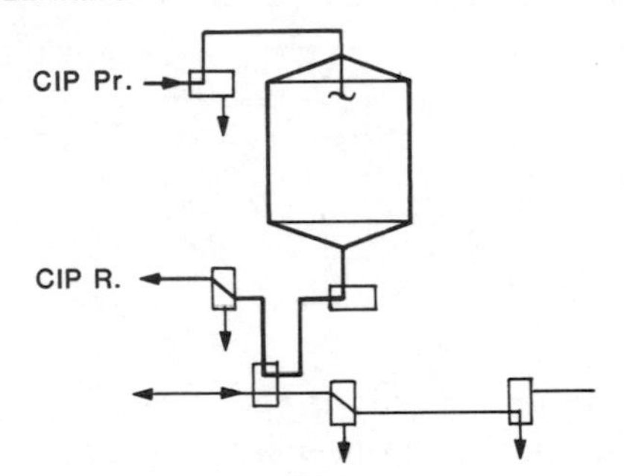

Fig. 5. Mechanisation level 3.

Control level 2

Two or more objects are controlled from *one* electrical or pneumatic switch (see Fig. 7). By using Control level 2, certain advantages are gained:

— fewer switches;
— fewer decisions for the operator;
— less risk of faulty commands.

Control level 2 can be used when switching on a CIP-circuit. (Pressure and return connection valves are operated together.)

Control level 3

Control of groups of objects linked together by their function in the plant (see Fig. 8). In a Control level 3 type operation, the advantages are the same as in Control level 2, but we have the possibility furthermore to

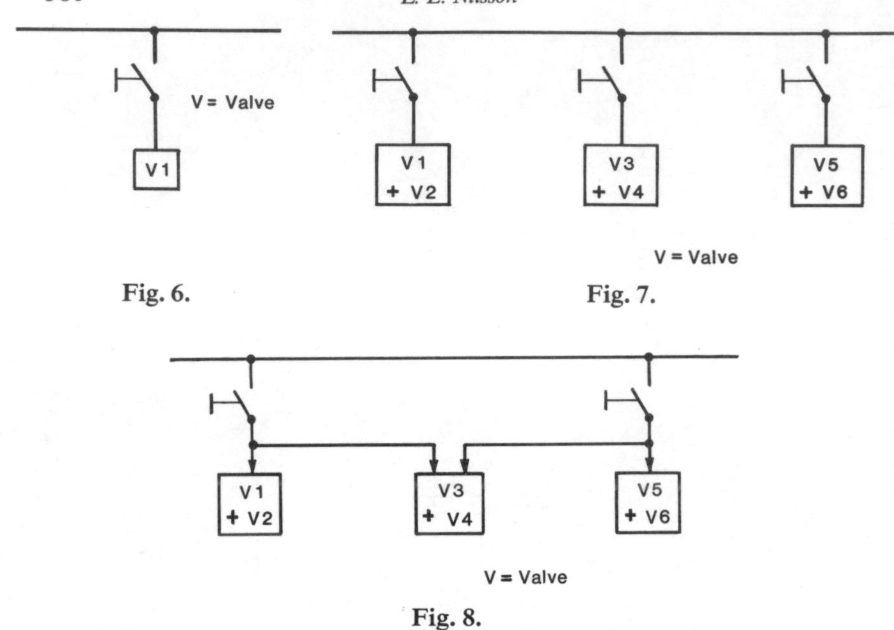

Fig. 6.

Fig. 7.

Fig. 8.

— start-up a complete function;
— control certain objects from *more* than one control switch.

Control level 4

Sequential control of objects. This means that the command to start a function is given by the operator, but the control of the actual objects is performed by

— time dependent step-by-step controllers;
— sequence operations depending on timer or process signals.

A typical process, controlled by a Control level 4 system, could be the *filling of a tank.* The valves are actuated and, after a preset time, the pump starts. The agitator is started when a certain level is reached and the function is reset by a signal from the HL- (full tank) electrode.

Control level 5

Sequential control of functions. This means that the commands by the operator (comprising several functions) are stored and then executed in a predetermined way. Examples of control level 5 options are:

— filling of several tanks in a preselected order;
— cleaning of several tanks in a preselected order;
— preselection of emptying and CIP of a tank.

Information Level

The information level tells us how much information is available to the operator, enabling him to read out the status of the plant. Two main information blocks must be considered:

The information about objects (feed-back).
The information about the process (temperature levels, etc.).

Information about objects without feed-back

— *Information level 1A*
The position of the activation switch indicates that a *command* has been given to a single object or function.
— *Information level 2A*
A visual indication informs the operator that a *command* has been given to a single object or function.

Information about objects with feed-back

— *Information level 3A*
A visual indication, based on the feed-back signals from the objects or function, informs the operator about the *status*.

This includes position indication of manually controlled objects (valves, swing bends).

Information about the process

Information level 1B
A visual indication informs the operator about the status of the process signal.

Information level 2B
A visual indication informs the operator about faults or changed status of process signals.

Information level 3B
Alarm is given for each new fault or changed condition.

Information level 4B
Protocol of faults and changed conditions is printed out.

Safety Level

If the information level selected includes the use of feed-back from the controlled objects, it means a considerable increase in plant safety. As will be seen in this chapter, feed-back is a must for obtaining a high 'safety level'.

The safety level is a measure of the ability of the control system to detect faults or faulty commands and take the necessary counter-measures to prevent damage or harm to the operator, products or plant.

Safety level 1
By suitable arrangement of switches and signs the chance of faulty commands is reduced.

Safety level 2
Interlocking of *start commands.* Protection against operator error.

Safety level 3
Continuous interlocking (not only *start* commands) achieved through monitoring of feed-back signals. Protection against process faults.

The handling of feedback signals can be done in two ways.

— Computation of important feed-back signals.
— Computation of all feed-back signals.

Safety level 4
Direct monitoring of processes, e.g. when filling an empty tank, the signal from the LLL-(empty tank) electrode must disappear within a certain time. Another example is in connection with flow meters, where a minimum number of pulses per timer interval is required.

Economic Aspects of Automation Levels

A very interesting exercise is to compare the total cost of a plant from different automation levels.

The plant, in the following example, is a milk-reception unit with two parallel lines, and a raw milk storage section including three tanks. The milk is pumped, de-aerated and cooled before entering the store. To simplify the example, it is enough to define three automation levels.

A. Manual system.
Manual valves, pumps, pipes, tanks, plate heat exchangers together with installation. No automation.

B. Semi-automatic system.
Manual controlled process and automatic controlled cleaning (CIP).
Automatic valves, pumps, pipes, tanks, plate heat exchangers together with installation.
A programmable controller (PC) plus sensors included.

C. Fully automatic system.
Automatic controlled process and cleaning (CIP). Equipment included as in B. Additional automatic valves, sensors and software-hardware for the PC.

The important factor is the relation between the costs, not the exact figures. Figure 9 shows that the total cost of the fully automated plant (C) is twice as high as the manual plant (A), and the cost of the semi-automatic plant (B) is about 65 per cent of the fully automatic plant (C); note that the calculations take into account the cost of the plant only, including installation, but running costs are omitted. In the semi-automatic plant (B), the control system makes up 15 per cent of the plant costs, and for the fully automatic plant, the same figure is about 23 per cent. The costs are broken down further in Figs. 10 and 11.

Conclusions

- The computer is a minor cost in a complete project. More costly is the process interface system (I/O system).
- Software takes up nearly half the cost of automation.
- It is expensive to automate too much, and the fully automatic plant costs twice as much as the manual one.

Please note that the plant in the above example could be either a small or medium-sized installation; the figures may be different for a large installation.

CONTROL SYSTEM HISTORY

Automation is a fast moving field. In the 1950s when the dairy industry began to adopt the system, equipment included a cleaning plant (CIP),

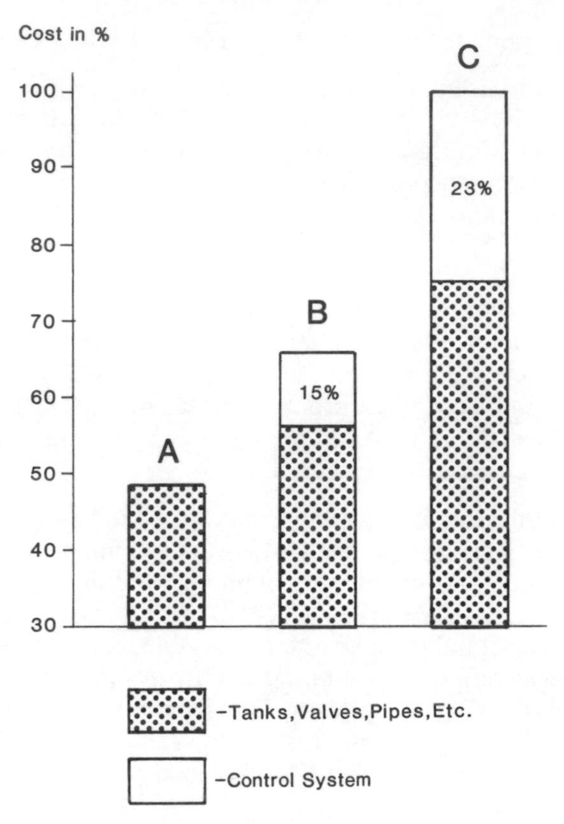

Fig. 9. Total cost of a manual and semi-automatic plant as a percentage of a fully automatic plant.

a pasteuriser and/or a filling machine which were controlled by an electromechanical relay system wired together in a logical pattern. Increasing demands for automation resulted by the early 1960s in the installation of large relay systems, and hundreds of miles of cable were used in systems that could include up to 10 000 relays.

The relay systems were replaced during the 1960s by hard-wired electronic systems; this was the new technology based on the development of the transistor. The systems were faster and more reliable because they contained no moving parts, but while the electronic systems needed less space than the relay systems, they were, like the latter, very inflexible.

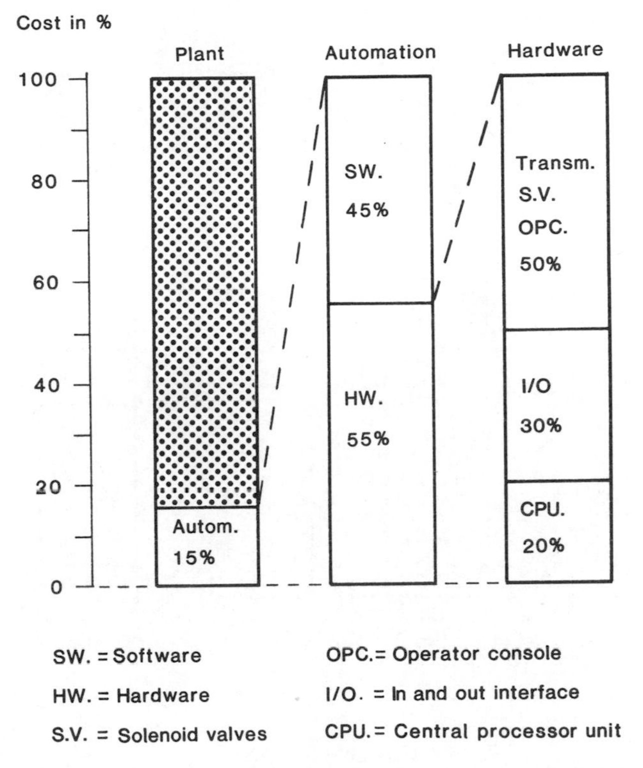

Fig. 10. Cost structure (version B) of a semi-automatic plant.

The next step was taken in the early 1970s, when computers were introduced as process controllers. In the computer based control system, the logic is expressed in data bits stored in the computer memory, and not in the physical arrangement of the wiring. This made it easier to change the program whenever necessary, as well as reducing the cost of the hardware.

In the mid-1970s, the computers were fairly expensive, so to utilise computer capacity in an economical way, centralised systems came into play. The concept of having one big computer running the entire plant was in vogue up until 1980, when the really low cost micro-processors came into their own (Fig. 12).

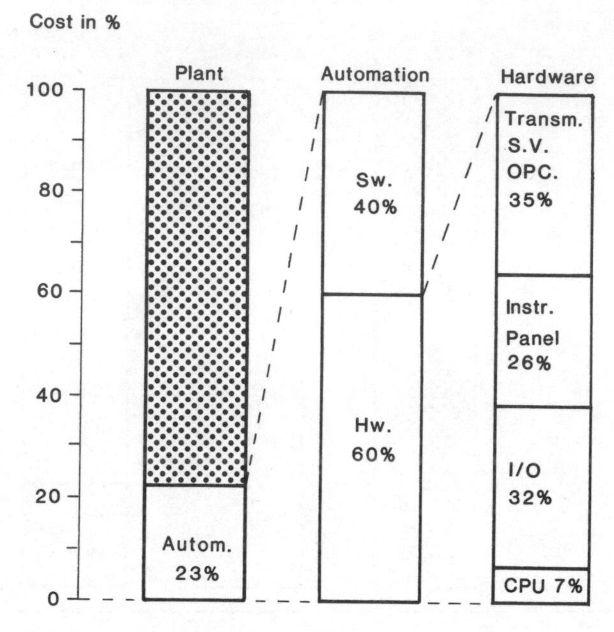

Fig. 11. Cost structure (version C) of a fully automatic plant.

Control Tasks

The control tasks of an automation system may be divided into the following different types:

1. On/off or digital control.
2. Analogue control.
3. Monitoring.
4. Reporting.

On/off control

On/off control is based on the fact that the controlled objects may be in two different states, on or off. For example, a motor may be running or shut off, and a valve may be opened, closed or in one of two positions, and on this basis, there could be a number of completely different automation levels:

A. Remote control, meaning that single objects are controlled from

Fig. 12. A compact control system for a large modern dairy.

a central panel, i.e. simply an extended arm of the manual control. This level should not, however, be considered as automation.

B. Group control, meaning that a group of objects are controlled at the same time, i.e. the valve battery under a tank.
C. Control of functions, i.e. opening or closing of product routings through the process or control of agitation.
D. Sequential control, meaning that functions are carried out, one by one, in a certain order. Examples of sequences are:
   — cleaning with different cleaning solutions in a predetermined order and at predetermined times;
   — preselection of product routes and filling levels;
   — starting up of a pasteuriser.

To take advantage of the capacity of today's control systems, the automation level is often D, i.e. the sequential control is used to a great extent.

Analog control

Analog control means that an object is regulated by means of digital signals from the control unit. Normally, this type of control is based on another (continuously varying) feed-back signal to the control unit. This type of control is, for example, used in the pasteuriser to regulate the steam or hot water supply. The feed-back signal to the control unit is given by the transmitter for pasteurisation temperature.

The analog control is very important for the function of the dairy processes. In the dairy industry, however, the analog control is often of a simple nature, and usually their control circuits are rather small. The most important applications are:

— pasteurisers;
— weighing systems, often including handling of formulas and blending;
— control of pumping capacities;
— standardisation of fat or dry matter.

Often the control includes both analog and on/off control simultaneously, so the two types of control are supplementary. The heating of a pasteuriser is continuously controlled analogically, while a temperature guard constantly monitors the temperature. The guard will sense any fall in temperature below a preset value and transmit a signal to the control unit, and the pasteuriser is then switched over to recirculation.

Monitoring

Monitoring means that the different process objects and process states are supervised, and that the system triggers an alarm if a fault occurs. Monitoring is based on feed-back signals from the object, and these can be designed in several ways:

— simple monitoring of certain critical objects only;
— simple registration of faulty conditions only;
— interlockings that prevent start or continue functions if faulty signals are received. Start of cleaning procedures may, for example, be prevented if the low level signal is not received from the tank to be cleaned;
— automatic restart of functions when the faulty condition is abrogated.

A very important part of the monitoring task is the continuous check that the control system is carrying out on itself — the so-called 'self-diagnosis'.

Reporting

The growing power of micro-processors now makes it a practical proposition to utilise them for improving productivity, not only on the shop floor but also at management level. They can be made to study and analyse the data they generate, and present it in a form on which rational management decisions can be based.

Modern systems available have this capability to a high degree, and a few examples of management routines are given here:

- Data logging, i.e. retrieval of data from the process.
- Production planning, where logged data is processed to provide records of material consumption, stock inventories, etc., as a guide to prediction of future requirements. Information can also be obtained on how much of the plant's available capacity is being utilised.
- Maintenance planning can be made much more efficient if the management has access to records showing how many hours each machine has run, and how many times each valve has operated since the last service.
- Optimisation of operations where computer-generated records of consumption figures, e.g. on power consumption, can be an invaluable aid to identify soft spots in operational economy.
- Quality assurance. A bad run can easily be traced to its source with the help of information from the computer.
- Total plant supervision. All information mentioned here may be logged, collated and printed out as periodical summary reports for the guidance of top management. When more extensive management routines are planned, it is often recommended that a separate management computer is used.

A GENERAL CONTROL SYSTEM

The process control system in a factory can be defined as the equipment installed between the operator and the process (see Fig. 13). The operator gives his commands to the process via the control system, which performs the required actions in a predetermined way. The control system monitors the process and sends all relevant information to the operator. The general control system has three parts, namely: input system; logic; and output system (see Fig. 14).

The first two are grouped together under the name I/O systems.

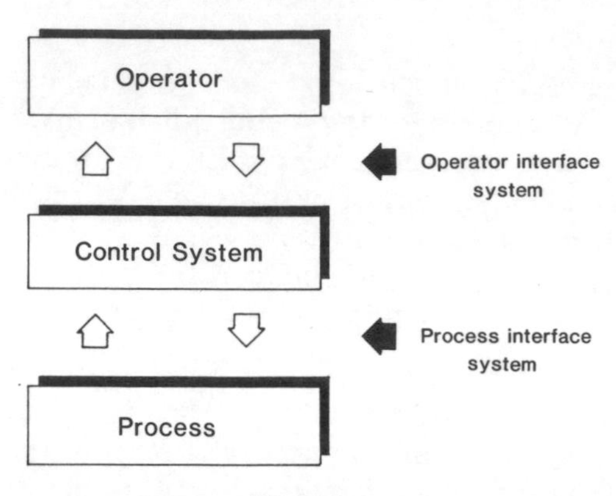

Fig. 13.

These have two separate functions: one communicates with the process and is called the process interface system, and the other communicates with the operators; it is called the operator interface system. Each will be handled separately in the text, depending on the way their profiles differ. The logic will consequently take care of some of the decisions and actions and ease the operator's workload. Parts of the process and production know-how are stored in the logic section, which, in a modern control system, is housed in a computer.

How Does the System Work?

Control is exercised by the logic which supplies output signals in a certain order actuating and shutting-off the various components in such a way that the logical conditions applying to the process are satisfied. The components send back acknowledgement signals confirming that the commands have been executed. These feed-back signals to the logic are used as conditions permitting the next step in the sequence to be actuated. The principal layout of the control system is shown in Fig. 15.

CONTROL SYSTEM

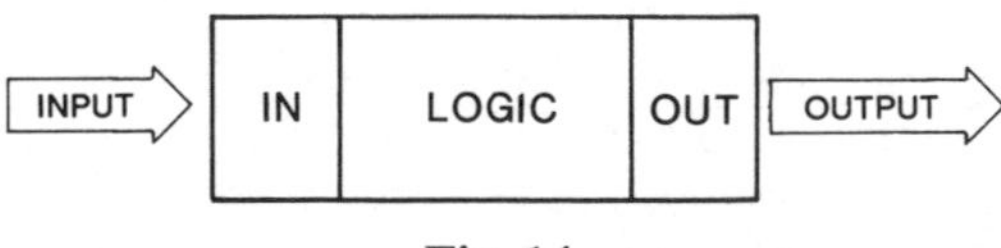

Fig. 14.

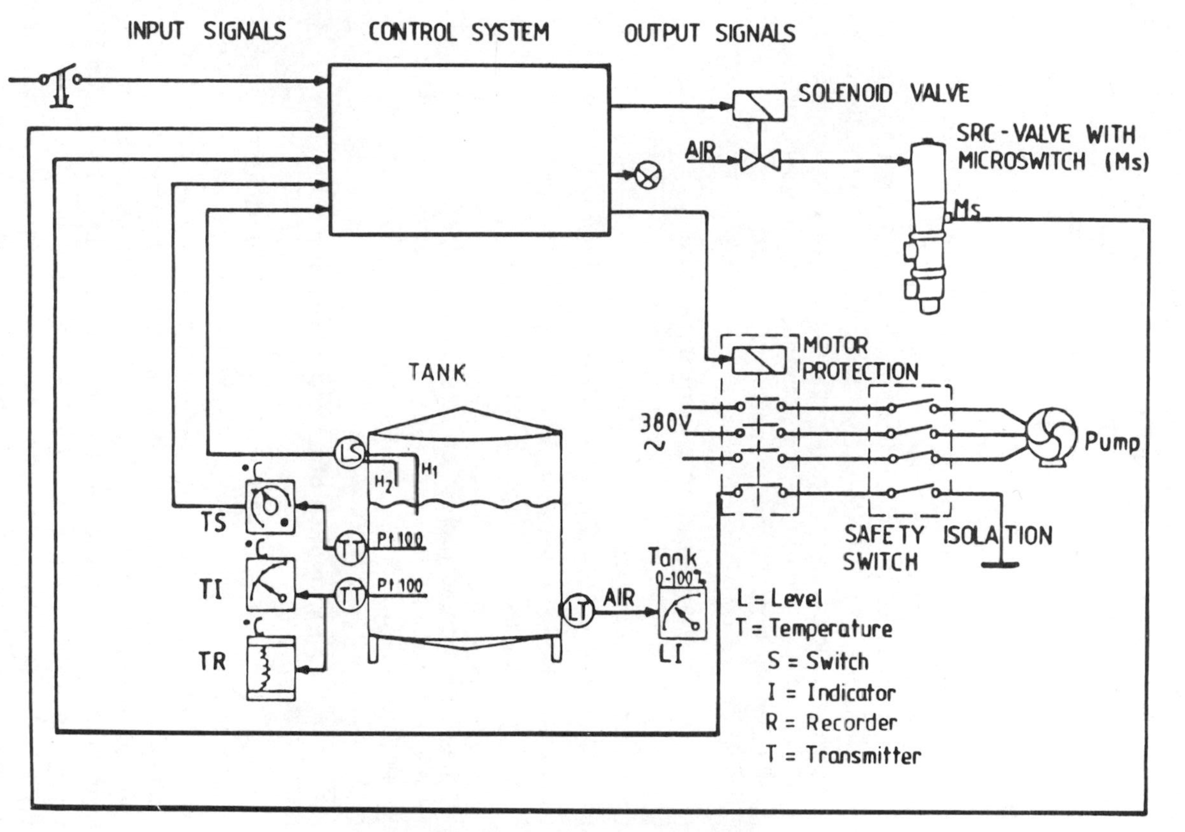
INPUT SIGNALS
CONTROL SYSTEM
OUTPUT SIGNALS
SOLENOID VALVE
SRC - VALVE WITH
MICROSWITCH (Ms)
Ms
AIR
MOTOR
PROTECTION
TANK
380V
~
Pump
H1
H2
LS
TS
TT
Pt 100
TI
TT
Pt 100
TR
SAFETY ISOLATION
SWITCH
Tank
0-100%
LT
AIR
LI
L = Level
T = Temperature
S = Switch
I = Indicator
R = Recorder
T = Transmitter

Fig. 15. A simple control system.

If no feed-back signal is received, an alarm signal may be actuated, and in this case, either the process stops or another part of the logic may be brought in to deal with the situation; this naturally assumes that the fault in question can be predicted. The more complicated the process is, and the stricter the demands made on operational security and economy, the more extensive the logic systems must be. All the transmitters and controlled objects in the process are connected to the logic, feeding information on temperatures, flows, pressures, and so on to the control system, and having absorbed these, the logic then transmits output signals to the control objects in the process.

In special so-called input/output units, the signals from and to the process are transformed to the correct form for treatment in the computer logic. All the necessary operator peripherals such as the operator console, visual display unit, printer terminal, separate keyboards and separate keyboard displays are connected to the logic.

Demands on a Control System

Of the different demands on a modern process control system, flexibility, reliability and economy are the most important. This means that

— the operator consoles should be comfortable and efficient;
— the system must be simple to expand;
— the programming language must be efficient;
— the system should include efficient electronics;
— the system should offer software for diagnostic test, modification and simulation.

Expanding of a Control System

There are plenty of automation systems available that are highly versatile and could probably be adapted to any production set-up. However, there is a much better chance of achieving top productivity with a plant if the control system is a specialised one. One of the most important demands on such a system is the potential for expansion, and it should be possible to build a system of any size, step by step, by adding standard components. A small controller installed to control a reception line for example, can later be expanded to control the milk treatment, the packaging, etc., by adding new control equipment from the same system. At the same time, management routines can be inserted into the existing processors or into a special management computer.

In the expanding process, it is very important that all system components between the operator and the process, i.e. from the remote sensor to the operator console itself, are part of the same system. An example of the expansion of a control system is given later on.

COMPUTER

In the latest generation of control systems, the designers have utilised the expanding capability, and diminishing cost, of micro-processors to distribute control functions to local units. This gives the system, as a whole, great flexibility and a very high potential capacity. The new processors can be used to control a single machine, or to build up a total control and management system to make the complete plant more productive.

Process Computer, Minicomputer, Microcomputer. What's the Difference?

The term 'process computer' expresses the function of the computer, i.e. that the computer controls the process, as opposed to carrying out administrative or scientific duties, and there is no basic difference between a process computer included in the control system of a dairy and a conventional computer for administrative duties. One of the advantages of the process computer is that it has a high capacity for using a large number of sub-programs simultaneously, each of which can be affected by the operator or the process.

The term 'minicomputer' merely denotes the capacity of the computer. In spite of the 'mini' prefix, this type of computer has a very large capacity, and many minicomputers are used as process computers in the dairy industry.

Finally, the term 'microcomputer' denotes an advanced electronic component, about the size of a sugar cube, which is included in the minicomputer. This component can also be used, in conjunction with memory and programming units, for assembling process panels for controlling individual items of equipment, such as a cheesemaking tank.

The Central Processing Unit (CPU)

This is the command centre of the computer, and its task is essentially to work down a sequentially listed program. It reads data located in specific locations in the various memories all the way down into its own working memory, carries out manipulations on this data according to a set instruction code, and then transfers the data to another store or displays it on a printer or VDU. On the sort of system with which we are involved, the CPU, once switched on, starts on an initialising phase. Thereafter, it operates on a round basis where it looks at all the tasks that it has been set, does all that it can at that moment in time, and then passes on to the next group in the sequence.

Take, for example, a bulk unit program which is in the middle of a caustic soda clean, and in which the computer can do nothing until the timer, which controls the length of clean, ends the session. In this case, it will pass on to the next task, and then after completing the last task, the CPU returns to the first one and goes through the cycle again (see Fig. 16). The cycle is regulated by a device called the Watch Dog Timer.

In going round this programmed loop of activities, the computer can also go round sub-loops if required. Generally, the time taken to go round the complete loop, the scanning time, is in the order of 0·5–1 s, which is almost instantaneous for most process plant purposes. However if very rapid action is required, as might occur on analog control of temperature, such action may be given an over-riding priority.

The Micro-processor

The performance of a micro-processor can be described in four words:

observe — memorise — calculate — act

It *observes* signals from the process, e.g. levels and valve status;
memorises the input signals and puts them into a memory;
calculates data according to input, stored parameters, etc.; and
acts by opening and closing valves according to calculations for example, or other instructions.

To manage all these tasks, the micro-processor needs an instruction set. It is unintelligent in the sense that it has to be advised to perform the simplest operation, and all operations are really extremely simple; a complex performance is always a series of elementary operations.

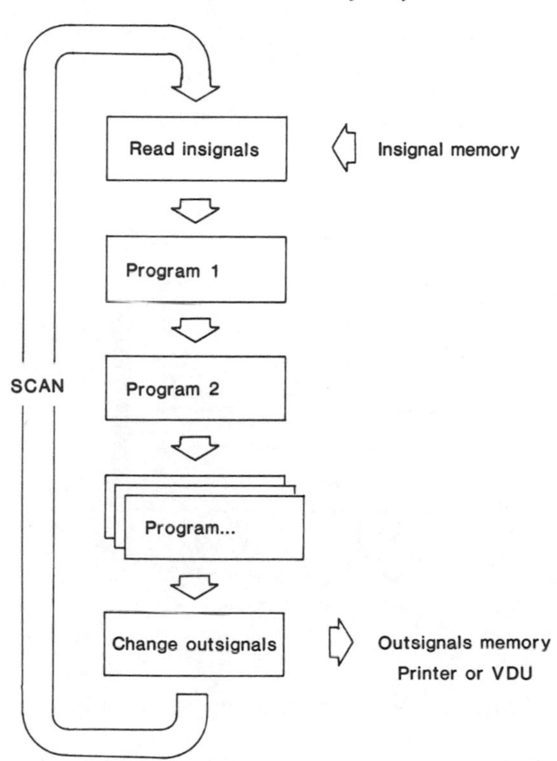

Fig. 16. Program scanning in the central processing unit (CPU).

The outstanding feature of the micro-processor is its speed. It can perform more than one million instructions in one second, and it always does the right thing. The micro-processor is totally reliable and never tires if it is given the right instruction. It is the obedient servant that always does its duty, but it is incapable of doing anything on its own.

Another stunning feature of the micro-processor is its size. The wafer containing the processor is no larger than a fingernail, but comprises several thousand transistors and other electronic components. A processor system is built around a central unit for logical and arithmetical calculations. The operations to be performed are put into a program memory and separate data, such as parameters, into a data memory (Fig. 17).

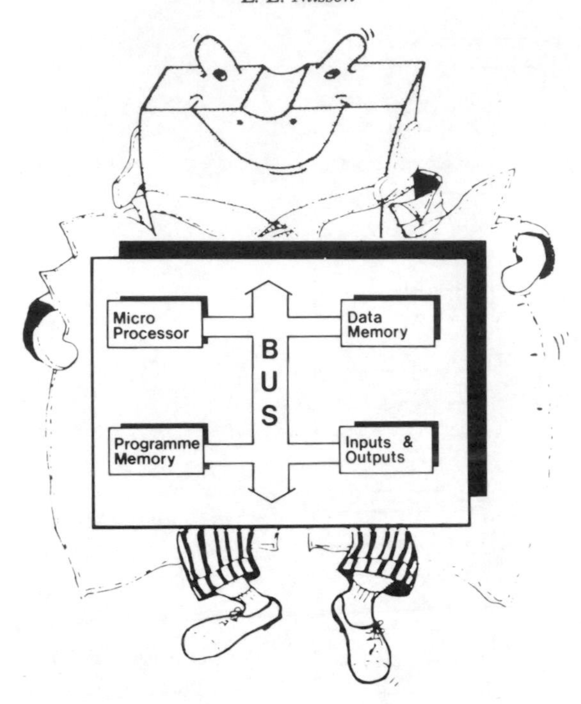

Fig. 17. Inside a micro-processor system.

Interfaces for input/output signals for external equipment, e.g. motors and valves must also be included in the system. The communication between the various components is done via an internal bus. At present, a micro-processor chip normally comprises the central unit and a limited data and program memory (Fig. 17). Additional memory and input/output units are put into separate chips. Together, all these chips form the micro-processor system. As manufacturing techniques grow more and more sophisticated, we can foresee the entire micro-processor system being comprised in one single chip.

The Programmable Logic Controller (PLC)

The programmable controller (PLC) was developed to obtain a flexible system in which the behaviour of the process could be altered easily. Instead of connecting and soldering wires, a number of instructions are entered into a memory just as with a computer. The PLC resembles a computer in that it consists of a central unit, an input/output unit for signals, a programming unit, and a memory for storing the program. The central unit scans the memory continuously and performs the stored instructions. A normal scan time is 10–20 ms.

The PLC is most commonly used for sub-processes like a pasteuriser section. Another typical application is the reception of milk in a dairy. In such a case, the milk is weighed and transported to a silo tank, where it is chilled and agitated. As standard routine, the pipes and tanks concerned must be cleaned every day. The PLC can direct the incoming milk to the right tank by checking level controls for spare space and opening the right valves. Afterwards, it can supervise the cleaning.

PLC systems suffer from one drawback — limited potential expansion. The present process, and future expansion, must be very well-defined, otherwise it could become necessary to change the whole system for a larger one; this can still be the case even when some of the more sophisticated PLCs are connectable to each other.

Therefore, after defining the process and level of automation, the number of suitable systems is very limited. It will be quite irrelevant to emphasise the existence of simple and cheap PLC systems, when there is a clearly stated need for a larger micro-processor system with a more powerful instruction set, better communication and more operator terminals. Flexibility is still, however, the major advantage of the PLC compared with a relay system, or an older electronic system. Control modification can be easily implemented by altering the program.

How Information is Stored

Kemper (1980): There are a number of means to store the large amounts of data fed into, or output from, computers. The earliest used cards or paper tape with round or rectangular holes punched in them according to a coded pattern, much along the lines of the music for the old fashioned pianola. The main series of memory devices in current use are electromagnetic. For example, 'core' memories have a large number of small ferrite cores which can be electromagnetised, that is to say magnetised by applying an electrical field. If one thinks of the soft iron core inside the coil of wire, as in the poles of an electrical generator, it is obvious that there are two possible directions for the magnet. Each end can become either a North or South pole. Thus, we can then take, by convention, each direction as being a 0 or 1, just as our paper tape had a hole or no hole. All our information is thus stored as binary arithmetic using only 0s or 1s; other examples of electromagnetic types of memory are magnetic tape, hard discs and soft or floppy discs.

Silicon chip memories are activated by applying voltages to the logic 'gates' so that, if the voltage is high enough (*ca.* 1 V), the gate will be induced to close. This again gives a binary type of data storage as previously discussed, i.e. closed or open.

One of the latest types of memory in our kind of application, which will probably replace the hard discs, is the bubble memory in which bubbles are formed in a film of synthetic garnet. The presence of a bubble is, conventionally, a 1 and its absence a 0.

The various memories can be made permanent and unalterable, i.e. ROM = Read Only Memory, which of course has first to be programmed, and hence is PROM = Programmable Read Only Memory. They can also be erased and reprogrammed, and they then become EPROM or Erasible Programmable Read Only Memory. In the case of a chip, the erasure is done by a beam of ultra-violet light shone through a window in the chip; the light discharges the voltages left on the transistors when the chip was originally programmed. With a magnetic tape, the data are erased as new data are recorded, i.e. the tape is remagnetised as the new recording proceeds as on a tape recorder.

Thus, each time we add a binary memory location, we double the amount of numbers that can be stored. Each Binary digIT or BIT is linked to others in a controlled way; 8 such BITs = 1 BYTE, and each Byte can store 2^8 different numbers or 256 in all.

We also talk of words which, unfortunately, can be composed of anywhere between 8 and 64 bits depending upon the computer. The PDP11 uses a 16-bit word, which is common, and gives a total of 65 536 binary numbers which can be stored.

Another form of memory is Random Access Memory or RAM. This is a form of memory which is only retained so long as the battery is kept connected, or the memory is not cleared by switching to a new calculation. For example, finding the square root of 2 on our hand calculator. There is a section of program held in ROM which is instructed how to take square roots which we enter by pushing the square root key. The part of memory used to hold 2 as we enter it, and the result of the square root, is RAM. A record of the calculation is lost when we move onto the next one. The RAM memory is sometimes described as a 'scratchpad'.

We can represent any decimal number in binary form as shown below for the number 239; this is expanded first in decimal form, and then in binary.

$239_{10} = 2 \times 10^2$ (or 200_{10}) + 3×10^1 (or 30_{10}) + 9×10^0 (or 9_{10})
$239_{10} = 1 \times 2^7$ (or 128_{10}) + 1×2^6 (or 64_{10}) + 1×2^5 (or 32_{10}) + 0×2^4 (or 0) + 1×2^3(or 8_{10}) + 1×2^2 (or 4_{10}) + 1×2^1 (or 2_{10}) + 1×2^0 (or 1_{10})
$239_{10} = 11101111$

All the calculations carried out in the computer are done by the very tedious process of repeated addition and subtraction. This is possible only because of the very high speed operation involved.

OPERATOR INTERFACE

The proper function of automation is not to make the operator superfluous, but to extend his reach and power. The more sophisticated the system, the fewer fiddling details he need concern himself with. The program should handle all the routine functions of the process, the tactics, while the human operator is responsible for the command decisions, the strategy.

The typical tasks will be:

— production planning in their responsibility area;
— line control;
— to act if the control system calls for attention, or if something abnormal happens;
— to check the functions by direct observation;
— to check by direct observation or with other means that the production fulfills predetermined quality demands;
— to be responsible for the daily maintenance, and in co-operation with the maintenance staff to plan for bigger maintenance tasks;
— to control and adjust the operator's part of the plant for better yield and efficiency, and for lower consumption of materials and service media.

Examples of actions that the operator is responsible for are: preselection tanks for production; start of CIP for different objects; changes of times; temperatures and other production parameters in the program; and decisions concerning measures for dealing with faults.

The human interface with the control system frequently involves operator workstations, colour graphic screens and functional keyboards. Operator instructions are entered via the keyboard, and routed to the appropriate programmable controller which responds interactively back to the operator via a mimic, a lamp matrix or a screen. The workstations will usually be arranged in a control room central to each production layout, and supplemented by a local control station where necessary.

The Operator Console

The operator console should be ergonomically designed for maximum comfort. The keys and indication lamps should be located on the console in such a way that the communication between operator and machine is made as simple as possible; Fig. 18 shows a modern operator console in a large control system. The keyboard is micro-processor based and has four groups of colour-coded keys, each group with a

Fig. 18. A modern operator console.

special relation to the process. Group 1 relates to process items, group 2 to numerals, group 3 to function and group 4 to command area.

The operator normally initiates a new operation by pushing one key in each group from left to right. The actuation of the keys is immediately confirmed on an alphanumeric process display unit. Information about the status of the controlled process is shown on the operator console; this can include more than 750 light-emitting diodes or lamps. It may be designed as a process mimic (graphic flow chart), or as a number of co-ordinate matrixes for indication of functions in operation and alarms.

The keyboard is also used for operator messages, special commands (e.g. altering times and overriding interlocks) and individual actuation of process items via the process controller. In addition, the micro-processor in the keyboard has a diagnostic program that monitors all the electronics in the operator console.

For smaller control tasks and for local control stations, there are special operator units available; one such system is shown in Fig. 19.

The Visual Display Unit (VDU)

When a black and white or colour VDU is connected to the control system, the information from the system may be obtained in several ways, e.g.:

- — direct readout from the system in figures and/or plain language;
- — a computer generated mimic or matrix display such as the one on the operator console.

The information on the VDU is formed as a number of pictures. The VDU, as an operator interface, has its greatest advantage when the volume of information is big, but only a part of it is relevant and to be activated at one time. One example is recipe handling in a cheese factory. There may be a great number of such recipes in the computer, but only one will be called up and displayed on the screen, and this can then be inspected and changed if need be. Such a recipe can include hundreds of parameters, such as time, volume, temperature and speed. It is, of course, possible to cut-in during the production and make temporary changes, such as prolonging or reducing the rennet time, changing the temperature, etc. Some components in the cheese recipe must be manually added, and an indicator at the cheese tank will show when, and the required quantity will be displayed on the screen.

Printer Terminal

The printer terminal has three main functions. The first is to supply hardcopy information from the process controller, such as fault reports to the operator or statistics for management purposes. The second is programming; instructions and modifications fed into the memory of the process controller from the keyboard of the printer are documented

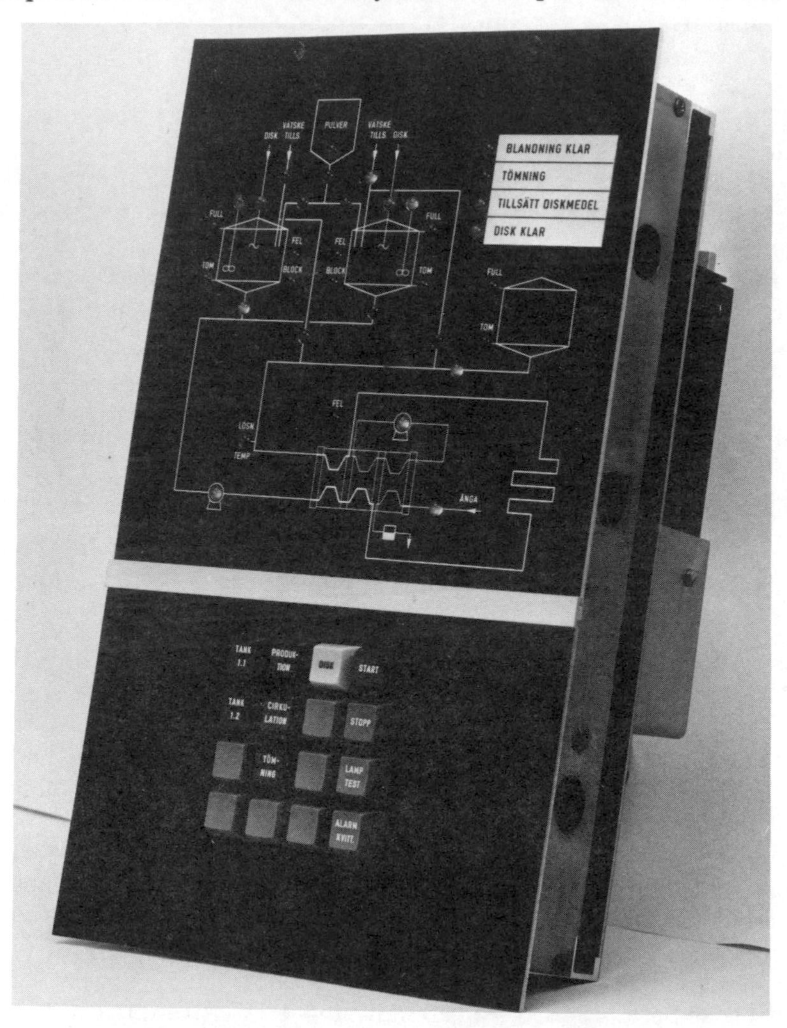

Fig. 19. Operator unit (translation: *blandning klar,* mixture ready; *tömning,* emptying; *tillsätt diskmedel,* add CIP detergent; *disk klar,* CIP ready).

on paper at the same time. Users and operators familiar with the process, and the programming language, can write and modify their own programs. The third function is as a back-up for the operator console.

Separate Keyboards and Process Display

Separate desktop keyboards are used in combination with a visual display unit or other terminals, or they may be built into the front of a control panel. The process display unit is sometimes installed separately, e.g. as the front of a graphic control panel.

PROCESS INTERFACE

The process interface, or the I/O system, is an important part of the process control system. Mechanically it uses units bigger than the computer itself, and the following are connected to the process interface:

- — process items;
- — instruments;
- — sensors;

and sometimes the control panel itself. Every type of item connected may, in certain cases, demand a specially adapted I/O unit. The types of signals involved are:

- — digital on-off;
- — analogue;
- — binary coded decimal (BCD) signals;
- — pulse signals;
- — special signals from temperature transmitters, proximity switches, level probes, etc.

I/O system suppliers often put high emphasis on the design of economical, easy-to-handle systems based on a few standard components with a minimum maintenance requirement.

Single Cable Distributed I/O Systems

The most modern, process interface systems are the single cable I/O systems. These are designed to handle process valves, motors, pumps and other digital on-off signals.

Running a dairy plant of any size involves keeping track of hundreds or thousands of valves, and operating them in different combinations and sequences.

Micro-processor control is a perfect way to remember an ideal combination for a given purpose and set-up that will perform in the shortest possible time. To do this, the control unit needs a channel for instant communication with all of the hundreds or thousands of valves, and this makes the installation costly.

It is to avoid this expense that a new system has been developed, which consists of valve units, one to each product valve, and I/O units with a number of digital input and output signals (see Fig. 20). The different units are connected to a common cable, and the valve units are connected to a common air line as well. The cable is connected to a modem, communicating with the control system. The installation is simplified and is, therefore, much cheaper than a traditional system.

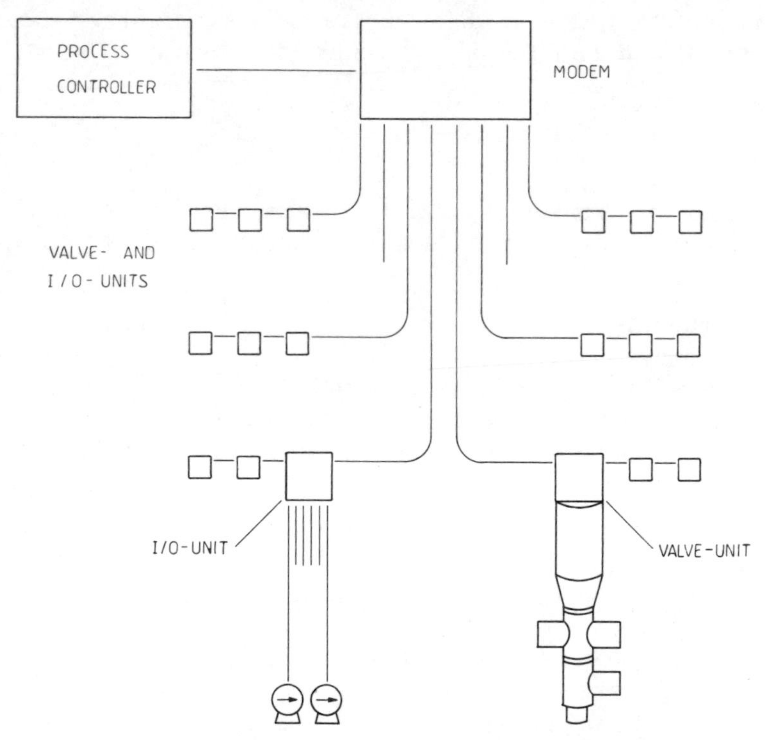

Fig. 20. Single-cable distributed I/O system.

It is possible to handle up to 64 valves or 72 pumps and motors via a single, coaxial cable. Eight coaxial cables can be connected to a modem, which can thus handle up to 512 two-position or butterfly valves, or a mixture with a number of pumps and motors. Another important advantage of the system is that it is two-way. When ordered to open/close, start/stop, the I/O units report back to the modem that it has done so. The modem can, therefore, scan the status of all valves and motors continuously and inform the process controller instantly of any malfunction. This makes fault tracing and repair work much quicker and easier too — especially as it is possible to disconnect individual I/O units without interfering with the operation of others in the system.

Sensors

Progress has been made in the field of sensors as well as in that of control equipment. Sensor technology includes devices for temperature measurement, strain gauge transducers for level, pressure and weight measurement, various ingenious techniques for flow measurement and so on. All these measurements are of fundamental importance in process automation.

Durham (1982): Moreover, the product structure is often complex and 'in process' measurements of attributes that can normally be measured physically is very difficult. For example, ice cream is a foam stabilised by ice crystals and fat crystals formed as air is beaten into an emulsion stabilised by protein and emulsifiers. The determination of which parameter to measure is not easy, let alone how to measure it. The taste of ice cream is also highly dependent on the melt-down characteristics which, in turn, depend on getting the correct physical structure.

Fortunately, instruments have now been developed, mainly for research purposes, based usually on some form of spectroscopy which are capable of being applied 'in line', and with outputs which can be readily fed into a computer system. The ability to use such instruments, however, does require considerable knowledge of product structure and process systems.

Example of characteristics that can be measured:

Physical characteristics:	Colour, viscosity, conductivity
Chemical characteristics:	Moisture content, sugar content, dry matter, water content, pH
Process parameters:	Temperature, level, flow, pressure, position

PROGRAM LANGUAGE

In the childhood of the computer, during the 1950s, all computer programs were built-up on '1' and '0'. The programmer used the computer's own language and machine code. This was the first generation of program languages.

Karlsson (1984): To write a program in machine code was very difficult, and the result hard to survey. That was the reason for developing the second generation of program languages, the assembler language. This consists of combinations of letters and decimal numbers and is considerably easier to handle. To make it understandable to the computer, it is necessary first, to translate the assembler language into machine code. This is done by a separate part of the computer, the assembler. The assembler languages were developed by the computer manufacturer and, therefore, tied to a certain type of computer. The mentioned difficulties disappeared with the next generation of program languages, the high-level languages. They are considerably easier to use, while they are closer to the human language and include ordinary words and phrases. The high-level languages must also be translated into a computer understandable form. This is done in a compiler. Examples of some of our most common high-level languages are Fortran, Pascal and Cobol. Each one with its own special advantages.

Experience from food process plants shows changes have often to be made in the operation of a plant. The user of the plant, the process operators, usually prefer to deal with the alterations themselves, rather than go back to the supplier. This requires the use of a simple language which can be understood and applied by people untrained in computer languages.

The fourth generation is an operator orientated, high-level language (Fig. 21). This means that the language makes it possible to describe the process instead of telling the computer how it should work. Thus, the denominations used in the flow chart can directly be used in the logic. Each part of the process will get its own module, e.g. tank filling and start of a pasteuriser. These modules have start and stop conditions to be activated by the operator, usually by means of a keyboard. Inactive modules can, in this way, be modified by any operator familiar with the process, and while other modules are running. It is not necessary to halt the computer, or stop processing in the plant, while changing parts of the program.

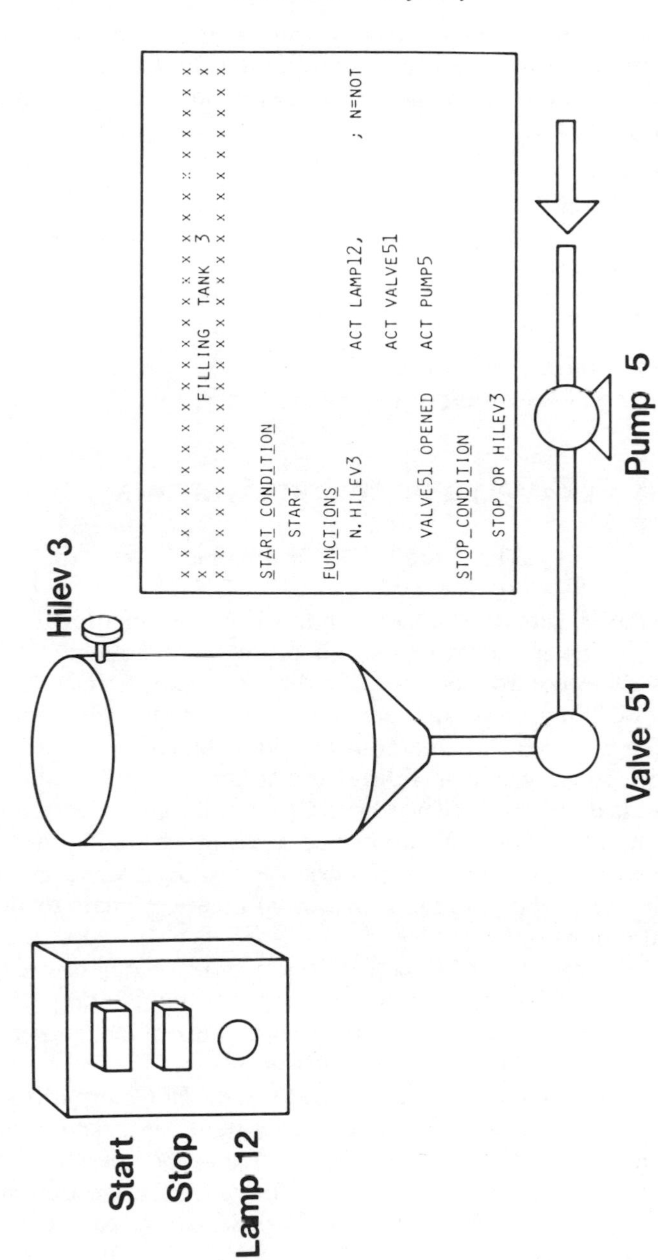

Fig. 21. Operator oriented high-level language.

The process program is normally made up of several standard packages from a software library, and added to these, are routines tailor-made for the specific plant. Some commonly used routines for dairy processing are:

— sequences for CIP, pasteurisers, etc.;
— analog control of temperatures, etc.;
— security interlocks, preventing, for example, detergent from being mixed with product.

Other commonly used routines are:

— valve position monitoring;
— handling of flow meters and weighing systems.

EXAMPLES OF CONTROL SYSTEMS

The Small PLC Controller

Figure 22 shows a small PLC controller for spot automation, i.e. local, automatic, programmed control of operations in a single machine or subprocess. This control could be of either milk reception, a plate heat exchanger working as a pasteuriser, or a cleaning system; other applications for the PLC could be material reception, batching, fermentation, sterilisation, cooking, carton filling, and so on.

The unit is a micro-processor based, programmable controller with up to 512 inputs and outputs connected to the process equipment. The inputs receive status signals (temperature readings, valve positions, etc.), while the output signals transmit command signals to pumps, valves and motors.

Figure 23 shows the PLC controlling a milk treatment area. The micro-processor in the unit constantly scans the inputs, comparing the current status of the process with the instructions in its program, and automatically taking whatever action is necessary.

In a stand-alone system as illustrated here, the PLC normally gets its instructions from an operator at a nearby panel, but having greater capacity than an ordinary PLC, this unit may be connected to an operator terminal, as in the figure, used for commands, programming or diagnostic messages. Alternatively, the instructions can come from another control unit.

Fig. 22. PLC controller for spot automation.

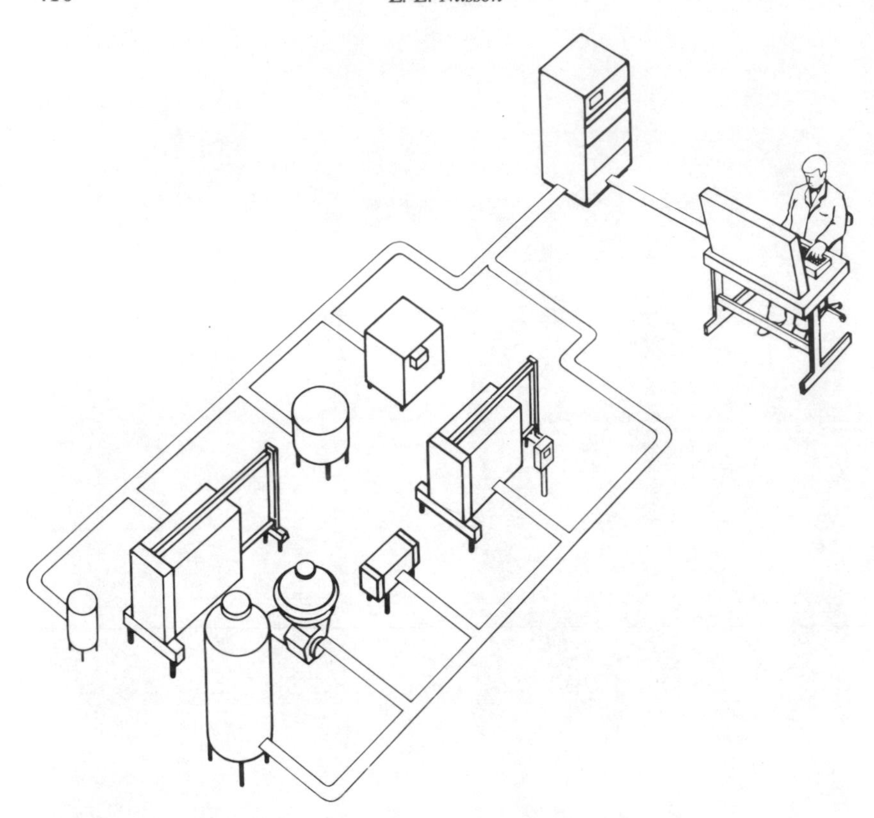

Fig. 23. PLC control of a milk treatment area.

The Process Controller

If the control system has to be expanded to take in a complete process line or, indeed, several process lines, more computation, communication and memory capacity is needed than the PLC can supply. However, the PLC does not have to be replaced; with the add-on capability of a bigger process controller, it can be incorporated into the expanded system for control of the whole process. In addition to the PLC, the system will be built-up of the following standard components:

— a process controller containing the central control unit for the process;

— one or more process interfaces for communication between the process controller and the process equipment;
— the process interface bus, a cable along which signals pass between the process controller and the interfaces. The original PLC is also connected to the bus, and so incorporated into the larger system;
— an operator console or some type of terminal for communication between the operator and the system.

The process controller is equipped with a more powerful micro-processor for process control and management. The program is normally written by an expert, but a skilled operator could change the program to adapt it to process changes. With the power of the micro-processor, and the extra distributed capacity of the process interfaces and the PLCs, a system such as this can control up to about a thousand process units (pumps, valves, etc.). In other words, it is perfectly feasible to control several milk treatment or comparable process lines from the same process controller with a single operator.

Distributed Control

The next step up will be a distributed control of all operations in a plant. Such a controlled system is shown in Fig. 24. In a distributed system, one operator is in charge of each section of the plant, assisted by a process controller that runs that section under his supervision.

The first distributed system came in when small, more or less standardised, process units got their own computerised control. A separator got its own control unit, a pasteuriser, a cheese vat, a milk-sterilisation unit, all got their own, often microprocessor based, control unit. What then happened was that conventional microcomputer process control systems were installed, and these had to link with the micro-processor controlled process unit to initiate operations and receive information.

It is now probable that distributed computer control systems could be installed a little more cheaply than central systems. Many of the costs remain the same, because there are the same number of valves and pumps, etc., in the plant, and hence things like wiring, interface, mimic diagram and programming costs are little altered. The micro-processors are probably a little cheaper than the minicomputers. The advantages lie in being able to automate plants in stages, without additional costs, to meet the growth in demand, and to avoid falling down on the job by trying to do too much in one go. It is easier and quicker to bring a plant

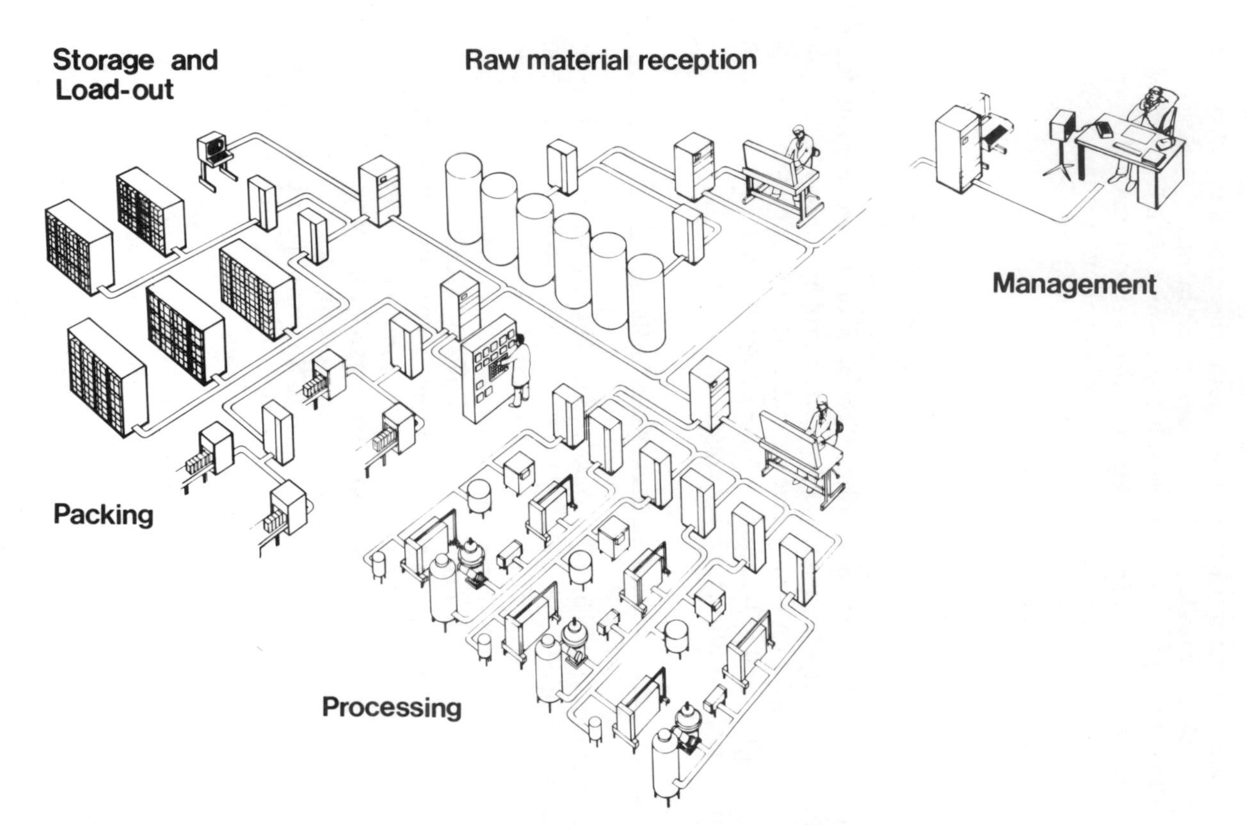

Fig. 24. Distributed control of all operations.

into service section by section, and the consequences of computer failure are more localised and can be dealt with by cheaper, stand-by systems.

Distributed control is a flexible system, and similar groupings of standard hardware components can be installed to control the raw material reception, processing, packing and handling of packed products. The difference is in the software; each departmental system is programmed for its own specific tasks.

MANAGEMENT SYSTEMS

The most important task for a control system is to run and supervise the actual process. The development of programmable systems, however, has opened up possibilities of their taking part in other duties like management and administration. There is considerable financial benefit to be gained if management can be informed immediately about such things as downtime and its reasons, waste and give-away in production. For this purpose, the plant manager or a senior operator can be furnished with a separate, supervisory control station. The equipment can be exactly the same as at the local stations, i.e. process controllers, but the management computer will be programmed for co-ordination tasks instead of for direct control. To do this, a management information link must be connected from the management computer to all its satellites to handle the communication.

We could define three different decision/information levels in a plant, each of which has its own specific tasks and demands (see Fig. 25).

Level 1: The Administrative Computer which takes care of things such as:

— invoicing;
— wage handling;
— order handling;
— accounting;
— etc.

The demands posed on an administrative system in respect to response time are not so critical. At the same time, it is exactly this response time which limits the use of an Administrative Computer as a Management System.

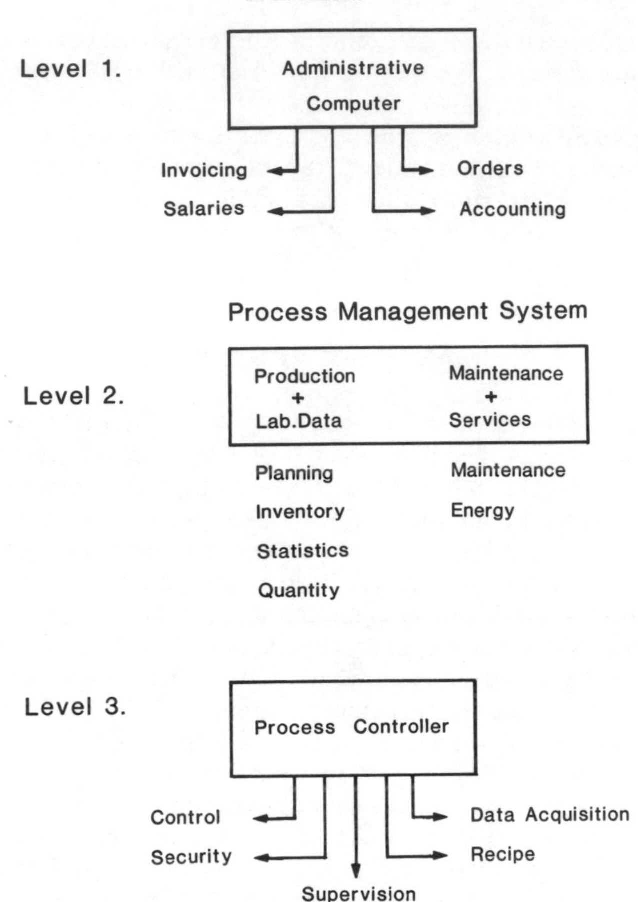

Fig. 25. Three decision and information levels.

Level 2: The Management Computer. The demands of a Management System are of such a nature that it should be considered to accomplish on a different level, such things as:

— production planning;
— statistic handling;
— inventory;
— process efficiency;
— quality;
— plant maintenance.

This system should have a response time that is slightly slower than that of a process control system, but still allows the possibility of keeping track of changes in the incoming data, when for instance monitoring a certain variable, e.g. the throughput of a pasteuriser.

Level 3: The Process Control System. This level has several tasks to perform, such as:

— control;
— supervision;
— security;
— data collection;
— running-time;
— recipe handling.

The Process Controller(s) should have a response time which is equal to, or less than 1 s.

The demands on the Control System are of such a nature that it should be able to handle the process control, the man–machine communication, as well as being able to communicate with a system on a higher level, i.e. the Management Computer.

In a modern plant, you very often find these three levels represented; an IBM Computer handles level 1 tasks, a DEC or IBM Computer handles the Management Control (level 2), and on level 3, we find a distributed Control System.

Management Duties

The management duties include substantial data acquisition and statistic analysis. The objectives are:

— production planning;
— optimisation of energy usage;
— quality assurance.

Maintenance may be an integrated part of a management system, but is here handled separately.

Data Acquisition

This implies the collection of information from the process, and can concern temperatures, levels, pressures, pH values, etc. These process values can then be used by the operator (without pretreatment), and presented on a display unit or, via a printer as a print-out. The same information is also passed on to higher level where it is stored. It can then be recalled, on request, and be subjected to statistical analysis.

Data acquisition is one of the main bases for a Process Management System (PMS), and plays a very important role.

Statistics Analysis

To get a better understanding and supervision of the process, statistical evaluations are of great importance. To be able to adjust and optimise production, the management has to have knowledge about their process. One way of doing this is to keep a manual record of what is going on, but this is very time consuming and, sometimes, not even possible. With computerised data acquisition and the right transmitters, a system which gives much wider possibilities can be created.

From a database in a process management system, selected process data are used for statistical analysis, and this could be done for a specific process line or machine over a predetermined time.

The process management system is able to calculate a number of statistical functions such as:

— totalising;
— mean values;
— standard deviations;
— minimum values;
— maximum values;
— correlation between different process variables;
— regression;
— significance tests according to the chi-square method.

The different results are displayed in a numerical way or as trend curves. This makes it simple to follow deviations of interesting process variables.

Production Planning

Production planning means prediction of raw material, services and packaging material usage as well as the right production volume or weight of each product. Substantial costs can be saved by keeping exactly the right stock both for incoming and outgoing products. Production planning also includes customer/supplier identification, and quantity accounting for paying suppliers and billing customers.

The initial data that are required are:

— order statistics;
— stock statistics;
— process know-how;
— plant lay-out and data;
— experience.

(see Fig. 26).

In order to achieve complete production planning, it is essential to have access to the above mentioned data, and while, however, it is quite

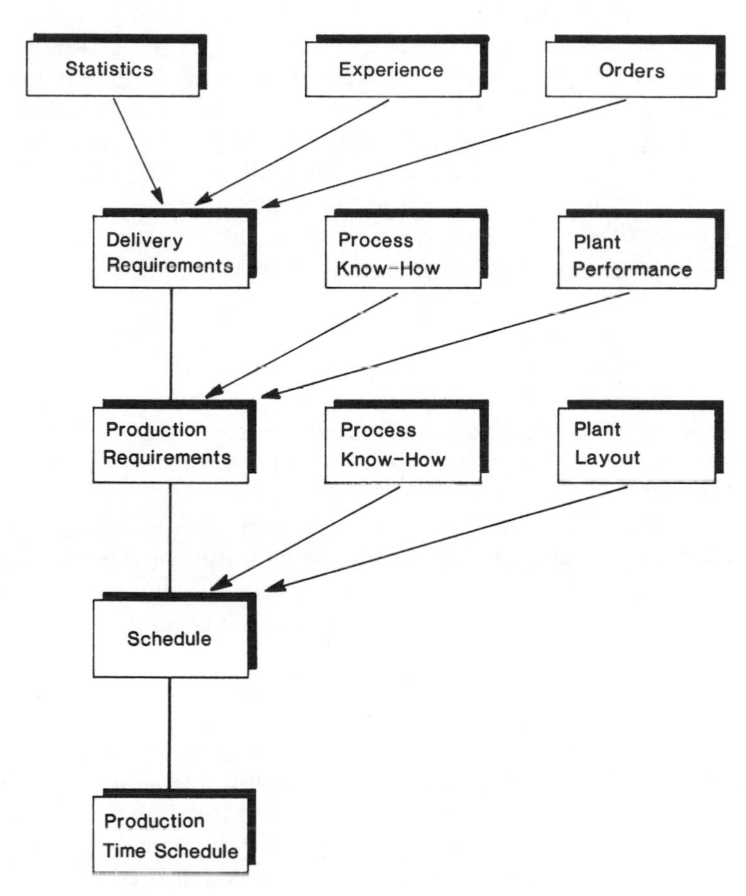

Fig. 26. Production planning.

possible that some tasks are performed by computer and others manually, the following steps can be recognised:

1. Prediction of required deliveries.
2. Required production per day.
3. Planning one day's production.

Prediction of required deliveries

The best way to make such a prediction is to combine statistics with experience. The order handling system should produce a statistical analysis for the last period, i.e. week, where it is possible to identify the quantities of each end-product both per day and totally. These data must then be combined with the experience of how the weather or holidays may affect the result, and it is desirable that this prognosis of deliveries is continuously compared with the actual incoming of orders. This comparison may be possible for those end-products where orders arrive before the actual production, but will not be possible if the production has to be started before the orders arrive.

The result of the prognosis is a plan of required end-products to be available for delivery each day during a period, e.g. a week.

Required production per day

Based on the information achieved by prognosis of deliveries, the actual production of each end-product can be planned. To compute this plan, there must be a complete model of the processing lines, in order to take into consideration the correlation between production of different end-products.

The model must also contain all process know-how so that consideration may be given to the processing methods: now is the time to cut down peaks by means of stock production, to cut down losses by planning long production times for each product, and to utilise the plant best by creating an even load.

Some end-products which are not very dependent on the date of production are more easy to plan, and can even be stocked, but others have to be freshly produced close to the delivery date.

The result of these considerations and computing of data will be a plan of which production tasks have to be performed on a certain day; some of them will result in an end-product on another day. When we know what production tasks are to be performed, a prediction of the usage of raw material can also be achieved and the corresponding quantities of powder, milk, etc., can be ordered; of course, this can also be done on a period basis.

Time schedule for a specific day

With daily production needs defined from the computing, it is possible to make a time schedule for one day, so that each basic element of the different end-products is already defined and planned for a certain day. We now have to consider such items as the time for starting the pasteuriser, or what tanks to utilise, and three steps can be recognised (see Fig. 27).

The first attempt can be either manual or computerised. The simulation is a program with a built-in model of the processing plant, which is then real-time simulated. All process items, such as tanks and pasteurisers, are switched in when needed.

The products are traced to identify how the volumes in, say, the tanks are increasing or decreasing. At this stage, considerable attention must be paid to experience, whether it is built in to the program or introduced

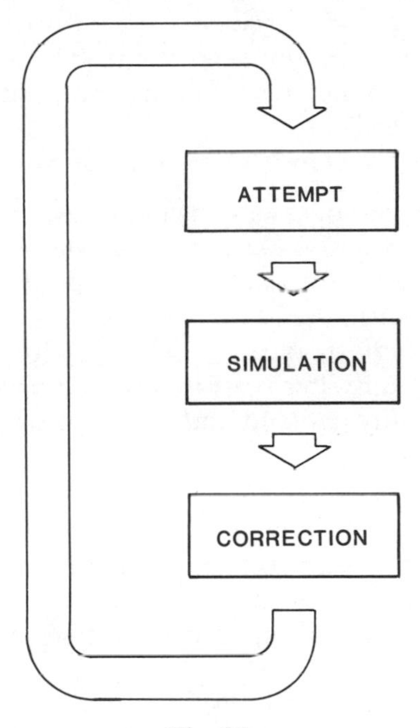

Fig. 27.

manually. The result of the simulation will show, in diagrams, how the tanks are being filled and emptied and so on, and if the attempt is failing, corrections will have to be made; they can be done either automatically or manually. The best way seems to be an inter-active system, where attempts and corrections are made manually in dialogue with the simulation program. During the real-time simulation considerations can be given to energy consumption, as well as to minimising down-times or utilising CIP installations. During a computerised attempt, evaluation and correction is harder to achieve. There can be difficulties in mathematically defining the optimum, and one can foresee that some of those tasks may best be carried out manually. The result of this computing and simulation will be a precise time schedule for each department on what to produce, when, and where, and an operation list can be produced for each department.

Optimisation of Energy Usage

Energy management in industrial operations is a complex task, and a computerised Energy Management System will make the energy managers' job easier and more efficient. The system will present information in easy-to-understand terms and produce easily readable results suitable for decisions.

With an energy management system it is easy to

— measure energy and water consumptions;
— establish energy and water norms for the plant;
— distribute the energy costs to different products;
— calculate effectiveness;
— discover bad performance;
— relate energy and water costs to different operating conditions;
— analyse the utility demand and how to produce at lower cost;
— provide records.

System description

Figure 28 shows the principle of data collecting for the energy management systems.

The system could be built up to cover different plant sections:

(I) Plant intake

Measuring of total input of electricity, water and fuels to the plant. Electricity input for refrigeration and compressed air production.

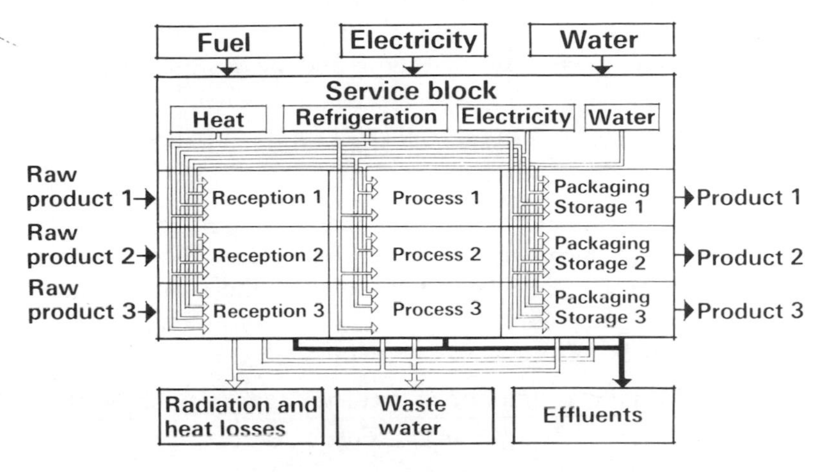

Fig. 28. Energy management.

(II) Consumption in the process section
Measuring of the different utilities input to the entire process section.

(III) Consumption in different process sections
Measuring of the different utilities to definite process areas.

Energy management software
If the data collecting system is designed in accordance with Fig. 28, the following information could be presented by the energy management system.

Energy intake/consumption reports
Intake of the different medias to the plant section I. Consumption of the different media in the plant section II, or in different process sections (III); e.g. consumption of fuel; usage of steam and water in the different CIP-stations.

Energy stock reports
— Products;
— quantity;
— usage in previous 24 h.

Efficiency reports

Efficiency of the different plant sections and the overall efficiency, that is:

(a) Energy and water usage per unit of production.
(b) Fuel usage per unit of steam.

As a measure of the efficiency with which the plant utilises different parameters, their measured value is compared with the predicted amount that would have been consumed if the plant operated as planned.

The Efficiency Index (EI) can be defined:

$$EI = \frac{(\text{Actual consumption} - \text{Predicted consumption} \times 100)}{\text{Predicted consumption}}$$

The efficiency index represents the variation from the predicted consumption, and a trend report of the efficiency index could be given on, say, a weekly basis to indicate areas of efficiency losses.

Forecasts

Utility load forecast: when the production schedule is given by the Production Management System, corrections can be made that permit a more efficient utilisation of the energy supply system, and avoid expensive peak loads.

Power demand control

A control system for energy flow in the plant is not truly a part of a management system, but often considered to be so. In such a control system, the consumption of power is measured continuously for the different energy carriers, and the costs of electric power consumption in a plant can be managed. The different consumers of electric power in the plant are then given priorities. Thus, when, during operation, the level of power consumption on which the tariff is based has been reached, the system will automatically stop certain consumers with low priority, such as refrigeration compressors and air conditioning equipment, until the power demand has again fallen under the preset level (see Fig. 29). Such a system can, in certain cases, give large cost savings.

Quality assurance

Process recording by data logging is one aspect of quality assurance. If a batch of bad quality is received, the flow of the product through the

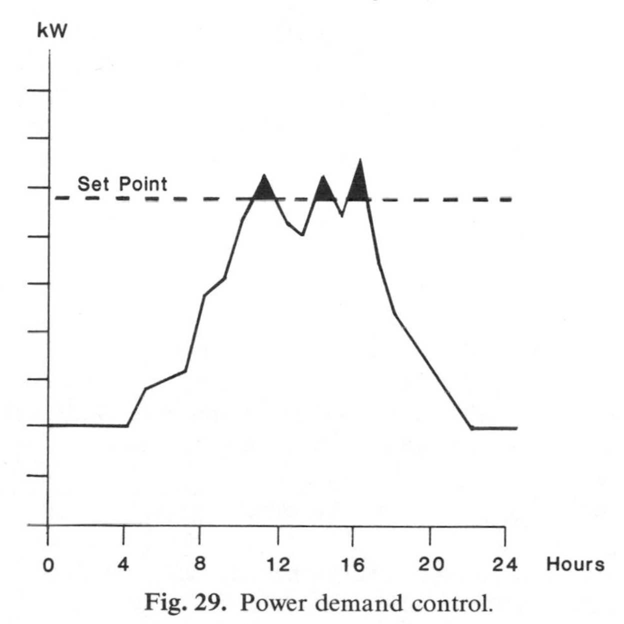

Fig. 29. Power demand control.

process can be traced, and the line where the quality of the batch was impaired can be identified.

MAINTENANCE

Electronics and computers exist today within all areas of society, and yet in spite of this, the sphere of electronics is still rapidly increasing. This spread means that the demand for maintenance also increases, although this demand will, of course, vary depending on the application area. Home and office electronics, currently the biggest expansion area, will not have the same need for maintenance as the electronics in the process industry, and for the reason that the cost of halting production in industry is very high; a single stoppage today could cost up to £10 000.

In what way then, can we maintain the electronics, and how can fault-finding and preventive maintenance be carried out?

Fault-finding

The maintenance of mechanical and electromechanical equipment has, very often, a short fault-finding time, i.e. it is easy to find that a winding motor is damaged, but it takes a long time to repair it. However, with electronics the situation is the opposite (see Fig. 30), in that it takes a long time to find the faulty circuit board, and an even longer time to find the fault on the board. On the other hand, the mean time to repair it is very short, the circuit board can simply be replaced.

New Problems when Applying Electronics in the Process Industry

When using electronics and computers without any connection to a process, fault-finding and maintenance are easy, but when the electronics are a part of the process, we can get faults, such as a motor failure, appearing as a fault in the computer. Fault-finding is made more difficult since all parts depend on each other, and there are no clear interfaces between the electronics and the process.

Self-diagnosis

The micro-processor is used to help the maintenance engineer determine which printed circuit card is faulty. A self-diagnosis system is incorporated into the process controller as well as in the operator console, conventional I/O systems, and single-cable I/O systems. In short, all electronic parts of the process control system have self-diagnosis, so that the maintenance engineer is able to determine exactly which card is faulty as soon as a fault on a printed circuit board appears. The fault diagnosis system includes alpha-numerical displays where the type of fault, together with the actual number of the faulty circuit board, will be displayed. A printout, together with the actual time and

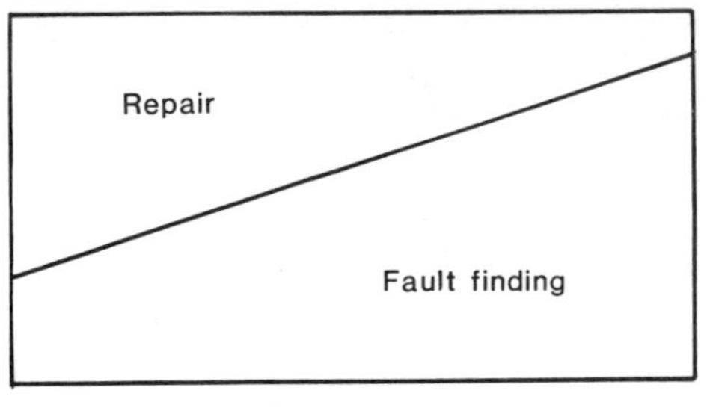

Fig. 30.

date, can be provided if a hard copy is required. The erring unit will be restored by replacing the faulty card with a spare one.

All faults are not faults in the electronics

Most of the faults in a process show up in the electronics, but most of them are not the fault of the electronics themselves. The electronics have a high availability, and by means of it, we are able to find faults in the process.

The above is leading us to develop a new strategy regarding the maintenance of electronics, and contains a great deal of preventive maintenance.

Preventive Maintenance

Many have claimed that it is not possible to implement preventive maintenance on electronics, and it is true that electronics do not wear out in such a way that you can discover weaknesses in the system. But preventive maintenance is a set of actions such as training, fault-finding, and documentation, which can increase the availability of the system.

Training

General product training does not normally include fault-finding, and a new type of training has to be designed which is tailored to the actual plant. The training, which should aim at fault-finding in the electronics or other equipment, should be run on site.

Documentation

Documentation is needed for fault-finding, and it may consist of circuit drawings, program listing and adjustment lists. All modern electronics are easy to alter and modify, but this flexibility creates two new problems. The first is that documentation has to be kept up to date with any alterations, and the second is that a badly thought-out alteration can affect another part of the plant.

Repair

Electronics repair is effected by replacing modules, which are then repaired centrally. Access to the right spare parts affects, to a great extent, the rate of repair, but at the same time, it is very costly to keep a big stock of spares. However, statistics show that more than half of all exchanged circuit boards are, in fact, correct, and the time for fault-finding is short and the cost of stockholding can be low — if the fault-finding procedures are carried out correctly.

Performance of the Electronics

While no electronic components may have been changed or worn-out, the performance of a plant can, nevertheless, change with accuracy of a transmitter; analog controllers being affected, for example. This is very often caused by some small adjustments that have been made, for several small, faulty adjustments can combine to give a worse performance. Even if every part in a plant is correct, a performance test still has to be made to check that the total system works according to specification.

Conclusion

For effective maintenance of the electronics and computer systems, the following five elements are required:

— fault-finding and self-diagnosis;
— training;
— documentation;
— spare parts;
— performance tests.

THE FUTURE

During the past ten years, developments in the automation of the dairy industry have forged ahead at a rapid pace from relay panels to computerised control. Greeves and Knott (1984): A new generation of chip technology is on the horizon, and it is only a matter of time before system manufacturers and software houses exploit the potential of these in offering more powerful, faster and more intelligent devices at lower costs.

We will, in the future, get programming languages that are even easier to learn and understand. Together with the coming generation of micro-processors, this will, maybe, make the computer understand normal human speech and, therefore, do what the user asks them to do. The increasing distribution of intelligence towards the plant interface and the sensors, where it can be used as a basis for higher levels of system diagnostics, is just one example of the way in which the industry is moving.

We can expect to utilise single-cable systems all over the factory. All sensors, motorstarters, valves and other process items, will be connected to the process computer by means of a single cable. The cable may even be replaced by a glass fibre, and the signal transmitted by a beam of light.

However, the control systems have always been matched to the process equipment which they are to control in the dairy. This will doubtless also be the case in the future. Future control systems will undoubtedly be able to operate in conjunction with yesterday's, today's and tomorrow's systems and machines.

REFERENCES

Alfa-Laval publications. The author acknowledges references to Alfa-Laval papers and publications too numerous to mention individually. These company publications may be obtained from: Alfa-Laval Co Ltd., Great West Road, Middlesex. TW8 9BT.

Durham, K. (1982). *Chemistry and Industry,* 363.

Greeves, T. W. and Knott, C. M. (1984). *The Role of Integrated Control and Information Systems in Manufacturing Productivity,* Bath, 10–12 April, Institute of Chemical Engineers, Series No. 84, 121.

Karlsson, U. (1984). *Ny Teknik,* **21**, 28.

Kemper, J. (1980). *The Microprocessor as Used in Food Plant Automation,* Alfa-Laval Co. Ltd., Middlesex.

Index